*M. Bard
and J.-D. Laredo
(eds.)*

Interventional Radiology in Bone and Joint

Springer-Verlag Wien New York

Michel Bard, M.D., and Jean-Denis Laredo, M.D.
Department of Bone and Joint Radiology
Hôpital Lariboisière, Paris, France

With 194 Figures

Library of Congress Cataloging-in-Publication Data: Interventional radiology in bone and
joint / editors, Michel Bard and Jean-Denis Laredo. XII, 273 p. 19.3 × 26.8 cm. Includes index.
ISBN-13: 978-3-7091-8950-4 1.Bones—Interventional radiology. 2.Joints—Interventional radiology. 3. Musculoskeletal system—Interventional radiology. I. Bard, Michel, 1925–   . II.
Laredo, Jean-Denis, 1951–   [DNLM; 1. Bone Diseases—therapy. 2. Biopsy, Needle. 3.
Chymopapain—therapeutic use. 4. Embolization, Therapeutic. 5. Intervertebral Disk Displacement—drug therapy. 6. Joints Diseases—therapy. 7. Steroids—therapeutic use. 8. Technology, Radiologic. WN 160 I 614] RC 930.I 58 1988 616.7'07572—dc 19

ISBN-13: 978-3-7091-8950-4   e-ISBN-13: 978-3-7091-8948-1
DOI: 10.1007/978-3-7091-8948-1

Foreword

The editors of this work, M. Bard, head of the Department of Bone and Joint Radiology of the Lariboisière Hospital, and his assistant, J.-D. Laredo, had the excellent idea of assembling in a single volume different topics concerning techniques of interventional radiology in bone and joint diseases. This includes all diagnostic and therapeutic procedures performed in departments of radiology which may be substituted for open surgery. These techniques have a number of advantages when compared with open surgery. They are more easily performed, less invasive, and less stressful for the patient. They require little or no hospitalization and are more cost-effective.

J.-D. Laredo, M. Bard, C. Cywiner-Golenzer, and J. Chretien (Hôpital Lariboisière, Paris, France) have extensive experience of percutaneous biopsy of the spinal column using a special trephine needle which J.-D. Laredo and M. Bard have designed for this purpose. Vertebral bodies and intervertebral discs from T 3 to L 5 can be biopsied under fluoroscopic control and local anesthesia. Percutaneous biopsy is especially useful in spinal infection. Histological and bacteriological examination of material obtained can be used to differentiate spinal osteomyelitis from other conditions which sometimes have a similar appearance, such as degenerative disc disease. In almost all cases of tuberculous spondylitis a definite diagnosis can be obtained by percutaneous biopsy (by the recognition of mycobacterium tuberculosis or typical tubercles). The technique can also be used to identify cases of pyogenic spinal osteomyelitis where the microorganism can be identified only in approximately 50% of cases. At the Viggo Petersen Rheumatology Center of the Lariboisière Hospital, percutaneous needle biopsy of the spine has been carried out almost routinely for several years in cases of spinal infection located from T 3 to L 5–S 1, resulting in the almost total absence of any need for open biopsy for diagnostic purposes, which is afar from benign procedure. Percutaneous spinal biopsy can also be used to detect metastases as well as other much rarer malignant vertebral lesions such as lymphoma, Ewing's sarcoma, solitary plasmocytoma, etc. A trephine needle similar to that designed for percutaneous vertebral biopsy can be used with fluoroscopic control to biopsy the sacroiliac joint when there is a suspicion of infection. Certain lesions of the limbs and girdles can also be biopsied after discussion with orthopedic surgeons. Recently the use of CT scan in preoperative assessment and biopsy guidance has proved to be especially useful in percutaneous biopsy of small deep osteolytic lesions and soft tissue tumors.

Percutaneous biopsy of the synovial membrane of the knee is commonly carried out without the aid of radiology. The same does not apply to other joints where fluoroscopic guidance is always useful and often necessary (J.-D. Laredo and M. Bard, Hôpital Lariboisière, Paris, France).

Provided certain contra-indications are kept in mind, chemonucleolysis is an alternative to surgery in sciatica due to a herniated disc and resistant to conservative treatment as described in this volume by J. Roucoulès, J.-D. Laredo, M. Bard and D. Kuntz (Hôpital Lariboisière, Paris, France). It is carried out under local anesthesia and requires only two to five days of hospitalization. Despite the rare anaphylactic reactions, papain remains the most widely used nucleolytic substance, though others are available or currently being evaluated. Due to the possibility of anaphylactic shock, chemonucleolysis requires the presence of an anesthesiologist fully prepared to deal with allergic reactions, as discussed by M.-C. des Essarts (Hôpital Lariboisière, Paris, France). Nucleolysis technique is relatively easy in the hands of experienced operators and is detailed by J.-D. Laredo, J. Busson, M. Bard (Hôpital Lariboisière, Paris, France), and M. Wybier (Hôpital Cochin, Paris, France). Under fluoroscopic guidance, the needle penetrates the disc laterally and its intradiscal position is confirmed by discography before injection of the nucleolytic substance. Several tens of thousands of chymopapain chemonucleolysis procedures for sciatic pain have been carried out throughout the world. Indications are the same as those for surgical treatment of sciatica due to a herniated disc. Results are considered to be excellent or good in 60–80% of cases. However, we feel that the precise evaluation in comparison with those obtained by surgery would justify even more extensive analysis than currently available.

Chemonucleolysis is not the only alternative to surgery for sciatic pain due to a herniated disc. Percutaneous discectomy using special instruments has also been carried out. Because of the topical nature of the subject and the interest which it arouses, several authors using different techniques have described their experience. P. Kambin (Graduate Hospital, Philadelphia, PA, U.S.A.) and G. Onik (Allegheny-Singer Research Institute, Pittsburgh, PA, U.S.A.) report satisfactory results with two different techniques of percutaneous discectomy through a lateral approach while W. A. Friedman and S. L. Kanter (University of Florida, Gainesville, FL, U.S.A.) had an unsatisfactory experience of percutaneous discectomy through a true lateral approach.

Low back and sciatic pain is sometimes due to posterior facet joint osteoarthrosis without disc herniation. CT scan provides a detailed analysis of degenerative changes within the facet joints revealing, in certain cases, a synovial cyst protruding into the spinal canal. CT also confirms the absence of any lesion of the intravertebral disc. Resistant sciatic pain due to osteoarthrosis of the facet joints may justify surgical release of the compressed root. By contrast, surgical treatment is generally not suitable for low back pain. The injection of corticosteroids during facet joint arthrography sometimes relieves sciatic or low back pain due to facet joint degeneration, as described by M. Wybier (Hôpital Cochin, Paris, France) and J.-D. Laredo (Hôpital Lariboisière, Paris, France).

In chronic low back pain there exists the possibility of percutaneous radio frequency denervation of the facet joints. B. Lavignolle, J. Senegas, J.-L. Houton, J. Guerin, and J.-M. Caille (Centre Hospitalo-Universitaire, Tripode, Pellegrin, Bordeaux) reviewed the results obtained in such indications.

Cervico-brachial neuralgia due to disc lesions or osteophytes usually recovers within less than two months but is often extremely painful. It is for this reason that G. Morvan, D. Mompoint, M. Bard, and G. Levi-

Valensin (Hôpital Lariboisière, Paris, France) have attempted to treat it with the intraforaminal injection of corticosteroids under fluoroscopic control, sometimes with good results but most often partial and transient. Certain cases of cervico-brachial neuralgia are sufficiently resistant to justify surgery aimed at removing the disc herniation or the osteophytic process responsible. In the presence of a herniated disc, an alternative to such surgery is cervical disc nucleolysis, experience of which is reported by Y. Lazorthes, J. Richaud, J.-C. Verdié, and A. Bonafe (Centre Hospitalo-Universitaire Rangueil, Toulouse, France).

Vascular radiology, in which there has been considerable progress, is essential in many areas of medicine. In bone and joint pathology it can be used to study the blood supply of some tumors such as angiomas. It is essential in demonstrating the blood supply of the spinal cord before surgery for tumors. However, it has other possibilities. Skilled specialists can embolize tumors with a particularly rich blood supply, e.g. certain vertebral angiomas and certain malignant tumors. It is also possible to obtain hemostasis in bleeding associated with skeletal trauma, as described by D. Reizine, S. Marciano, F. Gelbert, A. Aymard, and J.-J. Merland (Hôpital Lariboisière, Paris, France). S. Wallace, C. H. Carrasco, C. Charnsangavej, W. Richli, N. Jaffe, J. Murray, A. Ayala, A. K. Raymond, S. P. Chawla, and R. S. Benjamin (M. D. Anderson Hospital and Tumor Institute, Houston, TX, U.S.A.) have used intra-arterial treatment in osteosarcomas and inoperable giant cell tumors.

Until recent years, solitary bone cysts were treated by surgical curettage. Good results can now be obtained by the injection of corticosteroids into the lesion under fluoroscopic control, a technique which comes to us from Italy and which is described by M. Campanacci and R. Capanna from the Istituti Ortopedici Rizzoli (Bologna, Italy). A similar technique can be performed in the management of eosinophilic granuloma of bone (S. Wallace, C. H. Carrasco, C. Charnsangavej, M. Cohen).

Tendon calcifications due to hydroxyapatite crystals, often affecting the shoulder and sometimes the hip, wrist or elbow, may cause episodes of acute inflammation as well as chronic pain. The latter is sometimes so troublesome and resistant that it would seem to be legitimate to remove the calcification surgically. However, this is not always possible, even in the shoulder. A much easier therapeutic technique, which can be used in most chronically painful tendon calcifications, is that of puncture-aspiration of the calcification under fluoroscopic control with the in situ injection of a corticosteroid (C. Normandin, E. Seban, J.-D. Laredo, D. N'Guyen, D. Kuntz, M. Bard, Hôpital Lariboisière, Paris, France). In the hands of a rheumatologist as experienced as C. Normandin (Centre Viggo-Petersen, Hôpital Lariboisière, Paris, France), this technique offers approximately 60% good results and is worthy of wider use.

Close cooperation between rheumatologists and radiologists has been of enormous value in the management of bone and joint diseases. The honor which Michel Bard and Jean-Denis Laredo have bestowed upon me, as a rheumatologist, to write a foreword to their book is an indication of our long, friendly and fruitful collaboration. Radiologists provide for rheumatologists the images essential for the majority of diagnoses, the accuracy and precision of which have reached an astonishing level of quality with CT scan and magnetic resonance imaging. However, radiologists do not only offer increasingly refined imaging of bone and joint lesions. They have also conceived and developed diagnostic and therapeutic techniques replacing surgery where it was previously necessary. Radiology has become interventional and of the greatest therapeutic importance in rheumatology, as shown by this book, unique of its type, and which, I hope, will enjoy a fully-deserved success.

Antoine Ryckewaert
Professor of Rheumatology,
Paris

Acknowledgments

We are greatly indebted to Dr Jean Innes for her long and patient support in editing and translating the manuscript.

Gratitude is expressed to Professor Antoine Ryckewaert for writing the foreword.

We are much grateful to Joelle Goux for her expert work in manuscript preparation and secretarial assistance.

A special word of thanks is also expressed to Paula Schwartz, Dr Michael Prendeville, and to Stephanie Young for their help in manuscript translation and revision.

The editors

Contents

Percutaneous biopsy of bone and joints

Chemonucleolysis in the treatment of herniated intervertebral discs

Percutaneous lumbar discectomy

Facet joints percutaneous diagnostic and therapeutic procedures

Interventional vascular radiology in musculo-skeletal lesions

Direct steroid injection in the treatment of tumor-like lesions

Miscellaneous

Percutaneous biopsy of bone and joints

Percutaneous biopsy of musculo-skeletal lesions

J.-D. Laredo[1], M. Bard[1], Charlotte Cywiner-Golenzer[2], and J. Chretien[1]

Departments of [1] Bone and Joint Radiology and of [2] Pathology,
Hôpital Lariboisière, Paris, France

Modern treatment of skeletal disorders has become more complex and specific, increasing the demand for precise histologic and bacteriologic diagnosis. In many clinical situations, percutaneous biopsy (PB) of musculo-skeletal lesions can establish a definitive diagnosis without the disadvantages of surgery, making it a useful alternative to open biopsy. Although PB of musculo-skeletal lesions is an established technique, its potential contribution to the management of the patient with bone or joint disorders is not fully appreciated. Technologic advances in radiology have dramatically improved the capabilities of PB. With either fluoroscopic guidance or CT scan control, a biopsy of most lesions of bones, joints and muscles can be readily performed using a percutaneous approach. Preoperative CT scan is also a useful tool for determining the most adequate biopsy technique in each particular case. At the same time, progress in pathologic and bacteriologic capabilities have also improved the accuracy of PB results. Radiologists are in a unique position to perform PB of skeletal disorders [44] since they administer the multiple imaging procedures most likely to identify and evaluate the significance of the abnormality [39]. In addition, they have a dynamic three-dimensional approach to anatomy and spatial needle placement. With current levels of radiologic invasive diagnosis and therapeutic intervention, radiologists have the skills required for PB [39]. However, the radiologist must be a member of a team including the referring physician, a pathologist and a microbiologist. Biopsies should be planned and performed in a way that would not adversely affect subsequent definitive surgical procedures. Surgical advice concerning the biopsy approach should be sought each time surgical treatment of the lesion is considered.

In this text, needle aspiration biopsy will be distinguished from trephine biopsy. Needle aspiration biopsy consists of aspiration of fluid for cytological and bacteriological studies. Trephine biopsy, also called core, trephine [1, 35, 48], or needle biopsy [15, 34, 43, 50], is a process by which a core of tissue is obtained percutaneously for histopathological interpretation [37].

Background

In 1930, Martin and Ellis first reported successful pathologic diagnosis in a series of patients whose abnormal tissues were sampled by percutaneous needle aspiration using an 18-gauge needle. Their 65 successful biopsies included eight biopsies of the musculoskeletal system [33]. In 1931 Coley et al. reported biopsy experience with 35 consecutive bone tumors with a correct diagnosis rate of 91% [7].

Since these early reports, clinical experience and developments in biopsy needle design, radiologic technology, and bacteriologic and histopathologic diagnosis have improved the procedure of PB of the musculoskeletal system. The technique of needle aspiration biopsy was further developed by several teams [40, 41, 42]. In 1935, Robertson and Ball first devised a technique of vertebral needle aspiration biopsy, which was later refined by numerous authors [45]. In 1968, Valls, Ottolenghi and Schajowicz described a simple mechanical device using a 2 mm needle for aspiration biopsy of the lumbar spine; they reported that clinical decisions could be made in 80% of cases on the basis of aspiration biopsy results [51]. Ottolenghi, in 1969, reported a series of 1,050 aspiration biopsies of the spine, confirming the usefulness of this procedure. He also adapted this technique and instrumentation to the thoracic spine and reported 28 aspiration biopsies in the thoracic region, including the second through the ninth thoracic vertebra, with 27 positive results and no complications [42]. By 1976, Schajowicz and Hokama reported an experience of 7,165 aspiration needle biopsies of the skeleton, performed over a period of 33 years with an overall accuracy rate of 74% [46].

In the meantime, others developed the technique of trephine biopsy. In 1947, Ellis first introduced drill biopsy as a modification of aspiration biopsy and obtained better results (68%) than with standard aspiration biopsy [15]. Siffert and Arkin first used a trephine in 1949 to biopsy a lumbar vertebra [48]. This technique was further refined by Ackerman [1] and Craig [8] in 1956. Ackerman reported 223 uncomplicated biopsies of vertebrae and appendicular skeletons using a new trephine biopsy set which sampled a 1.5 mm diameter tissue core [1]. The same year, Craig described a larger bore trephine needle that could recover a 2 mm diameter tissue core [8]. Since then, Craig and Ackerman needles have become the most widely used bone trephine devices. In the following decades, the most important improvements in the closed trephine biopsy technique were the introduction of fluoroscopy guidance [24], and more recently, the use of CT scan in preoperative assessment and biopsy guidance of musculo-skeletal lesions. This is proving to be especially useful in the closed biopsy of small, deep osteolytic lesions [20] and soft tissue tumors [2, 36]. The trephine biopsy technique has been extended to additional skeletal sites. In 1981, Chevrot et al. [5] and Vinceneux et al. [52] independently reported a technique for percutaneous trephine biopsy of the sacroiliac joint using a posterolateral approach through gluteal muscles and iliac bone. We have previously reported a large series of uncomplicated percutaneous trephine biopsies of the thoracic spine from T 3 to T 12, using a special trephine set and fluoroscopic guidance [29, 30].

Equipment and instrumentation

The X-ray room

Radiologic evaluation of the needle position must be constantly available during PB procedure. This can be accomplished by using either fluoroscopic or CT scan guidance, depending on the biopsy site. A biopsy of most superficial lesions and sacroiliac joints can usually be achieved using single plane standard fluoroscopy. During biopsy of deep musculo-skeletal lesions, such as vertebral biopsy, biplane evaluation of needle position must be constantly available. This can be carried out with either biplane fluoroscopy, standard flurosocopy and portable C-arm unit, or standard fluoroscopy and cross-table lateral radiographs. CT scan is the method of choice for biopsy of soft tissue components and small, deep lytic lesions close to vital structures, such as vertebral lesions involving the neural arch.

Biopsy needles

A large variety of needles is available for both trephine and aspiration biopsy of bone and soft tissue components. We will cite here only the most frequently used needles. The specific needle used depends upon the nature, consistency (osteoblastic, osteolytic or mixed) and site of the lesion (Tables 1 and 2) [11].

Bone trephine needles

Bone trephine needles consist of round tubes of two basic types.

Ackerman, Mazabraud, Craig, Harlow Wood, and Laredo-Bard needles include both a serrated trocar that cuts the tissue, and an external cannula which is left in place against the bone while the serrated trocar is removed with the specimen. This allows for the sampling of several specimens during the same procedure without further manipulation. However, these needles differ in their insertion techniques and in their diameters. Ackerman and Mazabraud needles are directly inserted paralleling the anesthesia needle, using a sharply pointed stylet within the external cannula. The Craig needle is inserted over a blunt guide, while the Laredo-Bard needle is inserted in the track of the anesthesia needle using a procedure similar to vascular catheterization. Craig, Mazabraud and lumbar Laredo-Bard needles are large bore instruments, while Ackerman and thoracic Laredo-Bard needles have a smaller diameter.

In the Jamshidi and Tanzer instruments, an external cannula and a cutting trocar do not exist as two separate pieces. The external cannula, which has a cutting edge, is also used to remove the specimen. Both needles have a relatively large diameter. These instruments are technically very simple. However, they are completely withdrawn with each specimen. This is a big disadvantage for biopsies of deep bones. On the other hand, Jamshidi and Tanzer needles are convenient for biopsies of lytic lesions of superficial bones.

The Ackerman needle [1] (Slanco, Becton-Dickinson Co., Rutherford, NJ, U.S.A.) (Fig. 1). The Ackerman needle is a 12-gauge needle (external diameter) with a serrated trocar providing specimens 1.6 mm in caliber. The external cannula (B) and the sharp pointed obturator (A) are inserted together down to the lesion site parallel to the anesthesia needle (Figs. 1 and 2). The stylet is withdrawn and the external cannula is left in place. The serrated trocar (C and C') is then inserted through the external cannula and advanced into the bone with a twisting motion and a variable amount of pressure to take a specimen. The trocar is then withdrawn and the external cannula is left in place. The tissue is removed from the trocar with the probe (D and D'). The external cannula can subsequently be moved to a different location to take additional specimens.

Table 1. Characteristics of different trephine needles

Name	Manufacturer	Specifications: external diameter and length	Advantages	Limitations	Preferential indications
Ackerman	Becton-Dickinson	12-gauge, 140 mm	easy sampling of several specimens		thoracic spine
Mazabraud long short	Collin-Gentile	8-gauge, 240 mm 8-gauge, 95 mm	easy sampling of several specimens and easy handling	inserted in a separate pathway from anesthesia needle	lumbar spine sacroiliac joint and peripheral bones
Jamshidi	Travenol	8-, 11- or 13-gauge, 100 or 90 mm	easy use and handling	whole needle removed with each specimen	peripheral and flat bones
Tanzer	Collin-Gentile	7- or 8-gauge, 100 or 60 mm	easy use and handling		peripheral and flat bone
Craig	Travenol	10-gauge	railroading technique of introduction		lumbar spine
Harlow-Wood	Pembleton	6-gauge	railroading technique of introduction	blunt guide	lumbar spine
Laredo-Bard lumbar thoracic sacroiliac	Collin-Gentile	8-gauge, 210 mm 12-gauge, 140 mm 8-gauge, 110 mm	introduction in the same route as the anesthesia needle		lumbar spine, thoracic spine, sacroiliac joint and peripheral bone
MacLarnon	Needle Industries	6- or 10-gauge			lumbar and thoracic spine

Table 2. Conversion chart

∅ mm	Gauge	L mm	Inches
0.30	30 G	3	$^1/_8''$
0.35	28 G	6	$^1/_4''$
0.40	27 G	8	$^5/_{16}''$
0.45	26 G	10	$^3/_8''$
0.50	25 G	13	$^1/_2''$
0.55	24 G	15	$^5/_8''$
0.60	23 G	20	$^3/_4''$
0.70	22 G	22	$^7/_8''$
0.80	21 G	25	$1''$
0.90	20 G	30	$1^1/_4''$
1.00/1.10	19 G	35	$1^3/_8''$
1.20/1.30	18 G	40	$1^1/_2''$
1.40/1.50	17 G	45	$1^3/_4''$
1.60	16 G	50	$2''$
1.80	15 G	60	$2^1/_2''$
2.00/2.10	14 G	70	$2^3/_4''$
2.40	13 G	75	$3''$
2.80	12 G	80	$3^1/_4''$
3.00	11 G	90	$3^1/_2''$
3.50	10 G	100	$4''$
4.00	8 G		
4.50	7 G	110	$4^1/_4''$
5.00	6 G	120	$4^3/_4''$
		125	$5''$
		150	$6''$

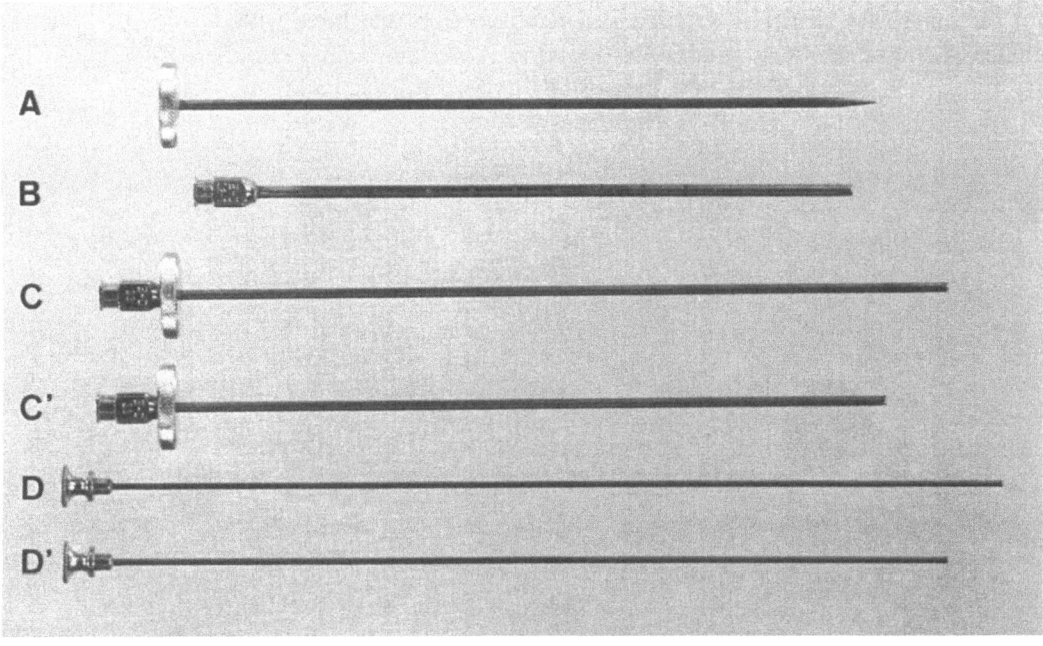

Fig. 1. Ackerman needle. *A* Sharp-pointed stylet; *B* external cannula; *C, C'* serrated trocar; *D, D'* probes

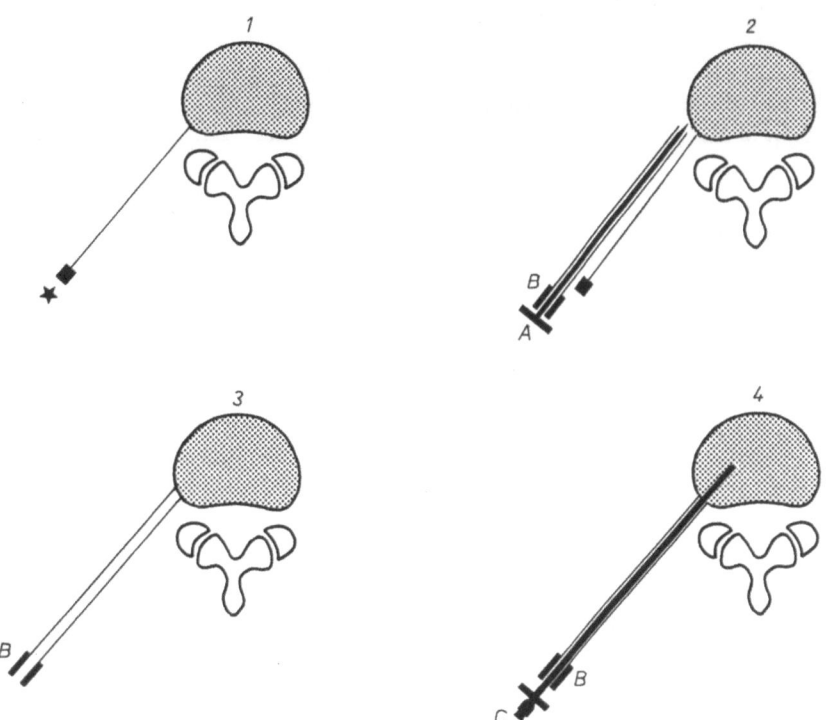

Fig. 2. PB technique with the Ackerman and Mazabraud needles. The trephine needle is inserted paralleling the anesthesia needle (★). The various parts of the trephine needle are labelled the same as those in Fig. 1

The Mazabraud needle [47] (Collin-Gentile-Drapier, Arcueil, France) (Fig. 3). The Mazabraud needle has a larger diameter (8-gauge) than the Ackerman needle and provides specimens 2.4 mm in caliber. It is available in two different sizes: 240 mm for lumbar spine biopsy and 95 mm for biopsies of the sacroiliac joint and peripheral bones. The Mazabraud needle has a very practical handle. Instructions for use are very similar to those for the Ackerman needle (Fig. 2), except that the serrated trocar must be advanced into the bone with clockwise rotation and removed from the bone with counter-clockwise rotation.

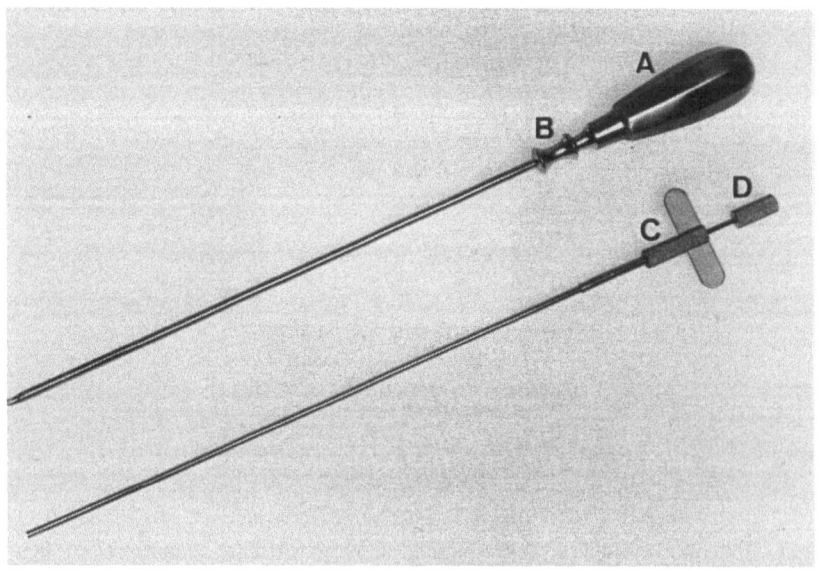

Fig. 3. Mazabraud needle. *A* Sharp-pointed stylet; *B* external cannula; *C* serrated trocar; *D* probe

The Craig needle [8] (Becton-Dickinson) (Fig. 4). The Craig needle is a large caliber needle (10-gauge) approximately 3.5 mm in diameter. The serrated trocar has a metallic handle which facilitates the rotation necessary for the introduction of the needle. The Craig needle provides specimens of a 2 mm bore. The 16-gauge blunt guide (A) is first inserted through a skin stab and advanced down to the biopsy site (Fig. 5). The external cannula (D) is slid over the guide and held firmly against the bone. The guide is removed and the serrated trocar (C) is inserted through the external cannula until the biopsy site is reached. With a twisting motion and a variable amount of pressure, the serrated trocar is driven into the bone to take a specimen. The rest of the procedure is identical to that using Ackerman and Mazabraud needles.

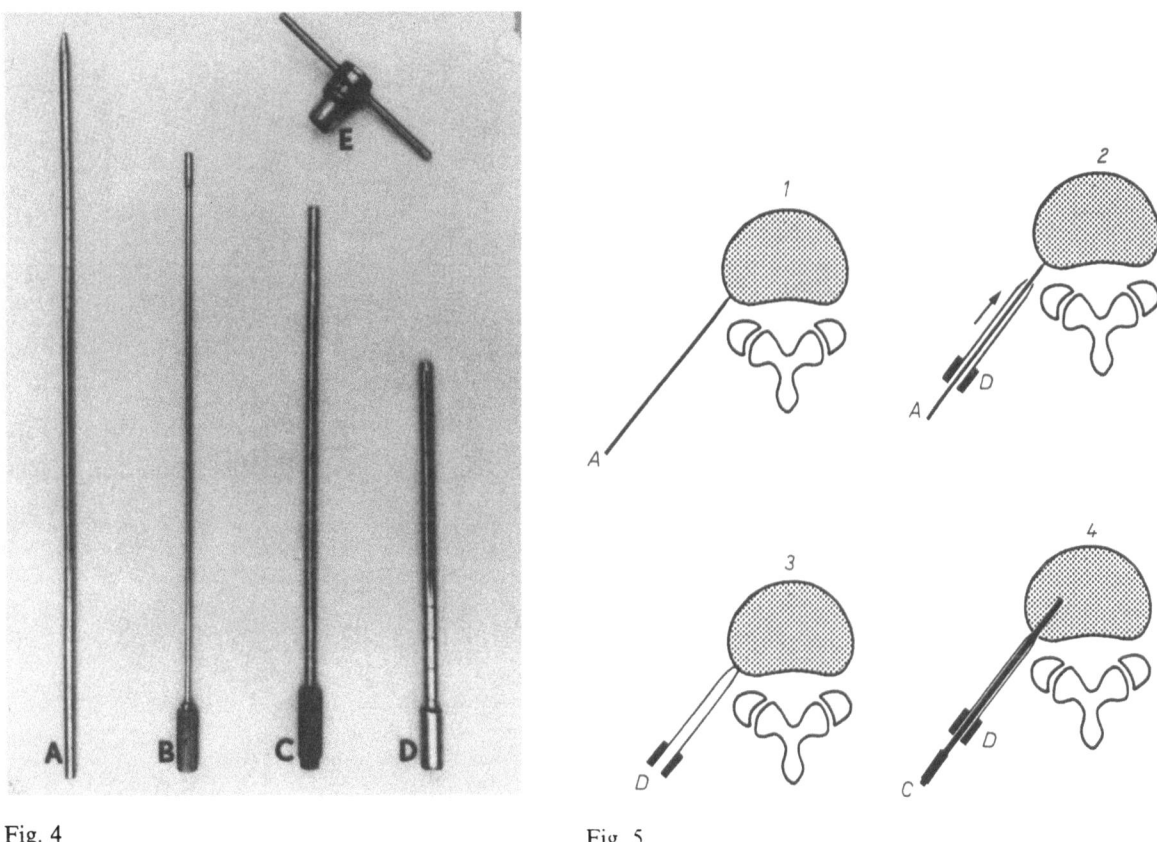

Fig. 4   Fig. 5

Fig. 4. Craig needle. *A* Guide; *B* probe; *C* serrated trocar; *D* external cannula; *E* handle

Fig. 5. PB technique with the Craig needle (or the Harlow-Wood needle). The various parts of the trephine needle are labelled the same as those in Fig. 4

The Harlow Wood vertebral trephine set [18]. This needle is inserted using a railroading technique similar to that of the Craig needle. However, the blunt guide is much thinner than that of the Craig needle and consists of a 1.5 mm caliber guide-wire 25 centimeters in length with a removable knob. The set also includes a tapered dilator, an external cannula with an external diameter of 5 mm, and a serrated trocar providing specimens of 3 mm bore, with an adjustable collet to control the depth.

The Laredo-Bard Trephine needle [25, 30] (Collin-Gentile-Drapier). This needle is used quite differently from the Ackerman, Craig, and Mazabraud needles (Figs. 6 and 7). The Laredo-Bard needle is placed following a pro-

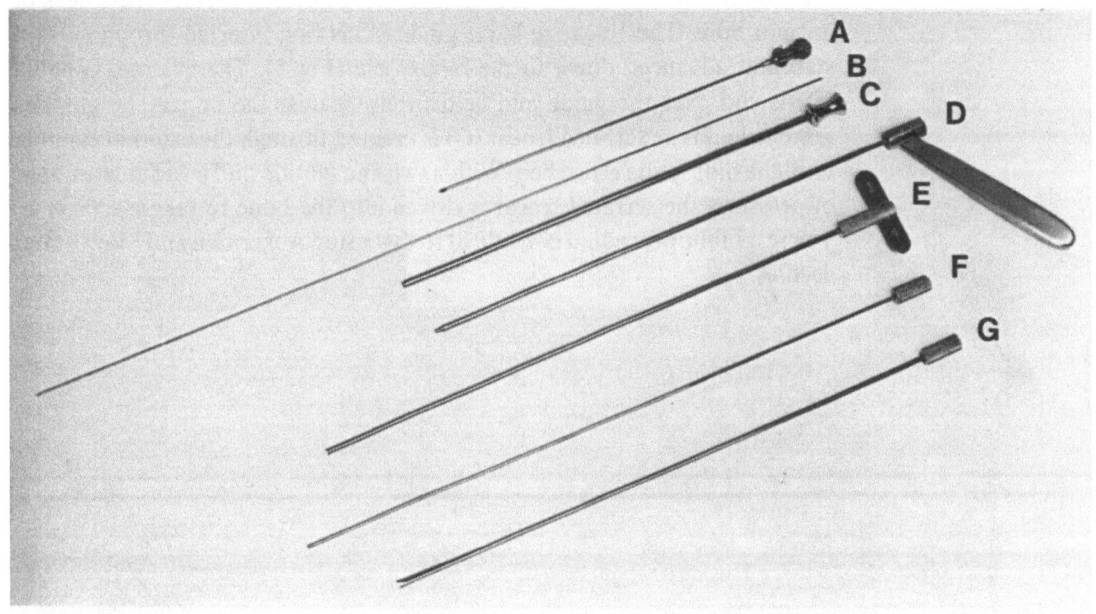

Fig. 6. Lumbar Laredo-Bard needle. *A* 12-gauge anesthesia needle; *B* guide-wire; *C* external cannula; *D* intermediate hollow piece with handle; *E* serrated trocar; *F* probe; *G* split cutting needle for soft tissue biopsy

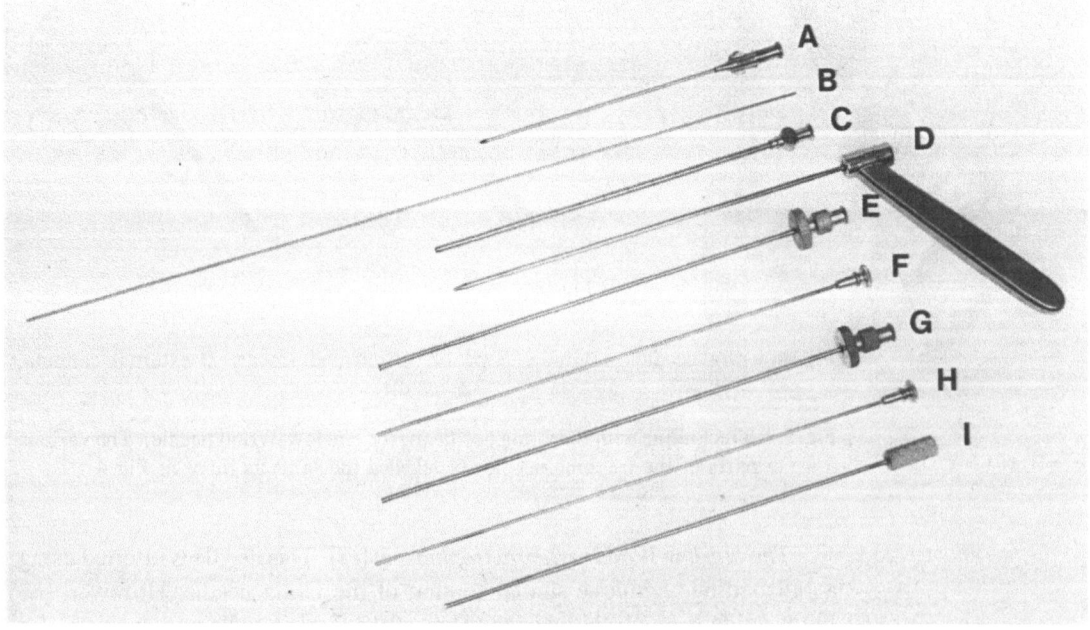

Fig. 7. Thoracic Laredo-Bard needle. *A* 17-gauge anesthesia needle; *B* guide-wire; *C* external cannula; *D* intermediate hollow piece with handle; *E, G* serrated trocars; *F, H* probes; *I* split cutting needle for soft tissue biopsy

cedure similar to that of vascular catheterization (Fig. 8) [25, 30]. Subcutaneous tissue down to the lesion is carefully anesthetized with a Xylocaine 1 per cent solution using a 20-gauge disposable needle. This needle is also used to control that the selected needle approach will avoid bony obstacles. A small stab wound is made and the 17-gauge anesthesia needle (A) included in the trephine set is inserted with its stylet through the skin stab and advanced to the bone under fluoroscopic guidance. When the bone has been reached, the stylet is removed and replaced by the guide-wire (B) (Fig. 8). The needle is removed and the guide-wire is left in place. The cannula (C) and the intermediate hollow piece (D) are slid together over the guide-wire and pushed gently through the intervening tissues to the bone with a rotating motion. Once the bone has been reached, the cannula is pushed firmly against the cortex and both the hollow piece and the guide-wire are withdrawn and replaced by the serrated trocar (E). Before cutting the bone, the position of the trephine is checked in both AP and lateral views using either fluoroscopic control or radiographs. The trephine is advanced into the bone with circular clockwise rotation, applying slight pressure to cut a bone core. The serrated trocar is then gently removed with the specimen while applying suction with a syringe connected to the needle. The cannula is left in place to take additional specimens. The specimen is removed from the serrated trocar using the flat-tipped probe (F). With this needle set, the same approach is used for local anesthesia and bone biopsy, and the trephine is advanced only after the safety of the puncture has been verified with the thin needle. If a nerve root is encountered, the thin needle can be easily adjusted without any complication, while crossing a nerve root with a large trephine needle

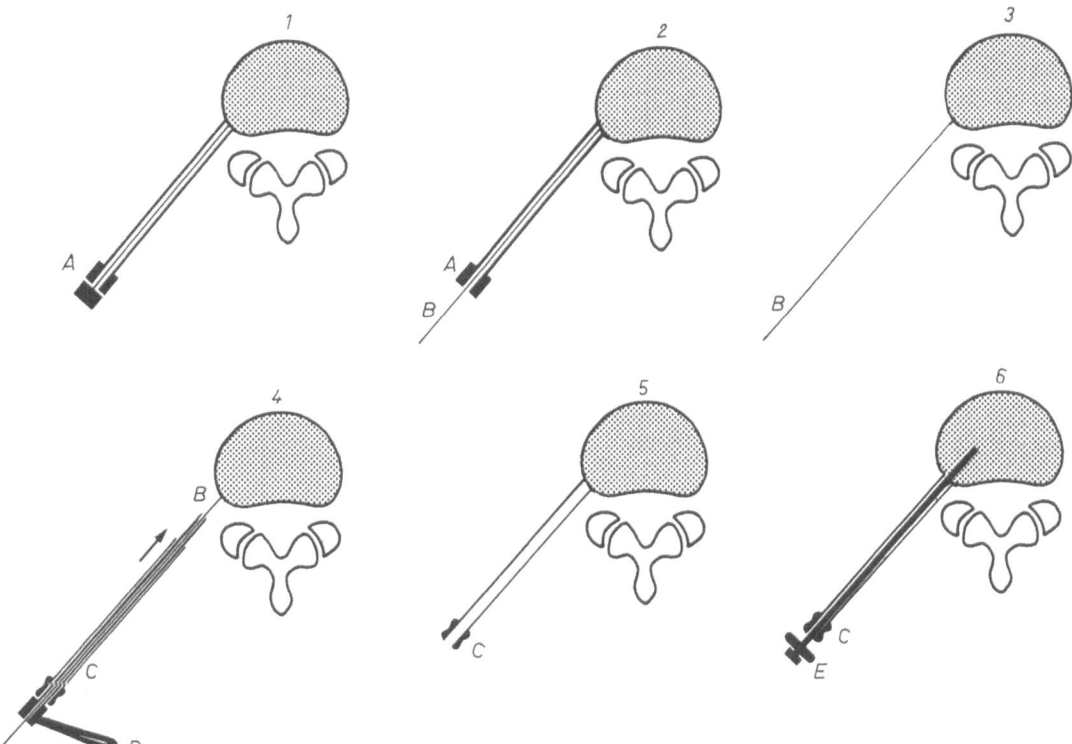

Fig. 8. PB technique with the Laredo-Bard (or MacLarnon) needle. The various parts of the trephine needle are labelled the same as those in Fig. 7

may be complicated by neurologic impairment despite immediate readjustment [37]. For these reasons, we now use the Laredo-Bard needle in all percutaneous biopsies of deep bones. Three different sizes and calibers of the Laredo-Bard trephine needle are available. The lumbar biopsy set is a large caliber needle with an external diameter of approximately 4 mm (8-gauge) which provides specimens of 2.4 mm in diameter. The thoracic biopsy set has a smaller caliber of 2.8 mm (12-gauge) and includes two separate serrated trocars of different lengths, providing specimens of 1.6 mm in diameter. A short set (110 mm) of large caliber (8-gauge) is also available for biopsies of more superficial skeletal sites such as peripheral bones and sacroiliac joints.

The MacLarnon needle [32] (Needle Industries, Medical Division, Redditch, Worcs., U.K.). The principle of this needle is similar to that of the Laredo-Bard needle. Two different sets of MacLarnon needles (for lumbar of thoracic spine biopsy) are available.

The Jamshidi trephine needle (Kormed Co., Minneapolis, MN, U.S.A., and Simmho S.A., Strasbourg, France) (Fig. 9). This needle is designed for bone marrow biopsy but can be effective in obtaining a core-biopsy of superficial and flat bones or relatively lytic lesions. Its external cannula (A), which is also used as a cutting trocar, has a conical, beveled tip which helps in retaining the specimen. The Jamshidi needle also has a very practical handle. The cannula is inserted with its stylet (B). Once the bone has been reached, the stylet is removed and the external cannula is advanced into the bone with a rotating movement. The needle is rotated slowly in a circular direction, gradually enlarging the circles to loosen the specimen. The entire needle is then removed with the specimen. With this needle, a new approach of the lesion site is necessary to take additional specimens. Adult (8- or 11-gauge) and pediatric needles (13-gauge) of different diameters are available. They allow removal of specimens of 3, 2, and 1.5 mm bore, respectively. The Jamshidi needle also exists in disposable form.

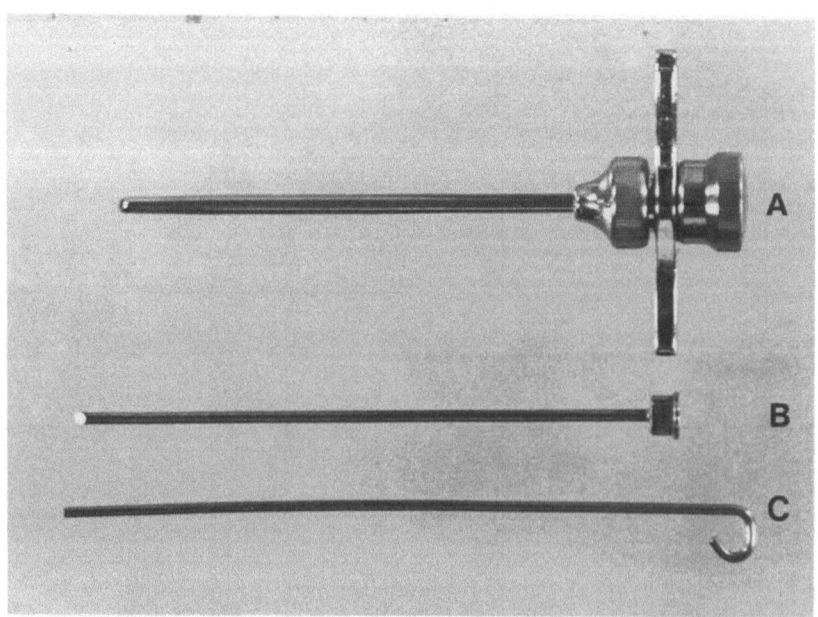

Fig. 9. Jamshidi needle. *A* External cutting cannula; *B* stylet; *C* probe

The Tanzer II trephine needle (Colin-Gentile-Drapier) (Fig. 10). The principle of the Tanzer needle is similar to that of the Jamshidi needle. Adult (7-gauge and 100 mm long) and pediatric needles (8-gauge and 60 mm long) are available. They produce specimens 3.5 mm and 3 mm in diameter, respectively.

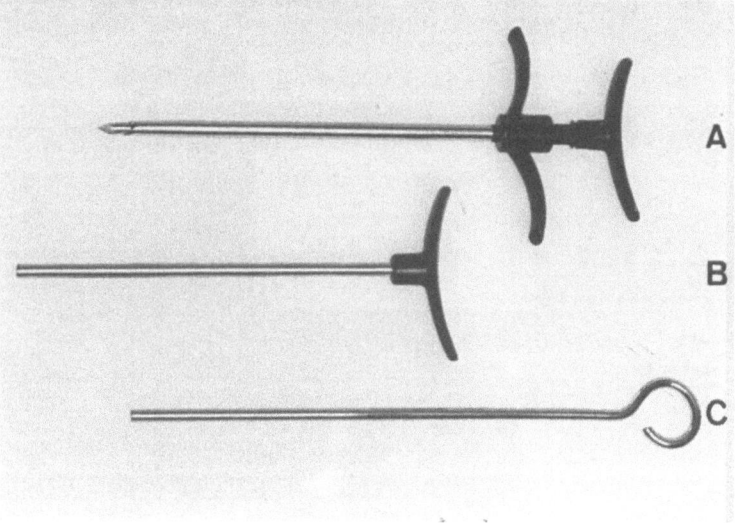

Fig. 10. Tanzer II needle. *A* External cutting cannula with long sharp-pointed stylet; *B* short flat-tip stylet used to push the cutting cannula into the bone; *C* probe

The hand-drill

Some practitioners use a pneumatic drill [6, 38]. Morrison and Deeley use a small pneumatic motor to rotate a hollow needle of 1.5 mm internal diameter at a speed of between 15,000 and 20,000 revolutions per minute, recommended only for lesions in which there is large bone destruction [38]. We have no personal experience with the hand drill. However, it may be useful in making a hole in long bones where the cortex is too thick for trephination and in sclerotic lesions, such as those of the sacroiliac joint. A trephine or a Tru-Cut needle can then easily be passed through the hole [6].

Instruments for soft tissue biopsy

The Tru-Cut needle (Travenol Labs., Deerfield, IL, U.S.A.). This is a very efficient needle for biopsy of soft tissues and osteolytic lesions (Fig. 11). The Tru-Cut needle includes an inner cannula with a 20 mm notch and a 14-gauge outer cutting cannula with a T-shaped handle. The needle is inserted through a skin stab with the inner cannula fully retracted to cover the specimen notch. The needle is advanced until the specimen notch is within the tissue to be biopsied. Without moving the inner cannula, the outer cannula is retracted to expose the specimen notch by pulling outward on the T-shaped handle. The T-shaped handle is then quickly advanced to cut the tissue which has been drawn into the specimen notch. Tru-Cut needles are available in 3 different lengths (75, 114, and 152 mm). These needles are completely removed with each specimen. However, multiple sampling of soft tissue through a single approach, can be achieved by using a 15,2 cm long Tru-Cut needle introduced into either a 9 cm long Jamshidi needle or a 6 cm long Tanzer II needle (see the second chapter of this volume, on biopsies of the synovial membrane). Also available is an aspiration biopsy

17-gauge Tru-Cut needle which allows fluid aspiration and multiple sampling
without removing the outer needle (Fig. 11).

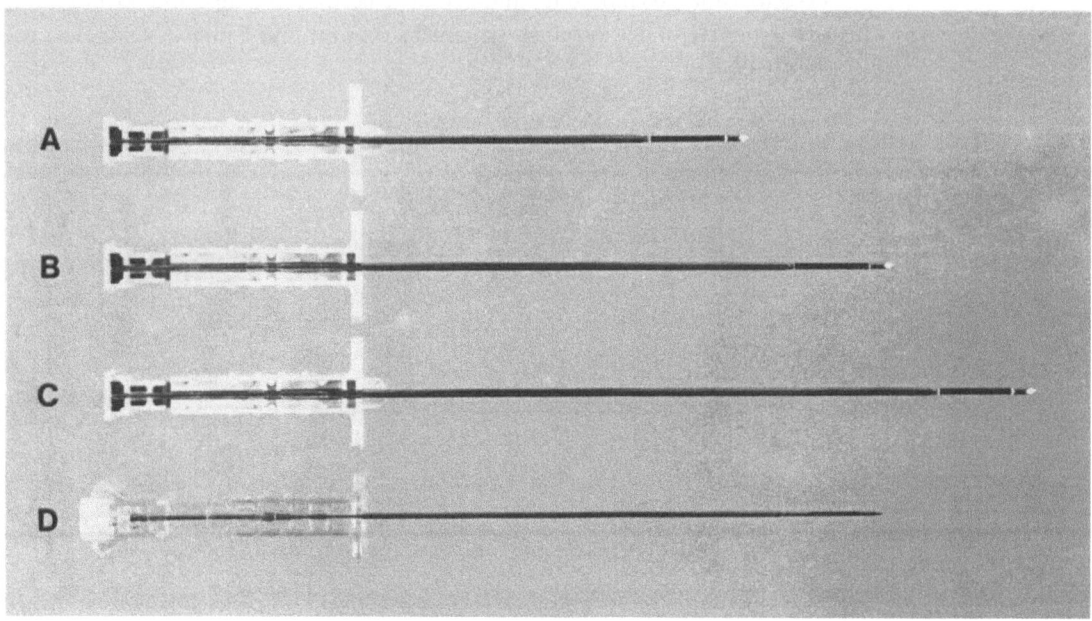

Fig. 11 a

Fig. 11 a. Tru-Cut needles for soft
tissue biopsy. *A* 14-gauge, 75 mm
Tru-Cut needle; *B* 14-gauge,
114 mm Tru-Cut needle; *C* 14-
gauge, 152 mm Tru-Cut needle; *D*
17-gauge, 152 mm aspiration Tru-
Cut needle

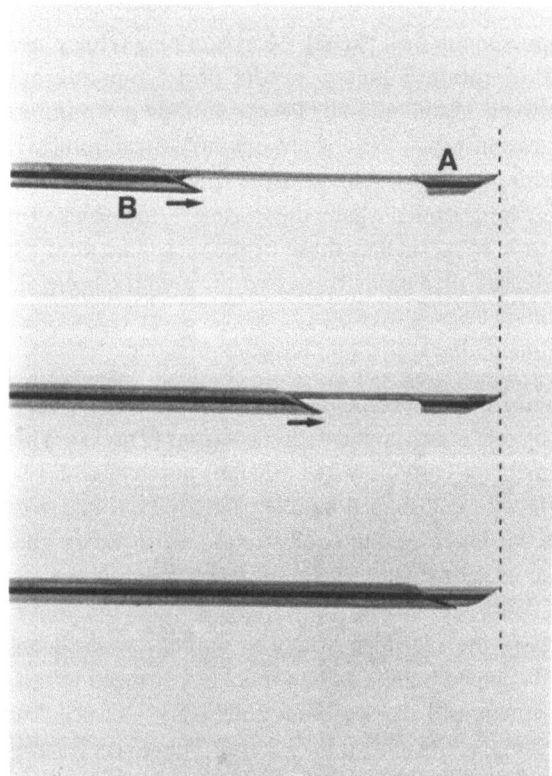

Fig. 11 b

Fig. 11 b. Directions for use of the
Tru-Cut needle: The inner part of
the needle (**A**) is held motionless
while the external cannula (**B**) is
quickly advanced

The Sure-Cut needle (The Sure-Cut needle (TSK Labs., Tokyo, Japan
and Ingenor, Paris, France). A thin 18- or 20-gauge needle is connected to

a syringe which allows sampling of microspecimens of soft tissue and osteolytic lesions (Fig. 12). The needle is inserted and advanced so that its tip
abuts the lesions. Suction is then applied to the syringe by withdrawing and
locking the piston. The needle is then quickly advanced within the lesion
and withdrawn with a to and fro motion. The needle is removed with the
specimen.

Fig. 12. Sure-Cut needle

Aspiration needles. Needles of different sizes are available, ranging from
relatively large needles to Chiba needles. Chiba needles (Cook Co., Bloomington, IN, U.S.A.) are 18- to 23-gauge thin-walled needles usually employed
for the biopsy of viscera (Fig. 13). They can also be effectively used in the
biopsy of soft musculo-skeletal lesions close to vital organs, or to obtain

Fig. 13. Chiba needles

cytologic material after the cortex has been perforated with any of the larger trephine needles. 18- to 20-gauge needles are the most frequently used. The needle with stylet is guided into the lesion under fluoroscopic control. The stylet is removed and a syringe is adapted for tissue aspiration. Rotation of the syringe may allow more adequate biopsy samples [44]. The material is placed on slides suitable for cytology and in laboratory containers for culture and bacteriologic examination.

Preoperative assessment

Prior to percutaneous biopsy, the patient's file should be sufficiently evaluated to determine whether non-invasive procedures might yield the desired diagnosis. Patient hemostasis must be checked in the days before the procedure. In all cases, radiographs in two projections and computed tomography of the lesion, as well as bone scintigraphy, are required prior to biopsy. The most accessible skeletal lesion and the proper anatomic approach are chosen [39] on the basis of all available imaging procedures.

CT is indispensable prior to PB. The usefulness of CT in planning fluoroscopy-guided percutaneous bone biopsies has been stressed by Hardy, Murphy, and Gilula [21]. CT helps locate and characterize skeletal lesions [39] and may influence the choice of both the biopsy site and the needle type. CT scan is especially helpful in determining the best route for biopsy. CT could also reveal additional information which may result in the cancellation of the biopsy [21]. Contrast-enhanced CT may be useful when a highly vascular lesion is suspected. Frank hypervascularization may call for needle aspiration rather than trephine biopsy.

The utilization of radionuclide studies in the localization of suitable biopsy sites must be stressed [44]. Bone scintigraphy may identify additional lesions that are more accessible to closed needle biopsy than the initially detected abnormality. However, closed biopsy of radiographically and scintigraphically positive lesions is more successful than biopsy of lesions that are apparent only on radionuclide scans [44]. In the case of a positive bone scan without roentgenographic confirmation, a CT scan of the positive bone scan area may demonstrate a lesion.

Immediate preparation

Biopsies are performed in the radiology department. Skeletal biopsies are usually done under local anesthesia, with the exception of children and restless patients, who are placed under general anesthesia or heavy sedation. The advantage of using local anesthesia is that the patient can communicate radiating pain due to nerve impingement so that the needle can be repositioned. Hospitalization for at least 24 hours is required following spinal biopsy. In most other locations, the biopsy can be done on an outpatient basis. The patient should not eat on the morning of the examination. Both a sedative (Hydroxyzine, 100 mg for an adult, one hour before) and pain medication are given prior to the examination. An intravenous catheter is placed and a saline solution is administered to keep the line open.

Technique of bone trephine biopsy under fluoroscopic guidance

General considerations

The examination is performed using either single plane fluoroscopy, single plane fluoroscopy with a second cross table X-ray tube, or biplane fluoroscopy, depending on the biopsy site (Table 3). The patient must be comfortable and stable. Radiolucent cushions are used for this purpose. The advantages and conditions of the procedure must be carefully explained to him at this time. In order to determine the point of skin puncture and the precise needle approach, a metallic marker such as a ruler or rod is placed upon the patient's side. This metallic marker simulates the approaching needle on the image intensifier screen. Its position is corrected until the optimal approach to the bone lesion, avoiding bone obstacles and vital organs, is determined. The best needle for a particular biopsy should be chosen according to the location and nature of the lesion (Table 3) [39].

Technique of approach according to the biopsy site

Lumbar spine

All lumbar vertebral bodies and discs are accessible by PB. The approach to lumbar vertebrae is posterolateral with the needle inserted at 7 to 10 centimeters from the midline of the spinous processes, depending on the biopsy level and the patient's build (Fig. 14). The point of skin puncture

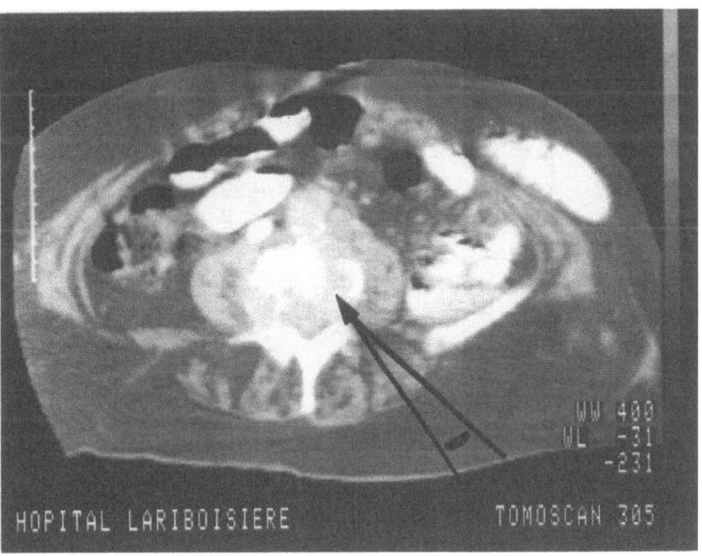

Fig. 14. Posterolateral approach for lumbar spine PB. Point of skin puncture is at 7 to 9 cm from the midline. The lumbar spine is approached at an angle of 40 to 60 degrees with the sagittal plane

must be close to 7 cm from the midline of biopsies of the upper lumbar spine and the thoraco-lumbar junction, and close to 9–10 cm for biopsies of the lower lumbar spine and the lumbo-sacral junction (Fig. 15). The lumbar spine is approached at an angle of 40 to 60 degrees with the sagittal plane. Either a right- or left-sided approach can be used, depending on the location of the lesion.

The patient is placed on his side on the radiolucent table. Knees, ankles and arms are protected by adapted positioning and foam cushions. A ra-

Table 3. Technique of percutaneous biopsy according to the biopsy site

Biopsy site	Type of anesthesia required	Approach	Guidance (minimal radiologic equipment)	Type of biopsy	Biopsy instruments
C1–C2	general	anterior, transoral	biplane fluoroscopy	trephine biopsy	Laredo-Bard (thoracic), Ackerman
C3–D1	local	antero-lateral	single plane fluoroscopy	aspiration biopsy	18 G needle, Sure cut
D2–D11	local	postero-lateral (5 cm from the midline)	single plane fluoroscopy + cross-table tube	trephine biopsy	Laredo-Bard (thoracic), Harlow Wood, MacLarnon, Ackerman
D12–S1	local	postero-lateral (7–11 cm from the midline)	single plane fluoroscopy + cross-table tube	trephine biopsy	Laredo-Bard (lumbar), Harlow Wood, MacLarnon, Craig
Sacroiliac joint	local	postero-lateral (through the iliac bone)	single plane fluoroscopy	trephine biopsy	Laredo-Bard (sacroiliac)
Long bones	local	perpendicular to bone cortex	single plane fluoroscopy	trephine biopsy	Tanzer, Jamshidi, Craig, Laredo-Bard (sacroiliac)
Small, deep and lytic bone lesions; soft tissue lesions	local	depends on biopsy site	CT scan	aspiration or trephine biopsy	Sure-cut, Tru-cut, Laredo-Bard (thoracic), Ackerman

diolucent block is placed beneath the flank in order to correct lateral deviation of the spine due to lateral position. During the first part of the approach, the X-ray beam must be strictly vertical, and orthogonal to the X-ray table. The image intensifier is positioned over the vertebra to be biopsied. Perfect lateral positioning of the patient is carefully checked so that a true lateral view of the vertebrae is obtained on the intensifier screen. Strict lateral positioning is crucial to a correct approach of the lumbar spine.

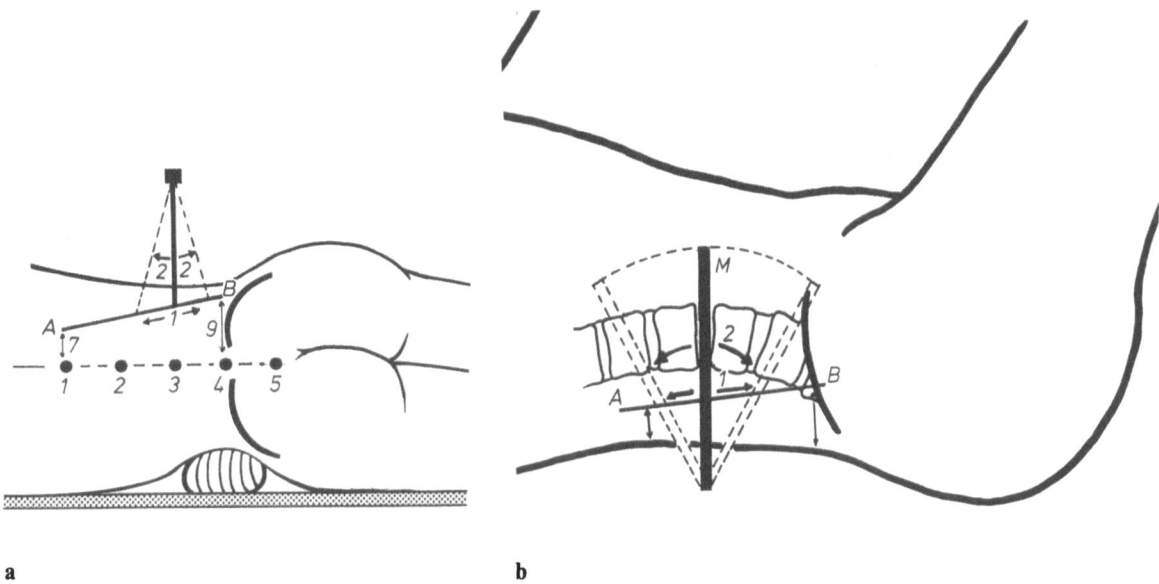

a   b

Fig. 15. Approach to the lumbar spine. Lateral (**a**) and upper (**b**) views of the patient placed on his side. The point of skin puncture is at 7 to 9 cm from the midline (line A–B). Both level of needle insertion (*1*) and angle of cephalad or caudal approach (*2*) are determined using a metallic ruler (*M*) under fluoroscopic control

The approach to the lumbar spine must avoid bony obstacles, namely, the transverse processes at all levels, and the 12th rib or iliac crest at the lumbar spine extremities. The exact level of needle insertion and the angle of cephalad or caudal approach are determined using a metallic ruler placed on the patient's side in order to simulate the approaching needle on the lateral view fluoroscopic screen. The ruler is tilted so that it projects as desired on the lumbar spine and avoids bony obstacles, especially the transverse processes (Figs. 15 and 16). In disc biopsies the approach must be as parallel as possible to the vertebral endplates (Fig. 17). Both disc and vertebral endplates are sampled in PB performed for infectious discitis (Fig. 16). On the other hand, vertebral bodies must be biopsied with a cephalad or caudal oblique approach in order to sample the entire vertebral body (Fig. 18). L 2, L 3, and L 4 vertebral bodies can be biopsied using either an ascending or descending approach, depending on the lesion site (Fig. 19). T 12 and L 1 are approached through an ascending route (Fig. 20) while L 5 and S 1 are approached in a descending direction.

After both the needle insertion point and the optimal approach have been determined, a surgical field is prepared and the skin and superficial planes are anesthetized. The needle is inserted and advanced under fluoroscopic control toward the lumbar spine at an angle of 40 to 60 degrees with the sagittal plane, depending on patient build and lesion site. The needle

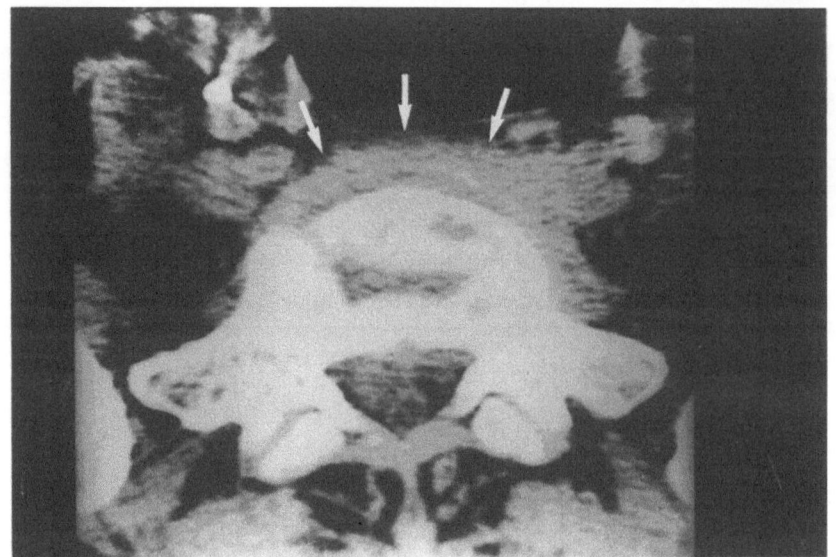

a

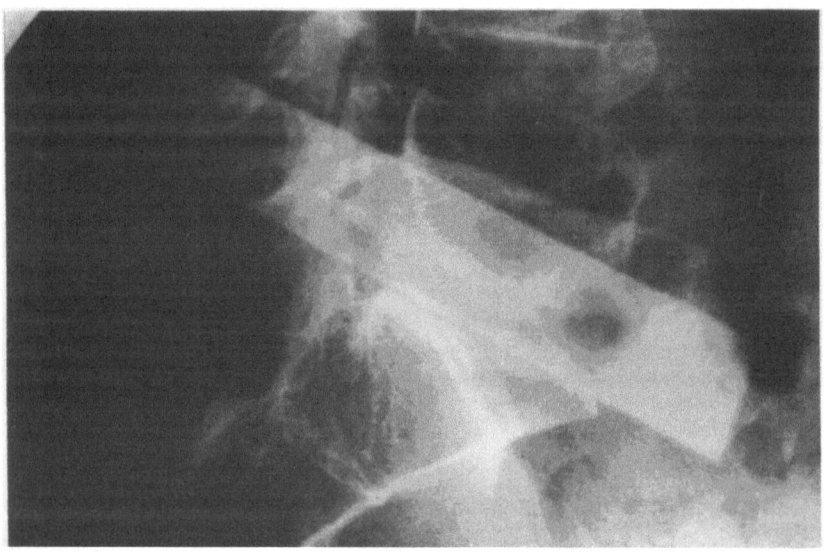

b

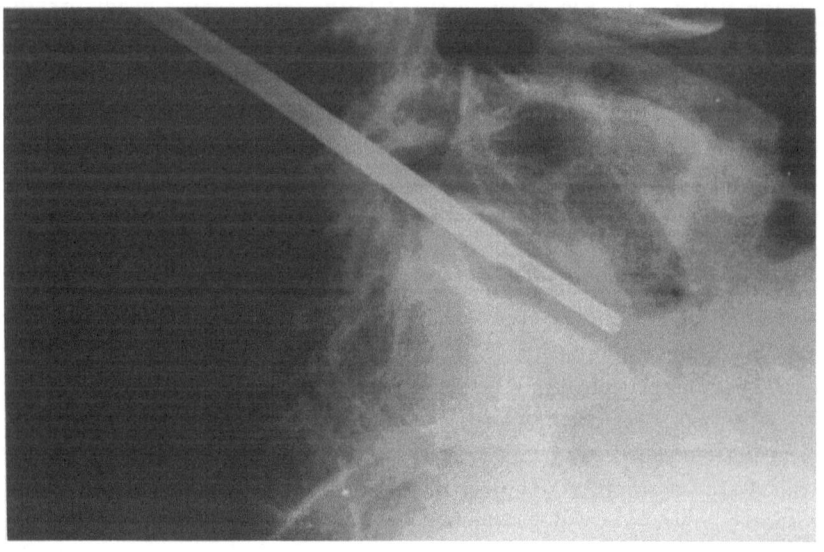

c

Fig. 16. Spinal infection: Post-operative contamination at L 5-S 1 interspace. **a** CT scan demonstrates a prevertebral abcess (arrows). **b** A metallic ruler is used to determine the appropriate level of puncture and caudal inclination of the approach. Lateral (**c**) and AP (**d**) roentgenographic controls during biopsy. Note that the approach parallels the disc space

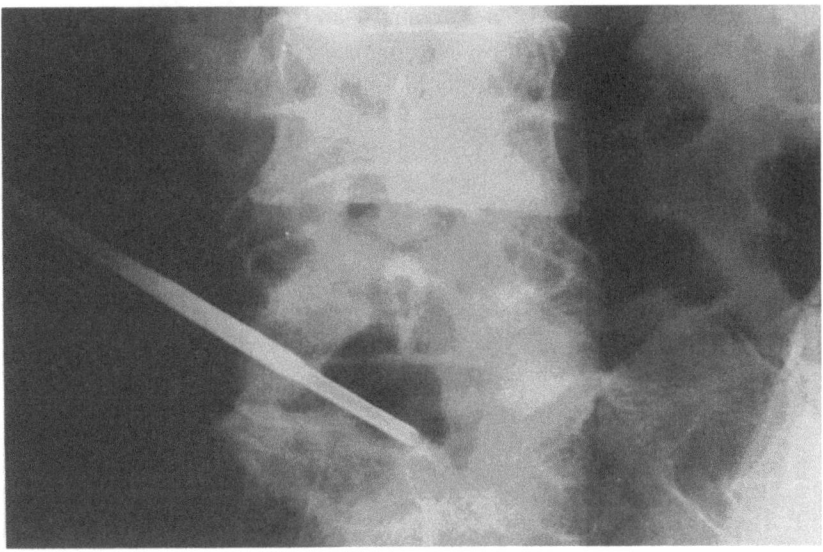

Fig. 16 **d**

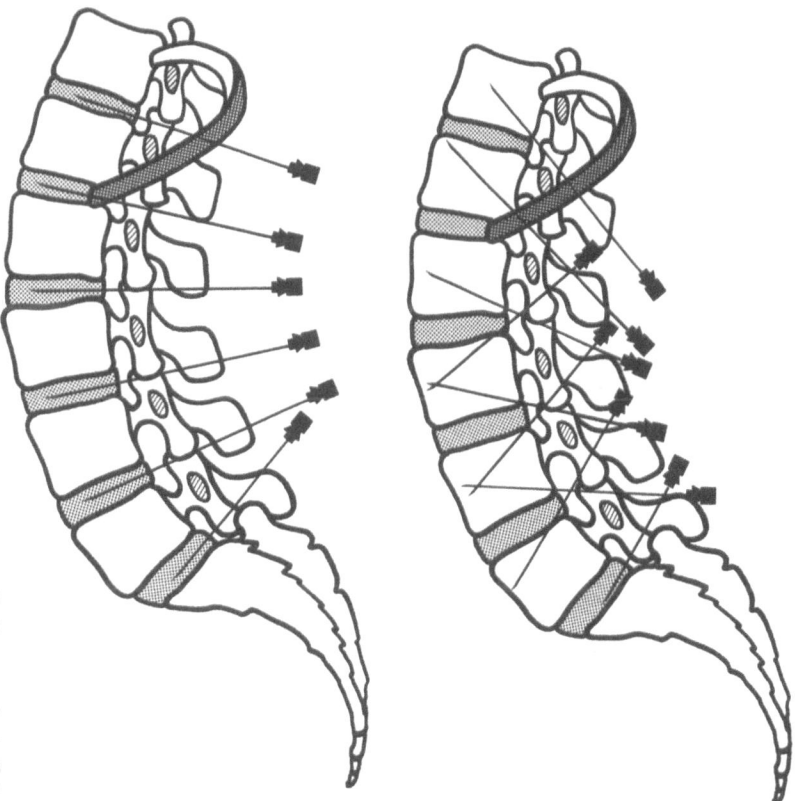

Fig. 17. Approach for lumbar disc biopsy should be parallel to the disc space

Fig. 18. Approach for lumbar vertebral body biopsy is oblique in order to avoid transverse processes and to sample the entire vertebral body

Fig. 17   Fig. 18

is also inclined in a cephalad or caudal direction, determined as mentioned above. The nature of any bony obstacle encountered can be determined on the lateral view fluoroscopy screen. Bony impingement at the level of the pedicles is probably due to transverse processes. In such cases, the needle must be moved in a caudal or cephalad direction. However, if needle place-

ment and approach have been determined and performed in accordance
with the above recommendations, the transverse processes are easily avoided.
Obstruction at the level of the lower half of the vertebral body or opposite

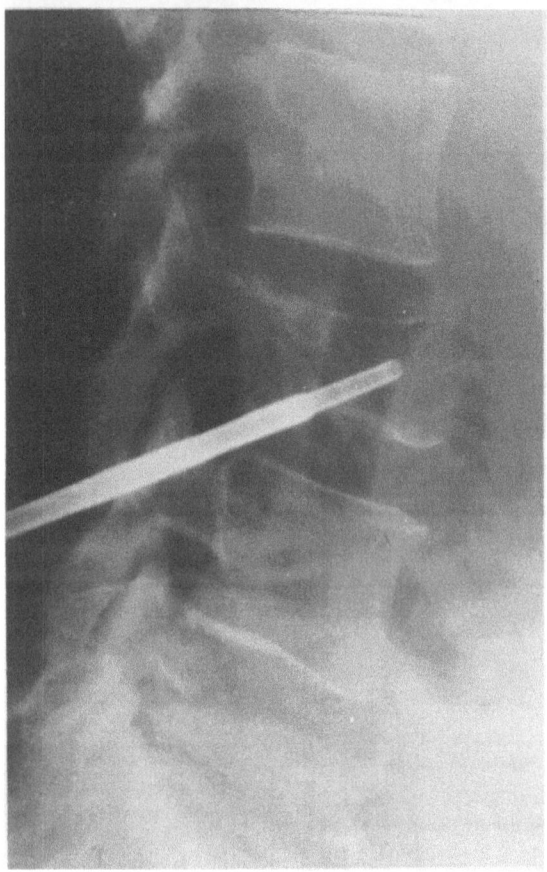

Fig. 19

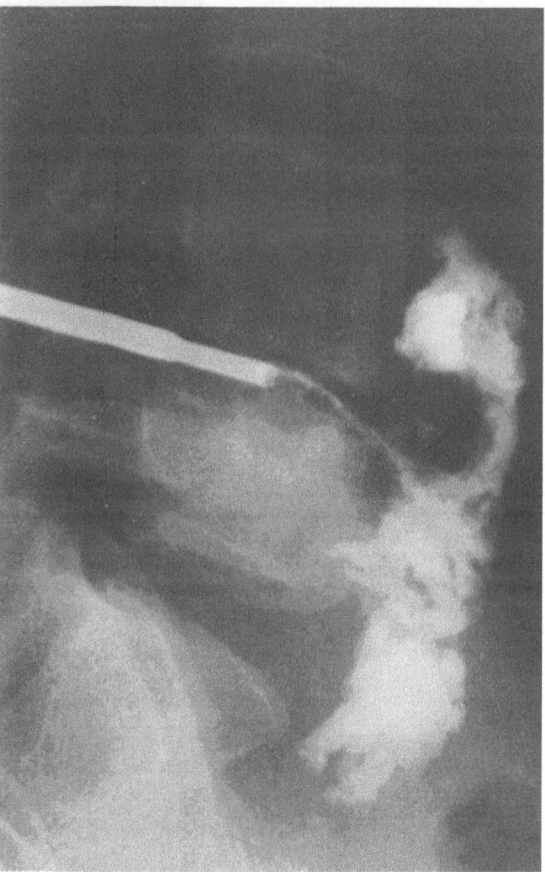

Fig. 21

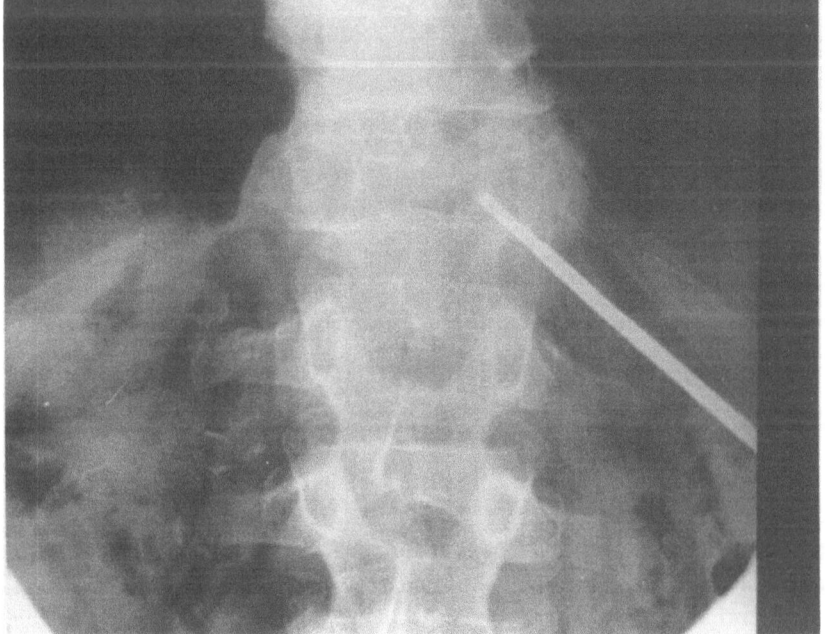

Fig. 20

Fig. 19. Biopsy of L 3 using
an ascending route in a patient with vertebral collapses
of L 3 and L 5. Amyloid deposits were found upon histopathological examination

Fig. 20. Ascending approach
to T 12 vertebral collapse. Biopsy results and clinical evolution confirmed osteoporosis

Fig. 21. Disc biopsy in spinal
tuberculosis at L 4-5 level with
opacification of a prevertebral
abcess

the disc space is usually due to the facet joint. In this event, the needle is withdrawn 2–3 cm and then advanced in a more sagittal direction. In the case of disc biopsy, the direction of the central ray can be tilted to better visualize the disc space after the needle has passed the line joining the transverse processes. In most cases, contact with the spine should be obtained when the needle reaches the posterior third of the vertebral body or disc space as shown on a lateral view. However, this depends on the lesion site. When proper placement of the needle has been verified on both antero-posterior and lateral views, the periosteum is carefully anesthetized and the trephine needle is inserted with the same approach used to perform the anesthesia. Throughout the procedure, exact positioning of the needles must be checked by using either fluoroscopic control or roentgenograms. Para-vertebral abcesses can be opacified at the end of the procedure (Fig. 21).

Thoracic spine

In order to improve safety, we have developed a well-codified technique for performing **PBPB** of the thoracic spine under fluroscopic guidance [29, 30]. Biopsy under biplane fluoroscopic guidance is recommended in those centers where it is available. However, the biopsy technique is the same with both single and biplane fluoroscopy.

The spine is reached via an oblique, posterolateral, intercostal approach at an angle 35° from the patient's sagittal plane, as recommended by Ottolenghi [42] (Fig. 22). We perform thoracic spine biopsies with the Laredo-

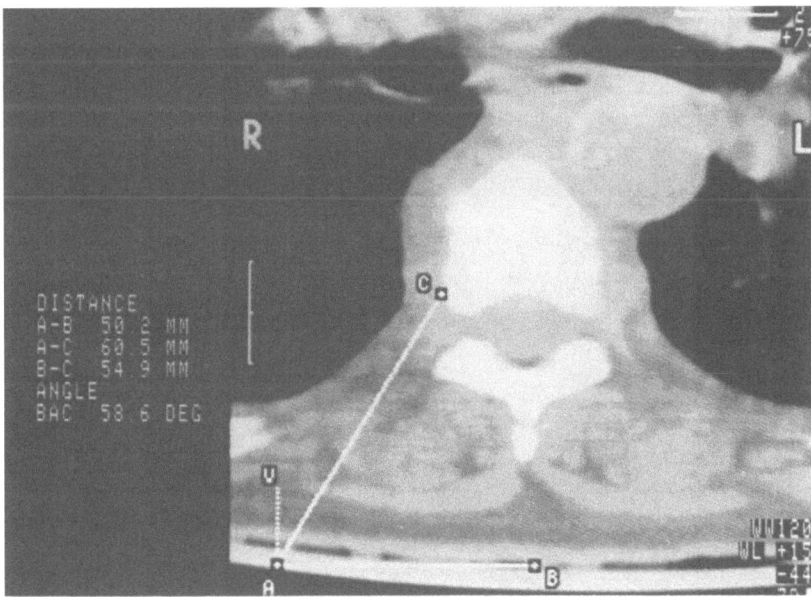

Fig. 22. Posterolateral approach to the thoracic spine drawn on a CT picture. Note that puncture site (*A*) is 5 cm from the midline, angle of approach (*BAC*) close to 55 degrees from the frontal plane and needle path between pleura and spinal canal

Bard trephine needle. Preoperative radiographs and CT scans determine the side of approach. The existence of a paravertebral abcess or mass facilitates the puncture. The patient is placed in a 35° prone-oblique (procubitus) position (Fig. 23). The puncture point is located at the level of the lesion, 4.5–5.0 cm from the midline, as indicated by the spinous process. A radi-opaque mark is placed at this point, and a radiographic examination of the

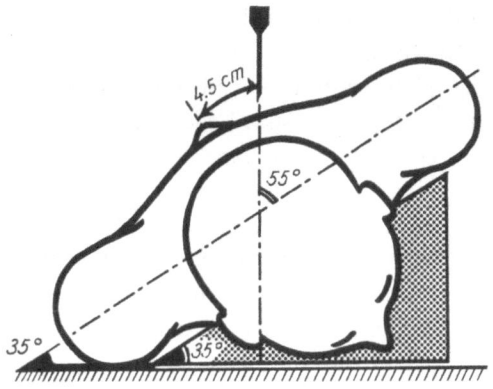

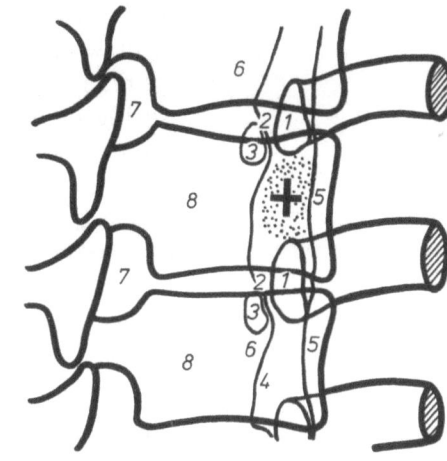

Fig. 23

Fig. 24

Fig. 23. Patient in 35° oblique procubitus position

Fig. 24. Diagram of anatomical relations in 35° oblique procubitus position. *1* Rib head, *2* costovertebral joint, *3* transverse process, *4* external edge of the articular process, *5* line of pleural reflection, *6* vertebral body, *7* contralateral lamina, *8* spinal canal. The puncture area is indicated by the dotted area

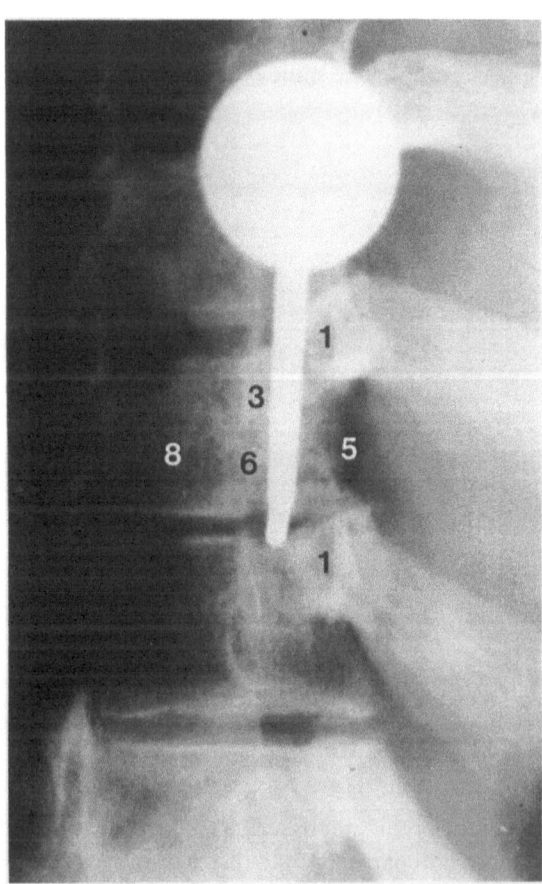

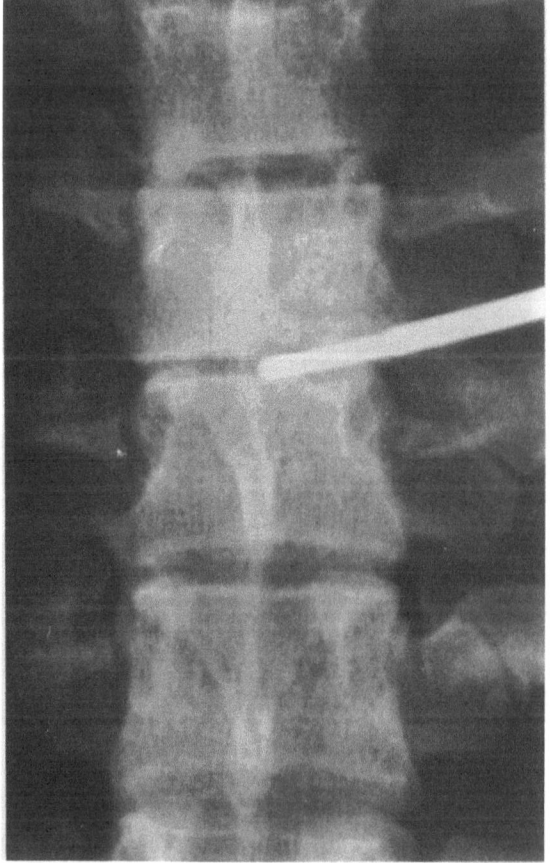

a   b

Fig. 25. Disc biopsy of a spinal tuberculosis at T 6-7. A thirty-five degree oblique view (**a**) demonstrates the same anatomic relations as those in Fig. 24. **b** AP view

thoracic lesion is made in this position. On oblique views, the heads (dorsal extremities) of the adjacent ribs define the direction of puncture (Fig. 24). The vertebral body or disc lesion can be reached through a special area (the "puncture area") bordered by the heads of the ribs, the lateral margin of the articular processes, and the lateral margin of the vertebral body or disc (Figs. 24 and 25). The pleural space and spinal canal are clearly visible and can thus be avoided during the procedure. Patient positioning and the level of the radiopaque mark are adjusted to bring the mark directly over the puncture area. Thus, the lesion lies directly below the puncture area. A 20-gauge needle is used initially to anesthetize the superficial planes, and the thin needle of the trephine set is then introduced through a small skin incision. The needle is advanced under fluoroscopic guidance and is vertically oriented to the puncture area in the direction of the X-ray beam. At the same time, this thin needle is used to anesthetize the deeper planes. The vertebral body or disc is usually reached at a depth of 6–7 cm [42]. After checking the fluoroscopic screen for good positioning of the needle tip and absence of bleeding, the thin needle is replaced by the trephine needle, which is introduced in the same track with the aid of a guide-wire as depicted in Fig. 8. When the trephine needle abuts the lesion, the patient is moved to a strict prone position. Two radiographs (a frontal view and a cross-table lateral view) are obtained to verify correct positioning of the trephine needle. Biopsies are then performed using the cutting cannula, and a new radiograph is obtained. Pain is usually mild or absent. If there is severe pain, a technical

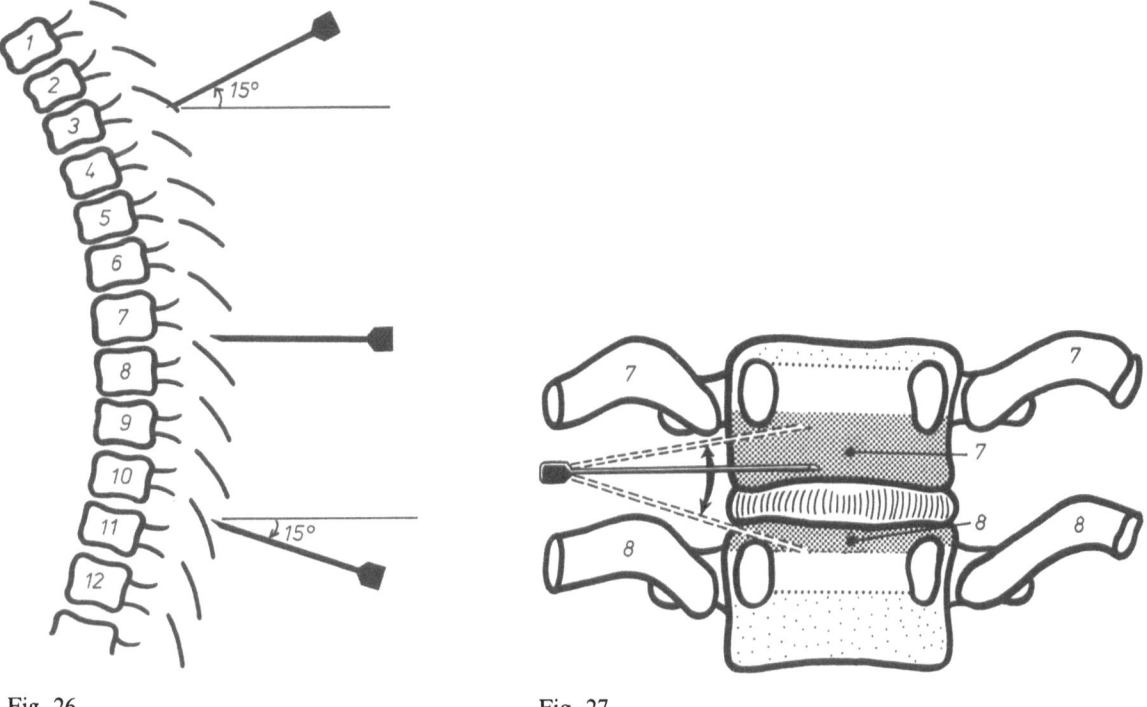

Fig. 26   Fig. 27

Fig. 26. Arrows show changes in angle of approach that must be made for various vertebral levels

Fig. 27. Shaded and striped areas show portions of disc and vertebrae that can be reached through an intervertebral approach. As an example, the T-7 and T-8 vertebrae and the intervening intervertebral disc are shown

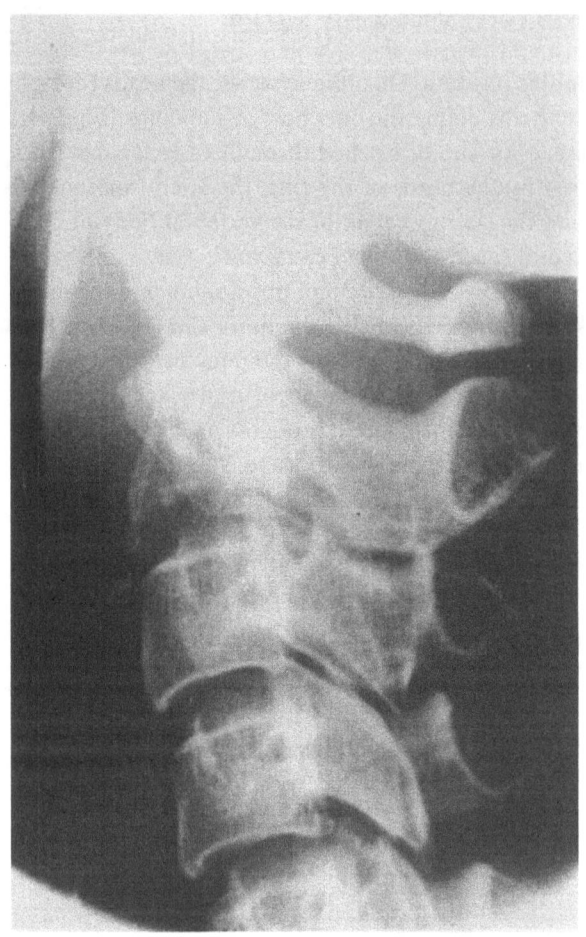

a

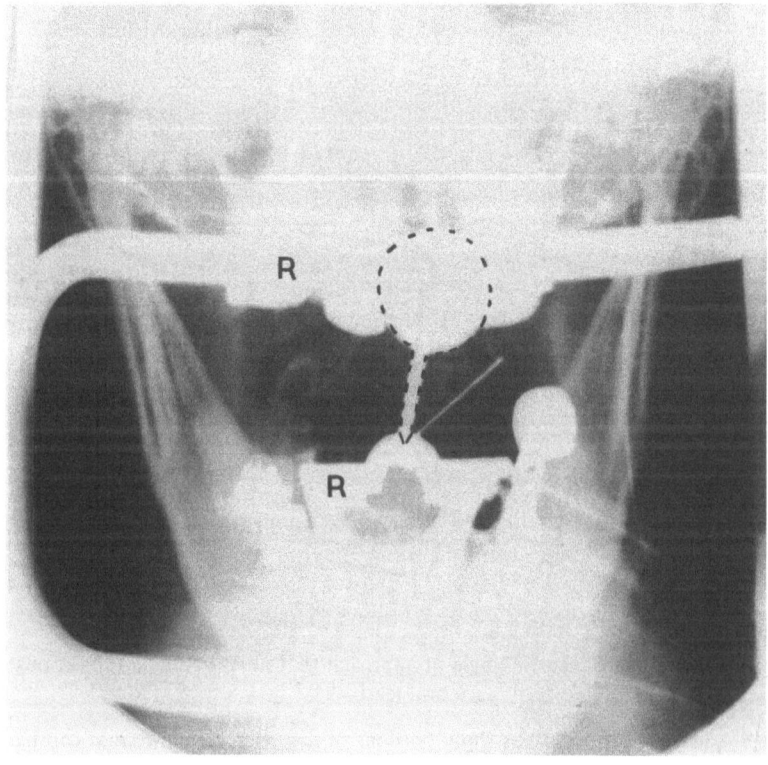

b

Fig. 28. Osteolysis of the anteroinferior aspect of C2 vertebral body (**a**). PB through an anterior transoral approach (**b**) using a retractor (*R*) and an Ackerman needle (dotted line) proves this lesion to be spinal tuberculosis

error in needle positioning must be sought. The entire procedure usually takes 60–90 minutes. A chest radiograph is obtained immediately after the procedure in order to verify the absence of pneumothorax. Vertebral bodies and intervertebral discs from T 2 to T 12 can be biopsied with this technique (Figs. 26 and 27).

The prone-oblique position appears to be the safest and easiest way to approach the thoracic spine, and with the patient in this position, single plane fluoroscopy is sufficient to guide puncture direction. Subsequently, when the patient is moved to the prone position for the biopsy itself, biplane fluoroscopy is more useful. However, AP and lateral radiographic controls are still necessary for verification of needle position, and for legal purposes.

Cervical spine

C 1 and C 2 can be reached through an anterior transoral approach under biplane fluoroscopic guidance and general anesthesia. The approach is greatly facilitated by the use of a retractor (Fig. 28). In our hospital, we perform these biopsies in collaboration with an ear, nose and throat surgeon. We have no personal experience with large needle trephine biopsy of the cervical spine from C 3 to T 1.

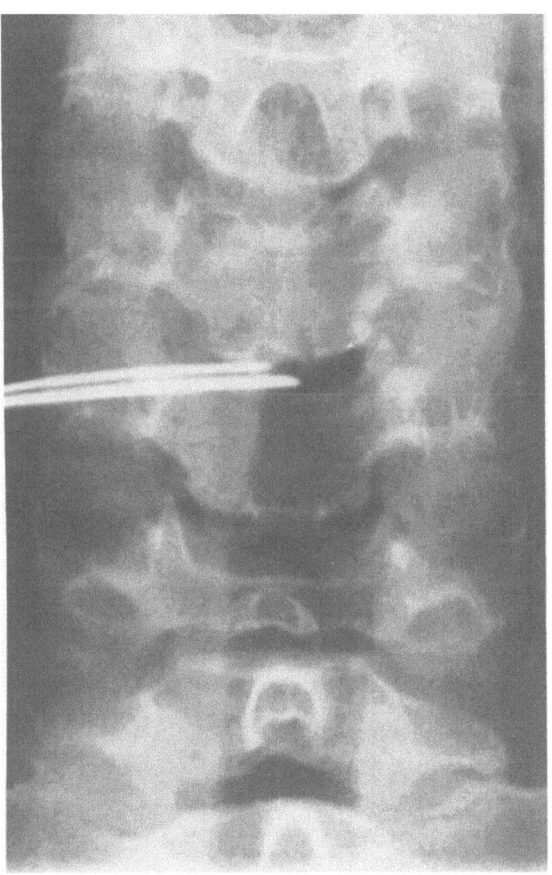
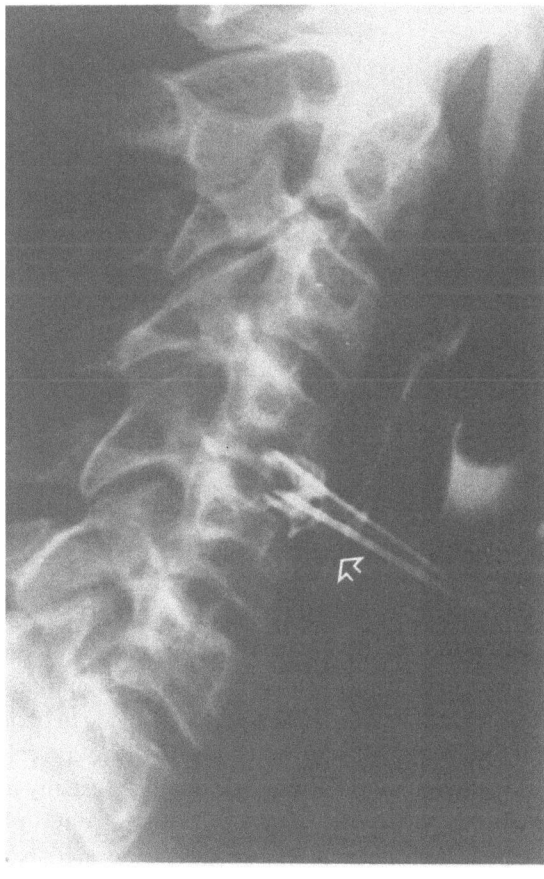

a

b

Fig. 29. Needle aspiration in a postintubation C 4-5 infectious discitis due to streptococcus A. Rinsing the disc space with saline solution may be easier using two parallel needles (**a**). **b** Injection of contrast demonstrates fistulization of the prevertebral abcess into the hypopharynx (arrow)

In this area, infectious discitis and lytic metastases can be sampled either by aspiration biopsy using a 17- or 18-gauge needle, or by a microbiopsy technique with a Sure-Cut needle (Fig. 29). The approach is anterolateral between the larynx and carotid artery, or lateral, behind the large vessels, at the posterior border of the sterno-cleido-mastoid muscle (Fig. 30). Sclerotic lesions cannot be sampled with this technique.

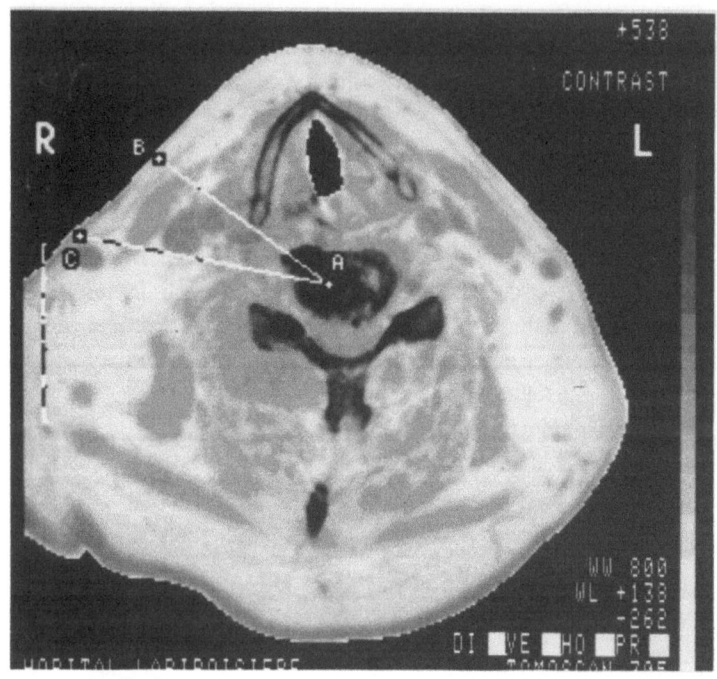

Fig. 30. Approach to the cervical spine is anterolateral (*B*) before carotid artery or lateral (*C*) behind the large vessels

Sacroiliac joint

Chevrot et al. [5] and Vinceneux et al. [52] have independently reported a new approach for the biopsy of the sacroiliac joint under fluoroscopic guidance. The sacroiliac joint is approached at right angles via a posterolateral route through the gluteal muscles and iliac bone. With this approach, cartilaginous portion of the sacroilic joint can be biopsied. The inferior portion of the sacroiliac joint is usually approached. At this level, the thickness of the iliac bone to cross prior to penetrating the sacroiliac joint is much less than at higher level. Either the short Laredo-Bard or the Mazabraud needle can be used. The exact technique of approach is determined on the basis of the preoperative CT scan using several measurement obtained on a CT slice located at one centimeter from the inferior margin of the joint and performed without inclination of the gantry (Fig. 31). The ideal approach is sketched on the CT picture and three measurements are calculated: the exact distance of the puncture point from the midline (usually around 10 cm) the angle of approach (usually 40 degrees from the frontal plane) and the point where the iliac bone must be reached (usually located in alignment with the external margin of the joint on an AP view) (Fig. 31). The patient is then placed prone on the radiolucent table. Single plane fluoroscopy is sufficient to perform the biopsy. A metallic ruler placed upon the patient is used to simulate the approaching needle and to determine the level of puncture (Fig. 31 c). The approach is horizontal at 1 cm above the inferior margin of

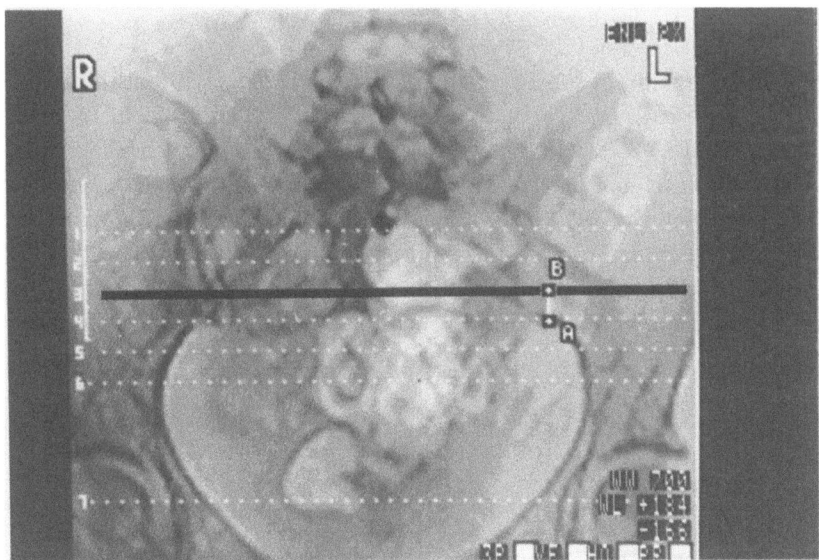

a

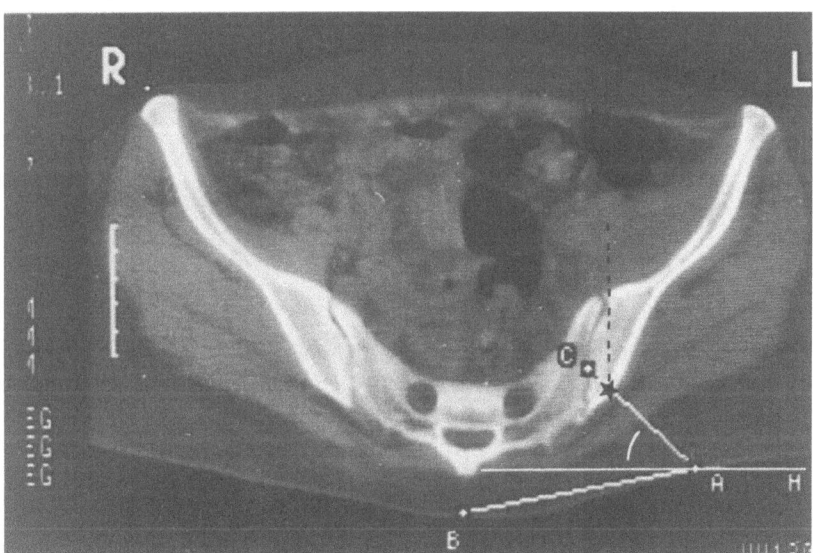

b

Fig. 31. Radiologic measurements prior to PB of the sacroiliac joint through a posterolateral approach. **a** Preoperative CT scan. Selection of a CT slice located at 1 cm from the inferior margin of the joint. **b** CT picture obtained without inclination of the gantry. Three measurements are calculated: distance of puncture point from the midline $(A - B)$ (approximately 10 cm), angle of approach (CAH) (approximately 40° to the frontal plane) and point where the needle must reach the iliac bone ($\star$) which is generally in alignment with the external margin of the joint on the AP view (dotted line). **c** Metallic ruler simulating the approaching needle used to determine level of puncture. Point of skin puncture ($\triangleleft$) and point where the needle must reach the iliac bone ($\blacktriangleleft$) are shown

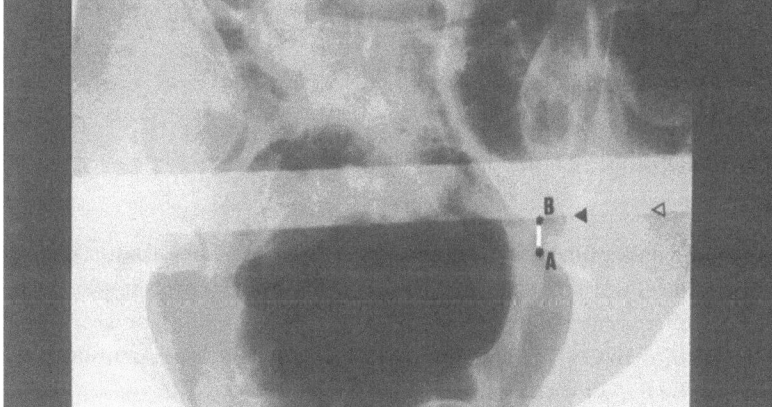

c

the sacroiliac joint, as shown on an AP view performed without tilting the X-ray beam (Fig. 31 c). The needle is inserted at 9–11 cm from the midline at an angle of 30°–40° with the sagittal plan and advanced to the iliac bone, which is reached immediately laterally to the external margin of the sacroiliac joint as located by CT (Fig. 31 b). The trephine is advanced 1–2 cm into the bone and the patient is moved to a prone oblique position by elevating his opposite side in order to control the exact position of the needle tip in relation to the joint space (Figs. 32 and 33). The trephine is then advanced

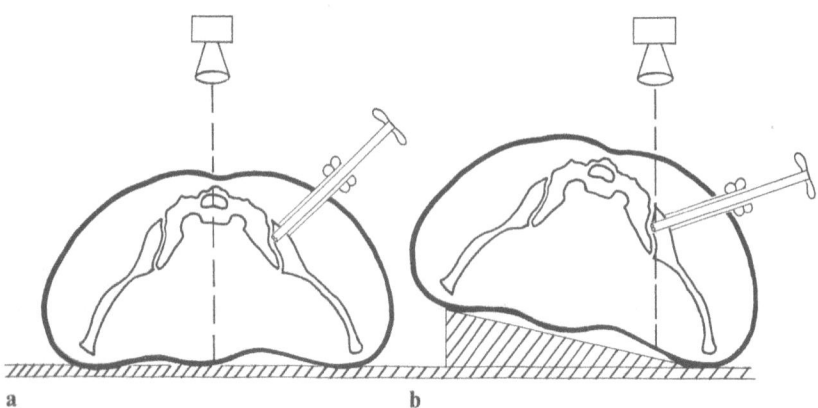

a   b

Fig. 32. Biopsy of the sacroiliac joint. **a** Approach and biopsy in strict prone position. **b** 15–20° prone oblique position to control needle placement

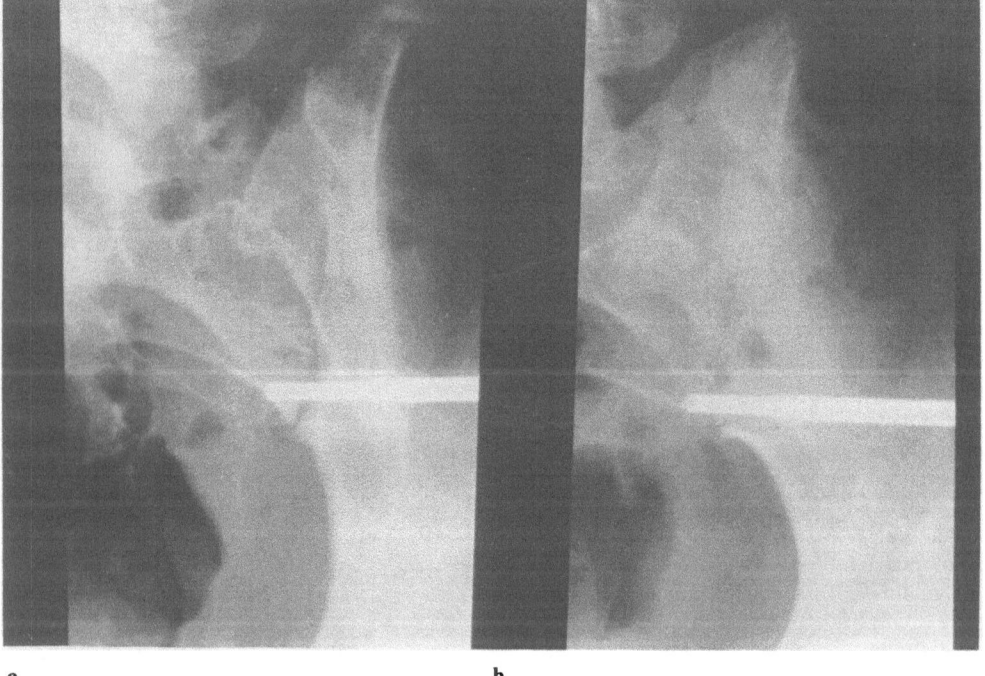

a   b

Fig. 33. PB of the sacroiliac joint. **a** 15° Prone oblique and **b** frontal views

into the bone under fluoroscopy guidance until it penetrates about 1 cm into the sacral margin of the joint (Fig. 33). To obtain adequate samples of the entire sacroiliac joint, the serrated trocar of a lumbar trephine set may be introduced into the external cannula (Fig. 33). This allows sampling of a longer specimen. At the end of the procedure, an arthrography is performed to verify that the joint space has been sampled (Figs. 34 and 35).

Simple aspiration of the sacroiliac joint using a thin needle can be also performed through a direct posterior approach as described by Hendrix et al. [22]. The patient is placed in a 10- to 30-degree prone oblique position by raising the side opposite the puncture. This allows direct entry into the caudal part of the joint with a vertically-oriented needle [22].

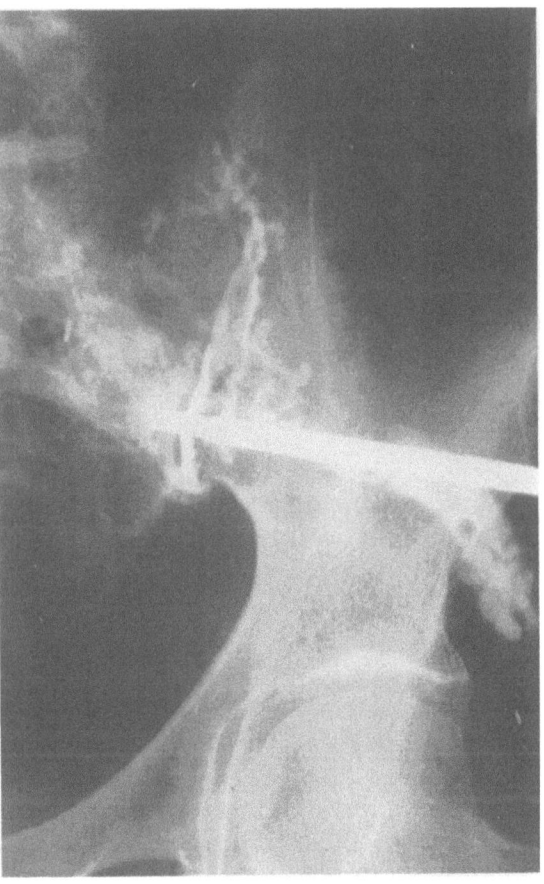

Fig. 34. Arthrography during PB of the sacroiliac joint

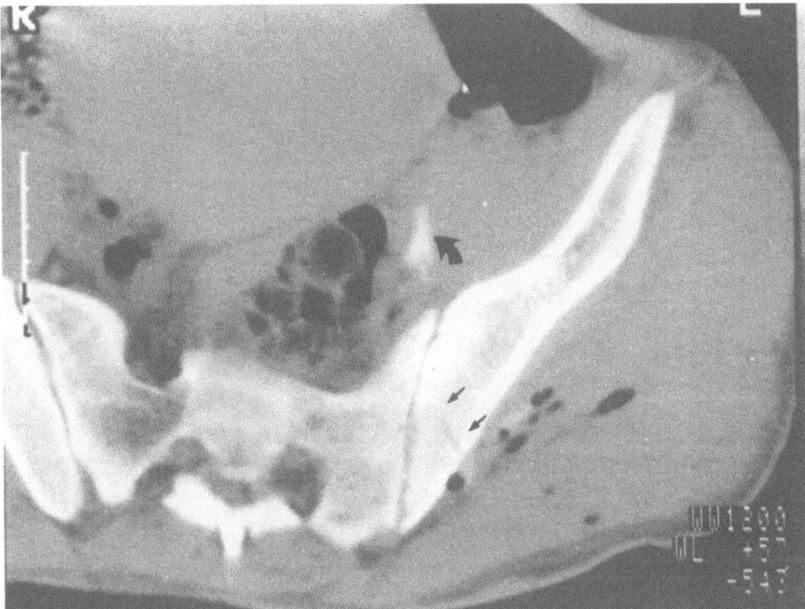

Fig. 35. CT scan control following PB olf thc sacroiliac joint. Note bore area within the iliac bone (small arrows) and contrast media within anterior abcess (large curved arrow)

Peripheral and flat bones

Most peripheral bones can be biopsied. A CT scan study and good knowledge of anatomy are useful in selecting the approach appropriate to the lesion site. For the biopsy of long bones, the needle is generally introduced in a trajectory perpendicular to the cortex of the bone to prevent it from sliding

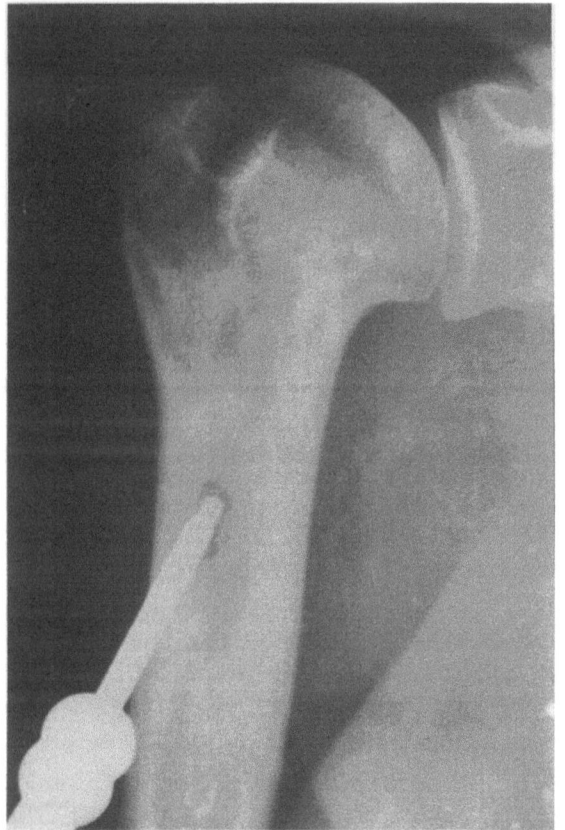

Fig. 36

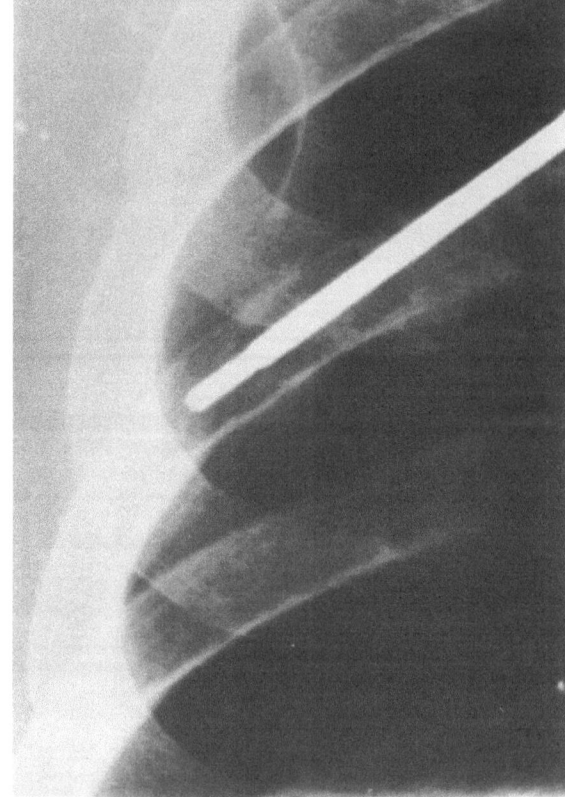

Fig. 37

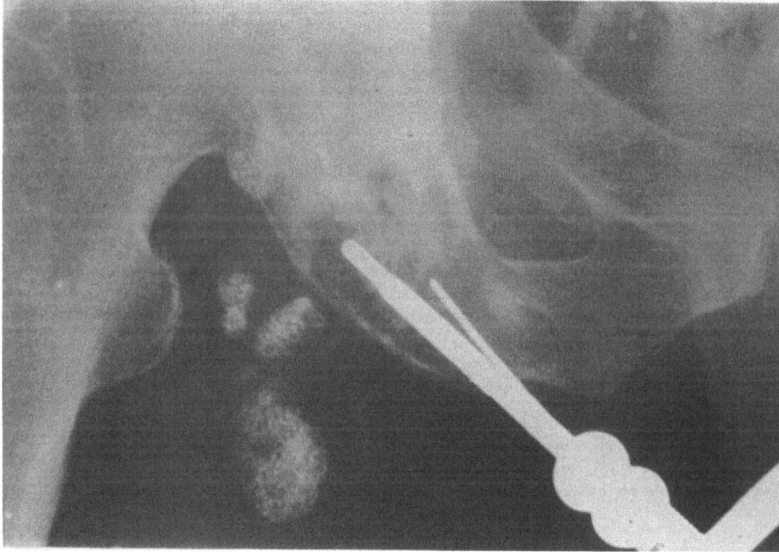

Fig. 38

Fig. 36. PB in humeral osteomyelitis (staphylococcus aureus)

Fig. 37. PB in rib metastasis. Approach is parallel to the axis of the rib

Fig. 38. PB of the ischium (Hodgkin's disease)

off the round cortex (Fig. 36). Roughening the periosteum with the edge of the sharp pointed stylet helps the needle cut into the bone [39]. The short Laredo-Bard needle or Ackerman or Craig needle is used first to obtain a core of cortical bone. This allows the passage of a trephine needle through the cortical defect into the marrow for aspiration. Penetration of the medullar cavity of long bones with the trephine is usually very painful, especially in the case of osteomyelitis, and may require additional anesthesia.

In the cases of flat bones such as the scapula, ribs (Fig. 37), and iliac bone (Fig. 38) an oblique approach is preferred in order to sample the maximum amount of bone possible and to avoid damage to the underlying structures [12]. In the case of biopsy of osteolytic lesions of the ribs, CT rather than fluoroscopic guidance should be used.

Sclerotic lesions

In sclerotic lesions, advancement of the trephine needle may become very difficult after only a few millimeters (Fig. 39). In such cases, excessive pressure on the trephine should be avoided by withdrawing the needle, removing the plug of cortical bone and replacing the needle in the same hole to continue the biopsy. This process may be repeated several times if necessary. It greatly facilitates the biopsy of sclerotic lesions such as chronic inflammatory processes involving the sacroiliac joint and the long bones. A hand-drill is also used by some authors to cut a bore area within the sclerotic bone in order

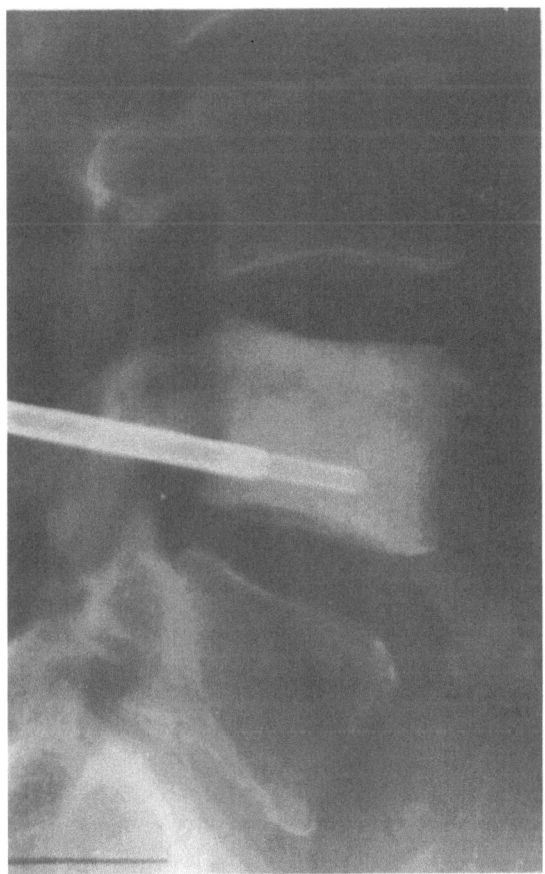

Fig. 39. PB in sclerosing metastasis from prostatic carcinoma

to create a passage for the trephine [6]. Preoperative CT scan evaluation is especially useful in sclerotic lesions. CT helps in locating lytic areas to biopsy within bone sclerosis.

Percutaneous biopsy under CT scan guidance

Computed tomography is useful in the assessment of bone lesions prior to biopsy. It may also be useful in the guidance of skeletal biopsy. In most cases of large lesions well-demonstrated on conventional X-rays, fluoroscopy is preferred to CT scan for the guidance of the biopsy procedure. In these cases, percutaneous biopsy under fluoroscopic guidance is most rapidly carried out and is more comfortable for the patient. However, CT guidance is of particular value in the biopsy of small lesions situated close to vital organs, such as lytic lesions of the vertebral neural arch (Fig. 40) and deep soft tissue lesions (Fig. 41). In 1981, Adapon et al. reported 22 cases of small-needle biopsy of the spine under CT guidance. Ten of these were in the thoracic region. Two of their biopsy specimens produced insufficient material for diagnosis [2]. Mick and Zinreich have reported six CT-guided biopsies of the thoracic vertebrae without complications using a Craig needle [36].

The patient is placed prone on the movable CT table and a digital radiograph (scout-view) is obtained (Fig. 42 a). The level of the lesion is identified and several cross-section slices are made through the lesion. It is better not to angle the CT gantry when doing these slices. The slice that best demonstrates the lesion is selected and the CT table is returned to the position in which the cut was made. The path that the biopsy needle should follow is drawn on the CT picture, and the distance from the midline, angle of insertion, and depth of needle penetration are calculated (Fig. 42 b). The localizing light is projected and a dermographic pen is used to mark the level of the CT slice on the patient's back [36]. The point of entry and the needle path down to the lesion site are anesthetized with 1% Xylocaïne. A

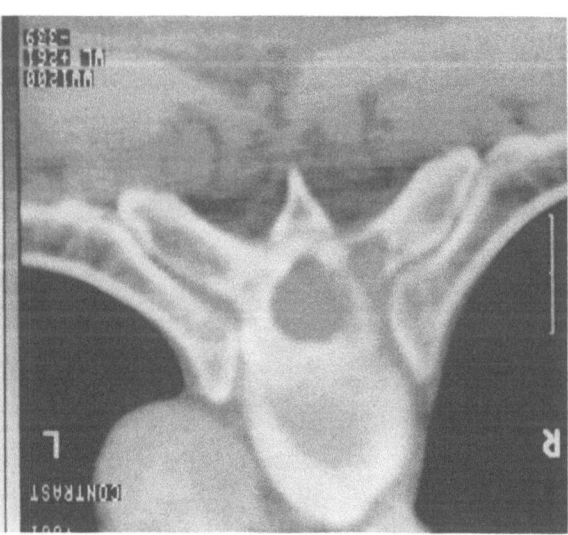 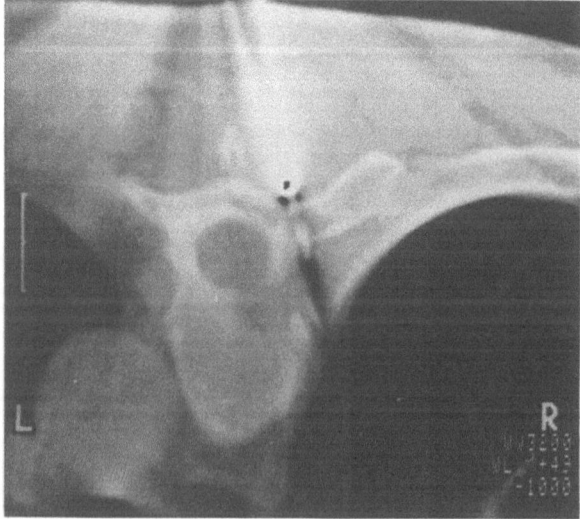

a   b

Fig. 40. a Bone metastasis from kidney carcinoma located at the right pedicle of T 6.
b Biopsy under CT scan control

small incision is made at the entry site to facilitate passage of the needle. The trephine needle is inserted in the plane of the gantry and advanced according to the angle of insertion calculated on the CT picture. Intermediary slices can be performed during advancement of the trephine to confirm

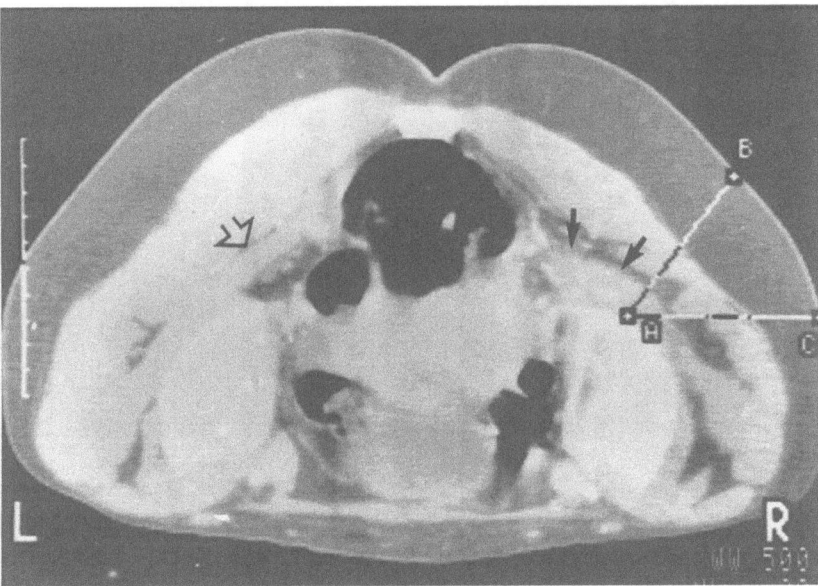

a

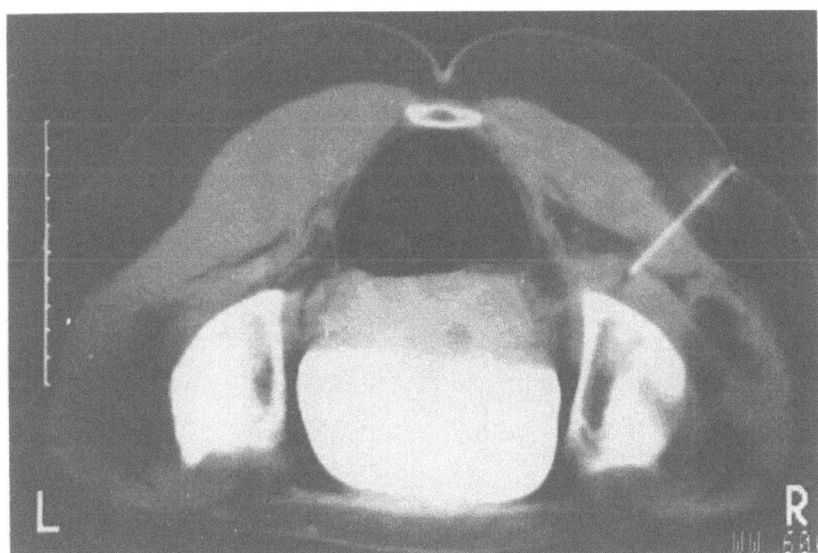

b

Fig. 41. PB of soft tissue under CT scan control in a 40-year-old woman with pain in the right buttock. **a** A soft tissue mass with contrast enhancement is present within the right piriformis muscle (arrows). Selection of appropriate approach (*A* − *B*). **b** Biopsy. Endometriosis was found upon histopathological examination

needle direction. Adjustments are made as needed. When the trephine abuts the lesion, a new CT slice is performed to check for correct placement (Fig. 42 c). Sometimes the biopsy needle is not placed in the exact plane of the gantry and the CT slice does not show the entire needle. Several slices may then be required to visualize the tip of the needle. The additional depth to which the cutting trocar should be advanced is calculated and a core is then taken. A third CT scan can be made prior to removal of the core to

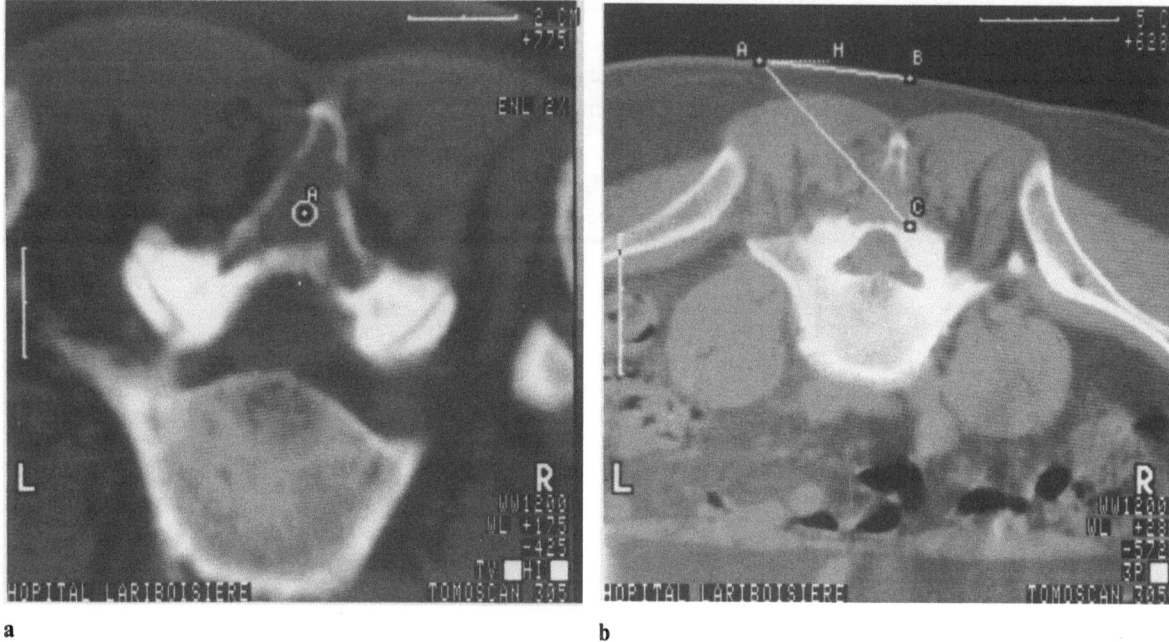

a   b

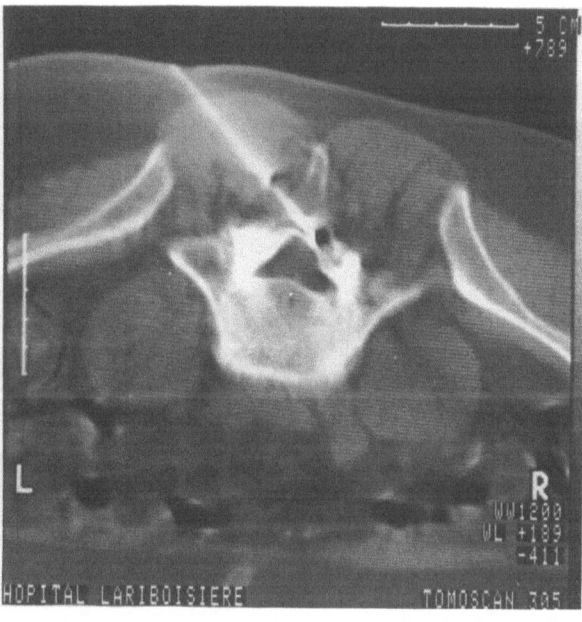

c

Fig. 42. PB under CT scan control of a myeloma located at the neural arch of L 5. Digital radiograph to localize the slice that demonstrates best the lesion (**a**). Preoperative measurements (**b**). Biopsy with a Tru-Cut needle (**c**)

confirm the exact site of biopsy. Several cores are then taken using slight alteration in the cannula single. Elderly patients may find it uncomfortable to remain motionless in a prone position for the duration of the procedure.

Specimen handling

Material should be obtained for both histologic diagnosis and appropriate tissue culture. Inspection of the core tissue obtained is helpful in deciding whether additional cores of tissue should be taken [37]. Core biopsies for histopathologic examination are fixed in 10% formalin, decalcified, mounted in paraffin block, sectioned transversally and stained. For electron microscopy the tissue must be preserved in buffered glutaraldehyde solution. Smears of the needle aspirate are prepared on frosted glass slides for cytologic examination and fixed immediately in Carnoy's solution or 90% alcohol. When an infection is suspected and no fluid can be aspirated, several millimeters of sterile saline solution may be injected and reaspirated. At least one tissue fragment should also be cultured.

Fluid aspirates should be transferred immediately to culture tubes or rapidly delivered to the laboratory. If a question arises as to proper specimen handling, the microbiologist or the pathologist should be consulted prior to the procedure. It is very important that the pathologist who reviews core-biopsy specimens be experienced in the methods and material involved. Close cooperation with the pathologist, who must be given all clinical information available, is imperative. The exact origin of core-biopsy specimens must be indicated in a diagram of the lesion.

Trephine or aspiration biopsy?

For most authors, trephine biopsy is preferred to aspiration biopsy in establishing a pathologic diagnosis [18, 39]. Trephine biopsy should be selected over aspiration biopsy each time there is no serious additional risk. However, in some locations such as the cervical spine from C3 to C7, cervicothoracic and cervicobasilar junctions, the lower and central part of the sacrum, and the ilioischial column, only aspiration biopsy or microbiopsy (Sure-cut) can be performed.

In infectious diseases of the bone, such as infection discitis, some authors perform aspiration rather than trephine biopsy. Simple aspiration of infectious processes may be sufficient to determine the nature of the causative micro-organism. However, bacteriologic examination may remain negative despite the aspiration of pus. Trephine biopsy allows for both histopathologic and bacteriologic evaluation of the infectious process. In bone tuberculosis, histopathologic results are obtained much more rapidly than those of bacteriologic cultures, allowing rapid institution of specific treatment. For this reason we favor trephine biopsy in most cases of bone infection, especially in the spine.

Adequate sample size

Most authors believe that the larger the sample of tissue obtained by biopsy, the easier the histologic diagnosis. Murphy et al. believe that specimen crush artifacts are minimized by using the larger internal diameter needles [39]. Fyfe et al. found that specimens obtained with needles and trephines of internal diameters of two millimeters or more almost invariably showed normal bone architecture with intact hematopoïetic tissue, while specimens obtained with small instruments showed considerable histological distorsion, especially with trephining; they reported a 50% accuracy rate with bone biopsies of less than 2 mm and a 90% yield with larger biopsies [18].

Complications

The overall complication rate of PB is low. In our recent experience of approximately 500 biopsies between 1979 to 1987, only one complication was encountered. This biopsy, of a lumbar lymphoma, was complicated by a hematoma which resolved spontaneously without sequellae. Moore et al. reported 531 closed biopsies with a one percent complication rate, primarily neural and mostly in association with vertebral lesions [37]. Murphy, in a large review of 9,500 percutaneous skeletal biopsies reported in the English literature, identified approximately 22 complications (0.2%) considered important enough to mention [39].

Not all reported complications are serious. Murphy et al. found a total of nine pneumothoraxes (including two in their own experience), many of which did not require a chest tube but were managed conservatively [39]. Other complications were either resolved on their own or resolved with appropriate therapy for the disease process (such as tuberculous sinus tracts). However, there were also serious complications such as transient or permanent paraplegia or quadraplegia, and other neural injuries, such as those with resultant footdrop. Five spinal cord injuries were recorded, including one case of meningitis and two with resultant deaths. Therefore, serious neurologic injury occurred in 0.08% of procedures reported and death in 0.02% [39].

The possibility of tumor spread along the needle in closed biopsy of bone metastases has been reported in three instances [16, 19, 53]. However, this is not a clinically significant complication in the reported experience of percutaneous biopsy of closed metastases. We did not encounter this complication in our personal experience of approximately 300 closed biopsies of bone metastases. In an experimental study of 25 adult rats with tumors of the flank, Burn, Deeley, and Malakar found the role of drill biopsy on tumor cell dissemination to be insignificant [3]. The risk of implantation of tumor cells along the biopsy approach is much more real in primary bone tumors such as chordomas and chondrosarcomas. In such cases, if closed biopsy is performed despite the risk, surgical advice must be sought concerning the puncture site and approach.

Results

Distribution of biopsy sites

The evolution of PB of the skeleton performed in our department since 1966 is depicted in Table 4. The number of procedures has greatly increased in recent years. The lumbar spine remains the most frequent site of biopsy. However, thanks to technical improvements, PB is being performed with

Table 4. Evolution of percutaneous biopsy sites (personal experience in 632 cases)

	No. of biopsies	Spine				Sacro-iliac joint	Other sites
		Cervical	Thoracic[a]	Lumbar[b]	Total		
1966–1982	205	2[c]	19[c]	178	199	2	4
1982–1984	161	6[c]	28	58	91	9	61
1984–1987	266	6[c]	59	111	176	30	60

[a] Including T12–L1
[b] Including L5–S1
[c] Aspiration biopsies

greater frequency in other skeletal sites, especially the thoracic spine, the sacroiliac joint and peripheral bones.

Overall results

The overall accuracy of PB of bone lesions reported in the literature varies from 66% to 96% [4, 10, 11, 27, 31, 37, 39]. However, the method of evaluation of skeletal PB accuracy varies greatly in the different series reported. Adequate evaluation of accuracy must take into account both true positive and true negative results. True negative and false negative results are difficult to confirm [39]. They require a second test or adequate follow-up [39]. Nevertheless, the accuracy of PB of skeletal lesions is reliably evaluated in some series. Debman and Staple reported that an accurate diagnosis was made or disease was excluded in 81% of patients or 74% of biopsy sites [9]. Murphy et al. found a 94% overall accuracy in 160 bone PB's. We carefully evaluated the results of 91 spinal biopsies performed in our department between 1982 and 1984. Overall accuracy was 83.5% (Table 5) [28].

The accuracy of trephine biopsy depends on many different factors:

1. The nature of the lesion, its location and radiological appearance.

2. The adequacy of preoperative radiologic assessment of the lesion.

3. Careful selection of the biopsy site, adequate instruments, approach and method of radiologic guidance.

4. Careful technical performance of the procedure, with emphasis on specimen adequacy [39]. This is ensured by:

   − taking several cores in different locations at each biopsy [37],

   − removing plugs of cortical bone from the trephine before advancing it into a soft lesion [37],

Table 5. Results of percutaneous spinal biopsies
(personal experience, 1982–1984 [28])

		True positive	False negative	Accuracy rate (percent)
Tuberculosis	21	20[a]	1	95.2
Pyogenic osteomyelitis	27	15[b]	12	55.5
Metastasis	12	12	0	100
Myeloma, plasmocytoma	2	2	0	
Primary malignant tumor	3	1	2	
Amyloidosis	1	1	0	
		True negative	False positive	
Osteoporosis	17	17	0	
Degenerative disc disease	5	5	0	
Other	3	3	0	
Total		76	15	83.5

[a] Based on bacteriological and histopathological examinations
[b] Based on bacteriological examination alone

- aspiration of blood from the lesion, which is included and treated as a bone specimen when a neoplasm is suspected [23],

- aspiration of fluid for culture, or rinsing of the lesion using sterile saline solution in the case of suspected infection when no fluid is aspirated.

5. The expertise of the pathologist and the bacteriologist.

PB is also useful in excluding a diagnosis. Murphy et al. obtained a 92% true-negative accuracy in skeletal biopsies performed when metastatic cancer was considered to be a possibility [39].

The accuracy of skeletal PB in each specific diagnosis varies greatly and is detailed below.

Detailed results

Bone metastases

PB has high diagnostic accuracy in bone metastases from 79% [37] to 96% [28] (Fig. 43). In patients without a history of previous malignancy, histopathologic examination of the biopsy material may help identify the primary malignancy. In particular cases such as bone metastases from hepatoma, the histopathologic examination may indicate the precise nature of the primary neoplasm.

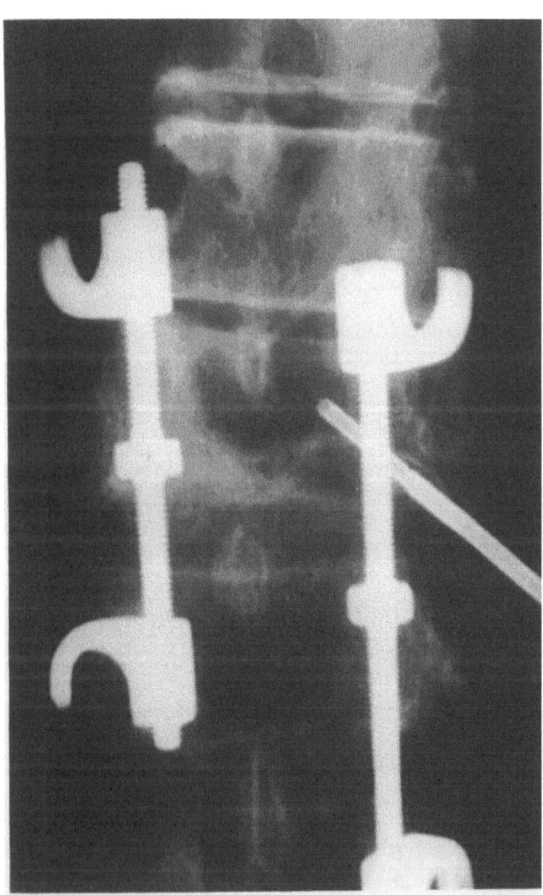
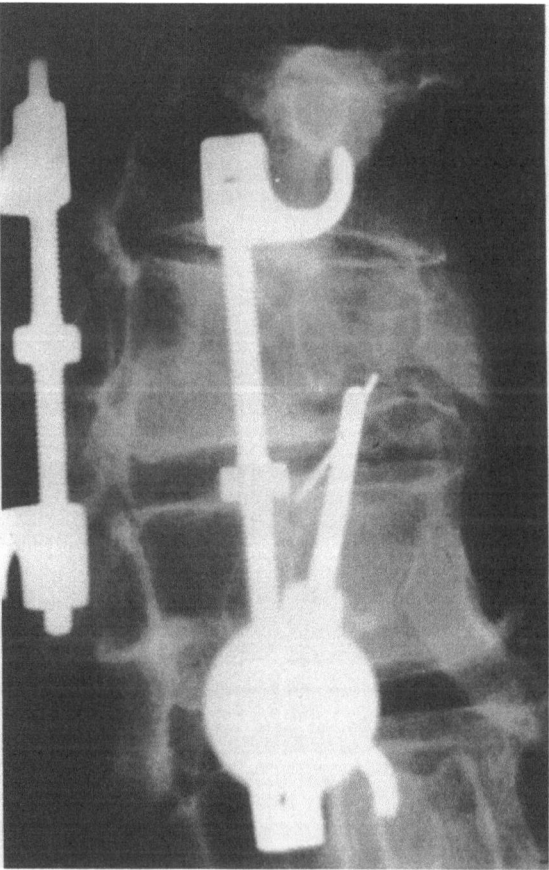

a   b

Fig. 43. AP (**a**) and oblique views (**b**) taken during the PB of a lytic metastasis of T 11 which was carried out despite posterior metallic fixation. Previous laminectomy failed to demonstrate the metastatic nature of the vertebral collapse

Primary neoplasm of bone

Most authors accept that some primary neoplasms of bone such as myeloma, plasmocytoma, lymphoma and Ewing's tumors can be accurately diagnosed by PB (Fig. 44). However, the value of PB in the diagnosis of other primary

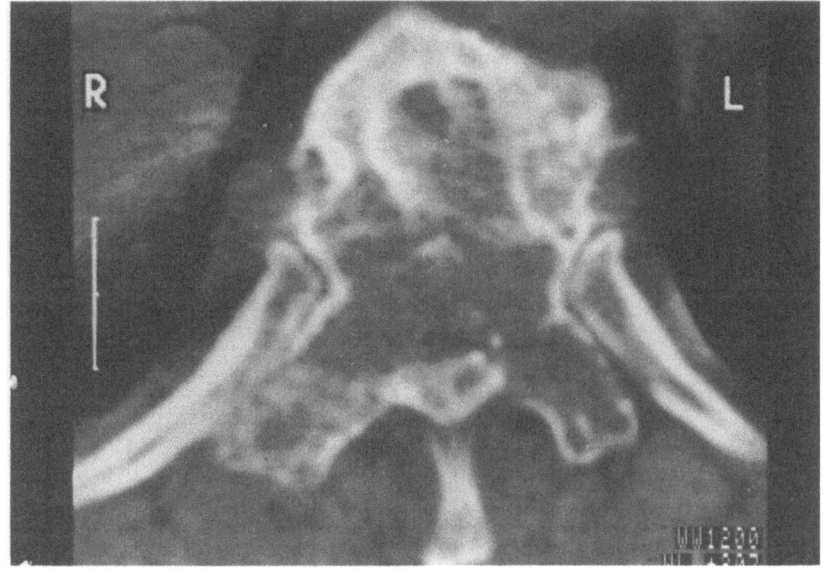

a

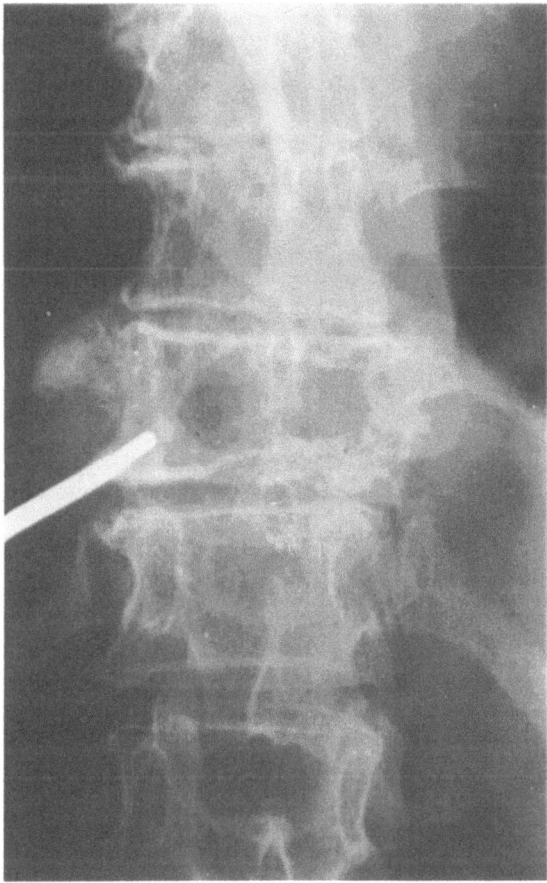

Fig. 44. Primary malignant lymphoma of bone of T 11. CT scan (**a**) and PB (**b**)

b

bone neoplasms such as bone sarcomas is still under debate. The accuracy of PB varies from 44% [49] to 85% [13]. In our experience, PB was not performed when primary neoplasms such as benign tumors and sarcomas were suspected. We believe that PB of bone may lead to erroneous diagnosis in cases of primary bone tumors with complex pathologic architecture such

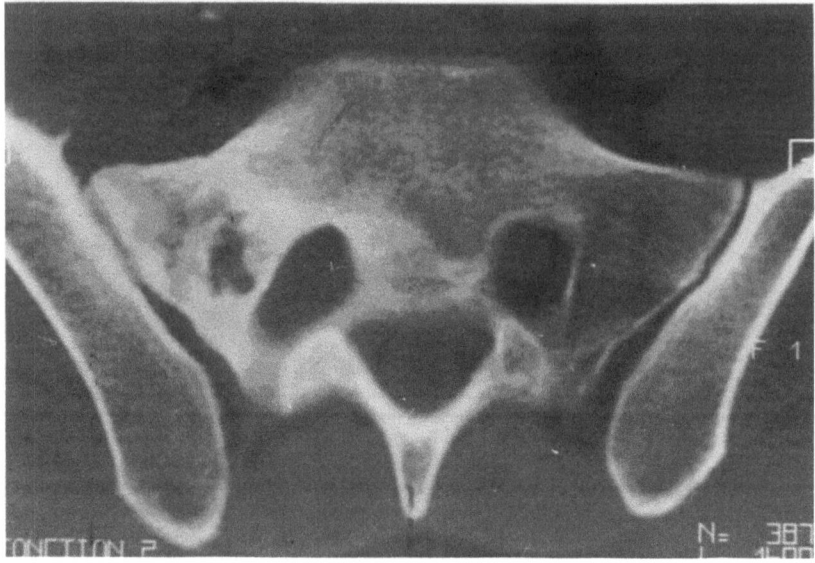

a

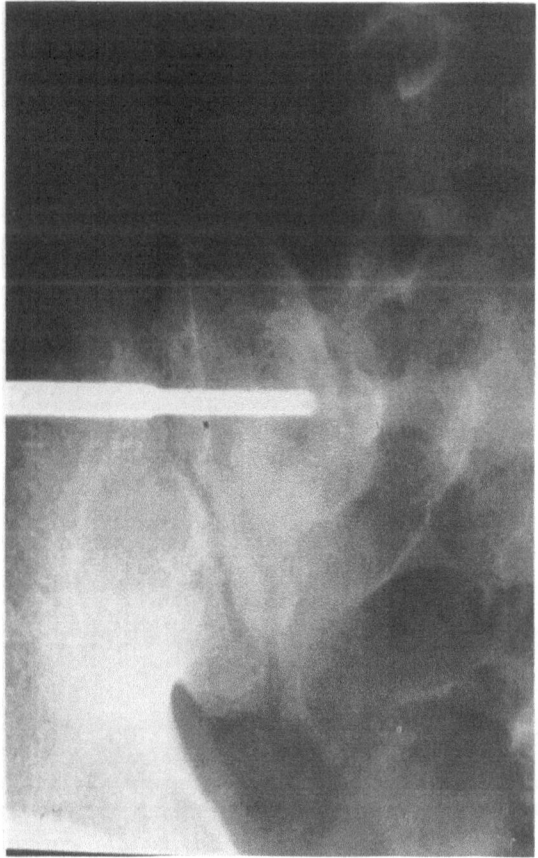

b

Fig. 45. Tuberculous osteomyelitis of the sacral wing at CT examination (**a**). PB through the iliac bone (**b**)

as giant cell tumors, aneurysmal bone cysts, osteoblastoma, and most sarcomas including osteosarcomas, chondrosarcomas, and fibrosarcomas. In these cases, we believe open biopsy, which samples a larger amount of tumor, to yield a higher accuracy rate.

Bone tuberculosis

In our experience, diagnostic accuracy of PB in skeletal tuberculosis is very high. In a series of 21 spinal tuberculoses biopsied in our department between 1982 and 1984, an accurate diagnosis was obtained in 20 cases (95.2%) [28]. This high rate of accuracy may be due to several factors: both histologic and bacteriologic examination may lead to a correct diagnosis; bone tuberculosis produces a large amount of pus; diagnosis is not impaired by antibiotic treatment prescribed without specific diagnosis. Bacteriologic examination may remain negative despite the aspiration of pus at biopsy. Therefore, lesion tissue for histopathologic examination must be obtained in all cases (Fig. 45). Even when positive, bacteriologic results may require a delay of four to six weeks, while results of histopathologic examination are obtained within one week. However, pathologic examination is sometimes non-specific, and may, in some cases, mimic those of a non-tuberculous pyogenic bone infection [4].

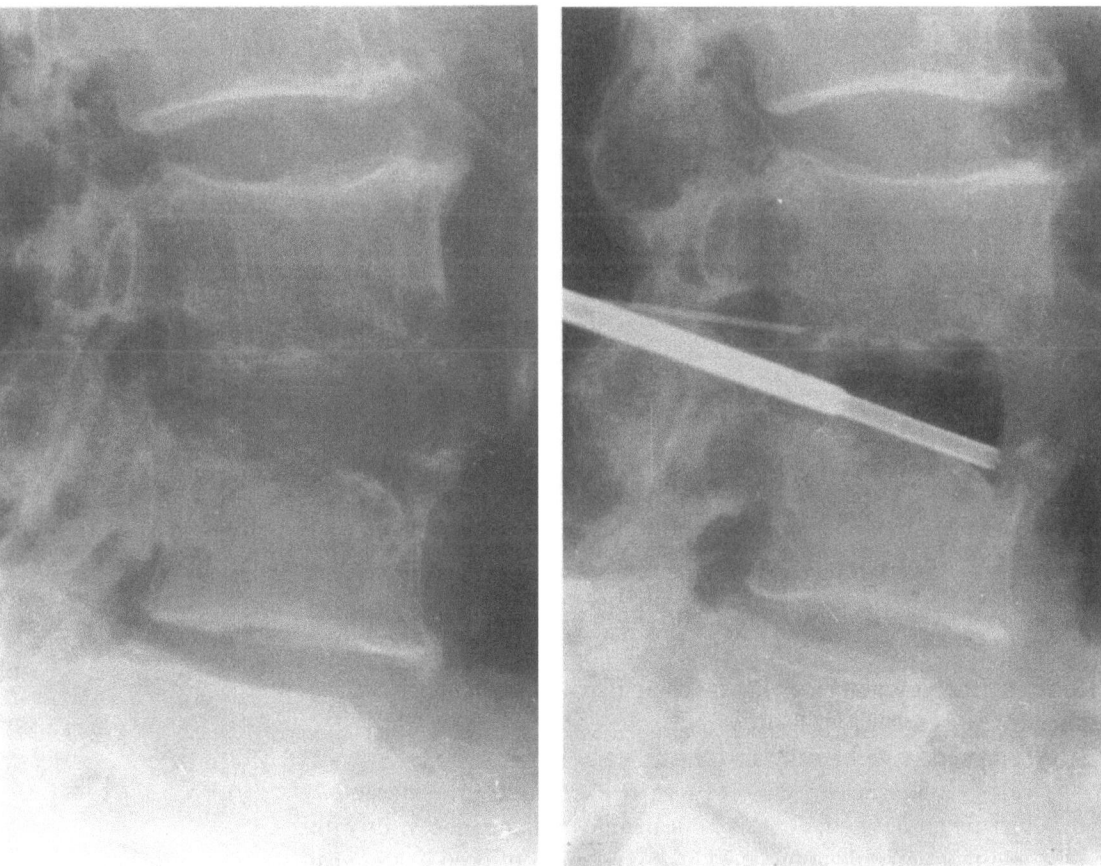

a   b

Fig. 46. Spinal infection due to Escherichia coli at L 3-4 (**a**). Note the vacuum phenomenon which appeared in the disc space after the aspiration of pus (**b**)

Pyogenic infection

Moore et al. obtained a 80% accuracy rate in 117 pyogenic bone infections [37]. The causative infectious organism was isolated in 15 (55.5%) of the 27 pyogenic discitis lesions biopsied in our department between 1982 and 1984 [28]; pathologic features suggestive of pyogenic infection were found in half the cases with negative bacteriologic examination. This relatively low accuracy rate of PB in pyogenic discitis (as compared to spinal tuberculosis) may be due to several factors: definitive diagnosis is only possible when the bacteriologic examination is positive (Fig. 46); the amount of pus is usually not as great; patients often received antibiotics prior to biopsy; spontaneous resolution with elimination of the microbial agent is possible. Microorganisms isolated in our experience of PB of spinal infection are shown in Table 6.

Table 6. Microorganisms found on bacteriological examination in 48 spinal infection biopsies (personal experience, 1982–1984 [27])

Mycobacteria tuberculosis	18
Gram positive cocci	
Staphylococcus aureus	4
Staphylococcus albus	3
Streptococcus	1
Hemophilus influenzae	1
Gram negative bacilli	
Proteus	1
Escherichia coli	2
Klebsellia pneumoniae	2
Brucella	1
Positive results	33 (66, 75%)
Overall number of biopsies	48

Miscellaneous

Our experience suggests that some lesions such as Paget disease and fibrous dysplasia can be accurately diagnosed by PB.

Advantages and disadvantages of percutaneous biopsy versus open biopsy

No useful comparison of results obtained by PB and open biopsy exists in the literature.
- However, the advantages of PB as compared to open biopsy can be summarized as follows [11, 44]:
  — The method is technically simple.
  — In almost all cases, closed biopsy is carried out under local anesthesia either on an out-patient basis as in peripheral bone biopsy, or with a minimal hospitalization of 24 hours, as in PB of deep anatomical structures such as the spine. This makes PB more cost-effective than open biopsy.
  — Complications of surgery such as infection, bleeding, thrombosis and nerve injury are very rare in PB.

— The small biopsy wounds heal rapidly. If radiotherapy is to be given, it can be initiated without delay.

— In bone, the defect made by closed biopsy is smaller than that made with an open biopsy, minimizing the chance of pathologic fracture.

● PB also has some disadvantages as compared to open biopsy:

— It is a relatively blind procedure. This disadvantage is minimized by careful radiographic study of the lesion using radiographs and CT scan.

— The biopsy specimen is small and may be not representative of the whole lesion. For this reason, large trephine needles are preferred to small aspiration needles and, if possible, several specimens should be taken in different zones of the lesion. Open biopsy is preferred in lesions with suspected complex or pleomorphic histology, such as most primary bone tumors, with the exception of round cell tumors.

Indications

(Table 7). Each decision is made in consultation with the referring physician and within the clinical context of the particular case.

Table 7. Indications of percutaneous biopsy. Comparison of preoperative and final diagnoses in 91 spinal biopsies (personal experience, 1982–1984 [27])

Preoperative diagnosis	No. of cases	Final diagnosis[a]
Osteomyelitis (undetermined causative micro-organism)	47	1 discitis of undetermined aetiology 46 osteomyelitis
Osteomyelitis or degenerative disc disease	7	2 osteomyelitis 5 degenerative disc disease
Metastasis	13	12 metastasis 1 angiosarcoma
Metastasis, myeloma or plasmocytoma	4	1 amyloidosis 1 myeloma 1 plasmocytoma 1 angiosarcoma
Osteoporotic collapse or malignant tumors	15	1 recurrent lymphoma 14 osteoporosis
Metastasis or radiation-induced osteonecrosis	2	2 osteoporosis
Infection or osteoporosis	1	1 osteoporosis
Other	2	2 other
Total	91	91

[a] Based on biopsy results and follow-up

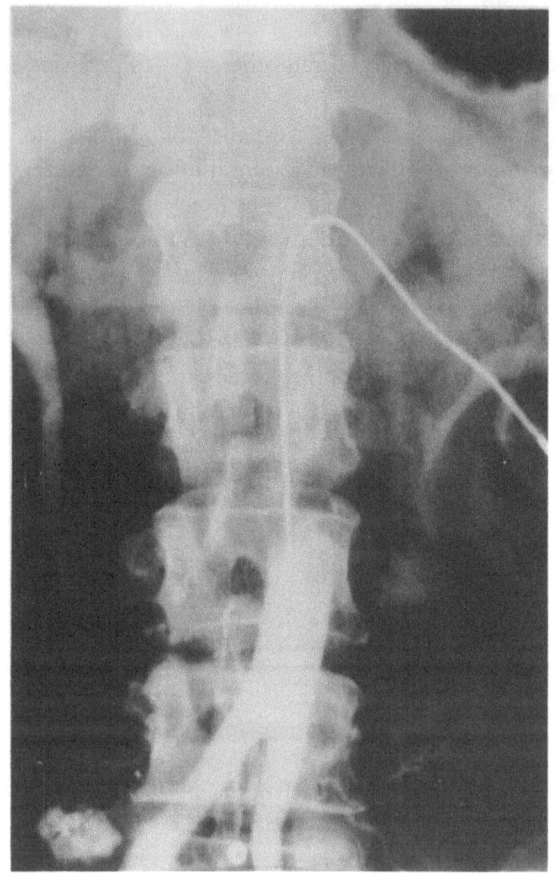

a

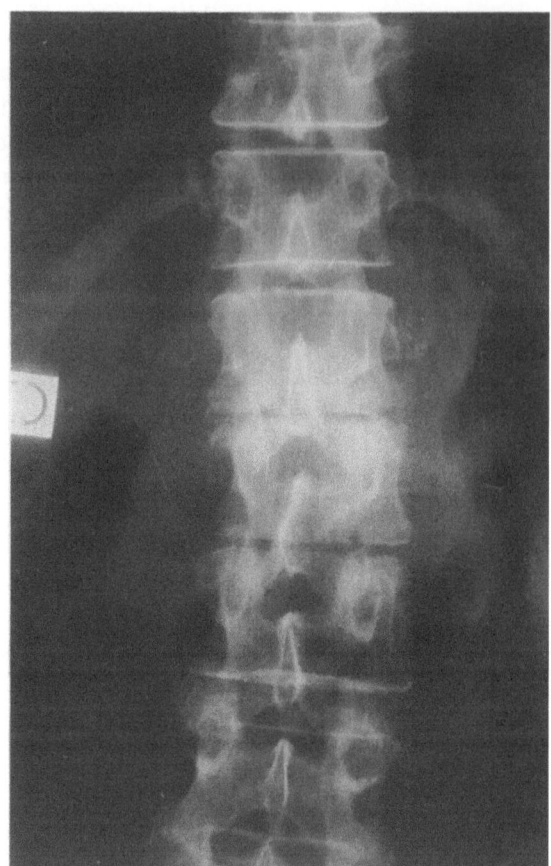

b

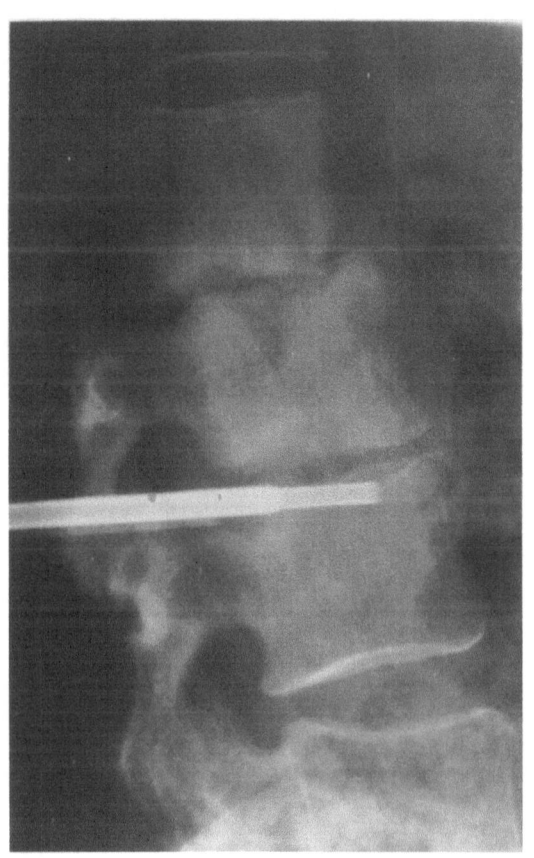

c

Fig. 47. One month after aortography using Seldinger's technique (**a**), this patient developed spinal infection at two levels, L 1-2 and L 2-3 (**b**). PB demonstrated that the causative microorganism was pseudomonas aeruginosa (**c**)

Bone metastases

PB may be indicated in many clinical situations:

Suspected bone metastasis in a patient with no known primary malignancy.

Single symptomatic lesions in patients with a previous history of primary malignancy when the cause and effect relationship may be unclear due to unusual clinical or radiologic manifestations [12].

New bone lesions in the presence of multiple primary lesions [12].

Apparently stabilized previously treated metastases, to determine activity [54].

Determination of the nature of bone changes in patients who have radiologic abnormalities after irradiation of the skeleton, and differentiation of radiation-induced necrosis from metastasis or local tumor extension [14].

Primary bone tumors

PB can be performed in cases of suspected round cell tumors such as Ewing's sarcoma, lymphoma or solitary plasmocytoma.

In primary neoplasms, where "en bloc" excision is indicated, such as chondroma, chondrosarcoma and chordoma, biopsy may complicate or prevent block excision.

Biopsy of histologically aggressive primary bone tumors may also spread tumor cells along the needle approach. If closed biopsy is performed, the puncture site and approach must be carefully selected to be excised "en bloc" with the tumor at the time of definitive surgical ablation. Deeley stated that there is possibly less risk of extensive spread of tumor cells after closed biopsy than with open biopsy because fewer lymphatic and venous channels are opened [10].

Infectious diseases

PB is indicated in skeletal infection when the nature of the causative microorganism has not been ascertained clinically (Figs. 47 and 48). PB is fre-

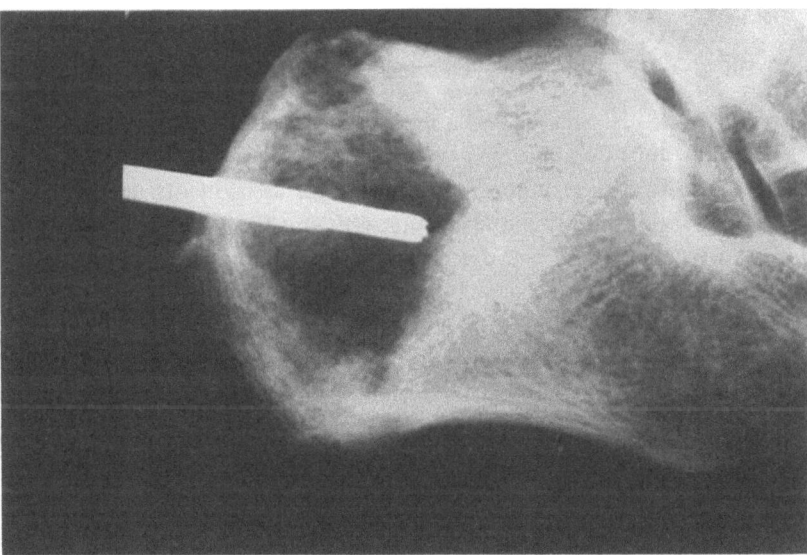

Fig. 48. Calcaneal lesion with mixed lytic and sclerotic appearance in a patient with multiple myeloma. PB demonstrated that the calcaneal lesion was atypical bone infection due to mycobacteria chelonei

quently indicated in order to differentiate infectious discitis from degenerative disc disease, and sacroiliac infection from unilateral rheumatismal involvement of the sacroiliac joint, such as Reiter's syndrome and psoriatic arthritis. PB is also frequently indicated to differentiate pyogenic osteitis from bone tuberculosis. In our experience, biopsy results proved that clinical and radiologic findings were often insufficient to differentiate bone tuberculosis from pyogenic osteomyelitis. When possible, PB must be performed prior to antibiotic treatment.

Miscellaneous disorders

PB of bone may be indicated for diagnostic confirmation in some disorders which may simulate bone neoplasm's such as fibrous dysplasia, eosinophilic granuloma and Paget's disease.

Vertebral collapse

PB is indicated in vertebral collapse when the clinical and radiologic manifestations do not show senile osteoporosis or trauma to be the cause.

Contraindications

There is no absolute contraindication for PB of bone. The risk of a particular biopsy is always measured against the risk from alternative diagnostic methods or against the risk of no specific diagnosis being obtained [39]. Biopsy is not performed if the result would not affect treatment or management [39]. PB should not be performed in patients with platelet counts below 50,000 or abnormal hemostasis unless they are adjusted prior to biopsy. Hemorrhage may be significant during PB of highly vascularized lesions such as giant cell tumor, aneurysmal bone cysts and bone metastases from kidney and thyroid carcinomas. These lesions may exhibit a rather characteristic roentgenographic appearance, with a bubbly lytic pattern with cortical expansion and bony septation. Skeletal metastases from kidney and thyroid carcinomas can be biopsied only if bleeding is not excessive during aspiration with thin needles. In case of important venous bleeding at biopsy, gelfoam can be injected through the trephine needle to obtain hemostasis.

Conclusion

PB of the skeletal system is not a recent technique. However, it has been dramatically improved during recent years by more careful selection of patients; refinements in preoperative radiologic assessments including detailed radiographic study, bone scintigraphy and CT scan; better adapted biopsy instruments; and the diversification of radiologic guidance, using either single plane fluoroscopy, biplane fluoroscopy or CT scan. Optimal results are obtained by close cooperation among rheumatologists, orthopedic surgeons, radiologists, pathologists and bacteriologists. Thanks to numerous improvements, PB of the skeletal system is a safe, accurate and simple technique which can be readily performed in the radiology department.

References

1. Ackerman W (1956) Vertebral trephine biopsy. Ann Surg 143: 373–385

2. Adapon BD, Legada BD, Limev, Silao JV, Dalmacio-Cruz A (1981) CT-guided closed biopsy of the spine. J Comput Assist Tomogr 5: 73–78

3. Burn JI, Deeley TJ, Malakar K (1968) Drill biopsy and the dissemination of cancer. Br J Surg 55: 628–631

4. Chevrot A, Godefroy D, Horreard P, Pallardy G (1980) Biopsie osseuse profonde disco-vertébrale au trocart sous contrôle de la radioscopie télévisée. Ann Med Interne (Paris) 131: 448–451

5. Chevrot A, Godefroy D, Horreard P, Pallardy G (1981) Biopsie osseuse profonde au trocart sous radioscopie télévisée dans les arthrites sacro-iliaques. Rev Rhum 48: 95–99

6. Cohen MA, Zornoza J, Finkelstein JB (1979) Percutaneous needle biopsy of long-bone lesions facilitated by the use of a hand drill. Radiology 139: 750–751

7. Coley BL, Sharp GS, Ellis EB (1931) Diagnosis of bone tumors by aspiration. Am J Surg 13: 214–224

8. Craig FS (1956) Vertebral body biopsy. J Bone Joint Surg [Am] 38: 93–102

9. Debnam JW, Staple TW (1975) Trephine bone biopsy by radiologists. Results of 73 procedures. Radiology 13: 215–224

10. Deeley TJ (1972) The drill biopsy of bone lesions. Clin Radiol 23: 536–540

11. Deeley TJ (1974) Needle biopsy. Butterworth, London

12. De Santos LA, Lukeman JM, Wallace S, Murray JA, Ayala AG (1978) Percutaneous needle biopsy of bone in the cancer patient. Am J Roentgenol 130: 641–649

13. De Santos LA, Murray JA, Ayala AG (1979) The value of percutaneous needle biopsy in the management of primary bone tumors. Cancer 43: 735–744

14. Edeiken B, De Santos LA (1983) Percutaneous needle biopsy of the irradiated skeleton. Radiology 146: 653–655

15. Ellis F (1947) Needle biopsy in the clinical diagnosis of tumours. Br J Surg 34: 240–261

16. Engzell U, Espositi PL, Rubio C, Sigurdson AZ (1971) Investigation on tumor spread in connection with aspiration biopsy. Acta Radiol 10: 385–398

17. Frankel CJ (1954) Aspiration biopsy of the spine. J Bone Joint Surg [Am] 36: 69–75

18. Fyfe IS, Henry AP, Mulholland RC (1983) Closed vertebral biopsy. J Bone Joint Surg [Br] 65: 140–143

19. Gagnerie F, Bonnard JM, Euler-Ziegler L, Commandre F, Ziegler G (1984) Ensemencement néoplasique du trajet d'une ponction-biopsie percutanée de corps vertébrale metastatique. Presse Med 13: 2322

20. Gatenby RA, Mulhern CB, Moldofsky PJ (1984) Computed tomography guided thin needle biopsy of small lytic bone lesions. Skeletal Radiol 4: 289

21. Hardy DC, Murphy WA, Gilula LA (1980) Computed tomography in planning percutaneous bone biopsy. Radiology 134: 447–450

22. Hendrix RW, Lin P-JP, Kane WJ (1982) Simplified aspiration of injection technique for the sacroiliac joint. J Bone Joint Surg [Am] 64: 1249–1252

23. Hewes RC, Vigorita VJ, Freiberger RH (1983) Percutaneous bone biopsy: the importance of aspirated osseous blood. Radiology 148: 69–72

24. Lalli AF (1970) Roentgen-guided aspiration biopsies of skeletal lesions. J Can Assoc Radiol 21: 71–73

25. Laredo JD, Bard M, Patrux CI (1984) Intérêt d'un nouveau matériel pour biopsie osseuse vertébrale percutanée. J Radiol 65: 297–300

26. Laredo JD, Bard M, Leblanc G, Lassau JP (1984) Repères radio-anatomiques pour la ponction-biopsie dorsale. J Radiol 65: 563–567

27. Laredo JD, Leblanc G, Bard M (1985) Ponction biopsie osseuse transcutanée au trocart sous amplificateur de brillance. Feuill Radiol 24: 418–425

28. Laredo JD, Chevrot A, Godefroy D, Auberge T, Leblanc G, Bard M, Pallardy G (1985) La ponction-biopsie disco-vertébrale radioguidée. Encycl Med Chir, (Paris) Radiodiagnostic I, 30660 A 10, 4.10.03, 10 p

29. Laredo JD, Bard M, Leblanc G, Folinais D, Cywiner-Golenzer C (1985) Technique et résultats de la ponction-biopsie transcutanée radioguidée du rachis dorsal. Rev Rhum 52: 283–287

30. Laredo JD, Bard M (1986) Thoracic spine: percutaneous trephine biopsy. Radiology 160: 485–489
31. Levernieux J, Huber-Levernieux C, Seze S de (1972) La ponction-biopsie du corps vertébral. Sem Hop Paris 48: 1029–1037
32. MacLarnon JC (1982) Biopsy of the spine using a needle with a rigid guide wire. Clin Radiol 33: 189–192
33. Martin HE, Ellis EB (1930) Biopsy by needle puncture and aspiration. Ann Surg 92: 169–181
34. Mazet R, Cozen L (1952) The diagnostic value of vertebral body needle biopsy. Ann Surg 135: 245–252
35. Michele AA; Krueger FJ (1947) Vertebral body trephine. Preliminary report. Public Health Rep 62: 1166–1167
36. Mick CA, Zinreich J (1985) Percutaneous trephine biopsy of the thoracic spine. Spine 10: 737–740
37. Moore TM, Myers MH, Patzakis MJ, Terry R, Harvey JP (1979) Closed biopsy of musculoskeletal lesions. J Bone Joint Surg [Am] 61: 375–380
38. Morrison R, Deeley TJ (1955) Drill biopsy: a technique using a high-speed drill. J Faculty Radiol 6: 287–289
39. Murphy WA, Destouet JM, Gilula LA (1981) Percutaneous skeletal biopsy 1981: a procedure for radiologists. Results, review, and recommendations. Radiology 139: 545–549
40. Ottolenghi CE (1955) Diagnosis of orthopaedic lesions by aspiration biopsy: results of 1061 bone punctures. J Bone Joint Surg [Am] 37: 443–464
41. Ottolenghi CE, Schajowicz F, De Schant FA (1964) Aspiration biopsy of the cervical spine. Technique and results in thirty-four cases. J Bone Joint Surg [Am] 46: 715–733
42. Ottolenghi CE (1969) Aspiration biopsy of the spine. J Bone Joint Surg [Am] 51: 1531–1544
43. Ray RD (1953) Needle biopsy of the lumbar vertebral bodies. A modification of the Valls technique. J Bone Joint Surg [Am] 35: 760–762
44. Resnick D (1981) Needle biopsy of bone. In: Resnick D, Niwayama G (eds) Diagnosis of bone and joint disorders. WB Saunders, Philadelphia, pp 692–701
45. Robertson RC, Ball RP (1935) Destructive spine lesions, diagnosis by needle biopsy. J Bone Joint Surg 17: 749–758
46. Schajowicz F, Hokama J (1976) Aspiration (puncture or needle) biopsy in bone lesions. Recent results. Cancer Res 54: 139–144
47. Seze S de, Levernieux J, Mazabraud A (1957) La ponction-biopsie du corps vertébral. Rev Rhum 24: 501–508
48. Siffert RS, Arkin AM (1949) Trephine biopsy of bone with special reference to the lumbar vertebral bodies. J Bone Joint Surg [Am] 31: 146–149
49. Tehranzadeh J, Freiberger RH, Ghelman B (1983) Closed skeletal needle biopsy: review of 120 cases. Am J Roentgenol 140: 113–115
50. Turkel H, Bethell FH (1943) Biopsy of bone marrow performed by a new and simple instrument. J Lab Clin Med 28: 1246–1251
51. Valls J, Ottolenghi CE, Schajowicz F (1948) Aspiration biopsy in diagnosis of lesion of vertebral bodies. J Am Med Assoc 136: 376–382
52. Vinceneux P, Lasserre PP, Grossin M (1981) Technique de ponction-biopsie percutanée au trocart de l'articulation sacro-iliaque pour le diagnostic bactériologique et histologique des sacro-iliites. Rev Rhum 48: 93–98
53. Von Schreeb T, Arner O, Skovsted G, Wikstad N (1967) Renal adenocarcinoma: is there a risk of spreading tumor cells in diagnostic puncture? Scand J Urol Nephrol 1: 270–276
54. Zornoza J (1982) Needle biopsy of metastases. Radiol Clin North Am 20: 569–590

Percutaneous biopsy of the synovial membrane

J.-D. Laredo and M. Bard

Department of Bone and Joint Radiology, Hôpital Lariboisière, Paris, France

Arthritis is a manifestation of a wide spectrum of systemic and local diseases. In selected cases, examination of the synovium may provide precise diagnostic clues or useful information about the nature of the articular process. Biopsy of the synovium can be performed through open arthrography, percutaneous biopsy, or as part of an arthroscopy procedure during which the biopsy area can be visualized.

Percutaneous biopsy of the synovium (PBS) permits removal of specimens from several regions of the joint lining while causing a minimum of trauma. However, since laboratory examination of the synovium is useful for diagnostic evaluation in a limited range of diseases, the precise indications of PBS must be well known.

PBS was first introduced by Forestier in 1932 [4]. In 1951, Polley and Bickel reported a large series of PBS of the knee performed with a specific instrument of 5 mm in diameter [10]. In 1963, Parker and Pearson described a small-caliber synovial biopsy needle which permitted removal of synovial

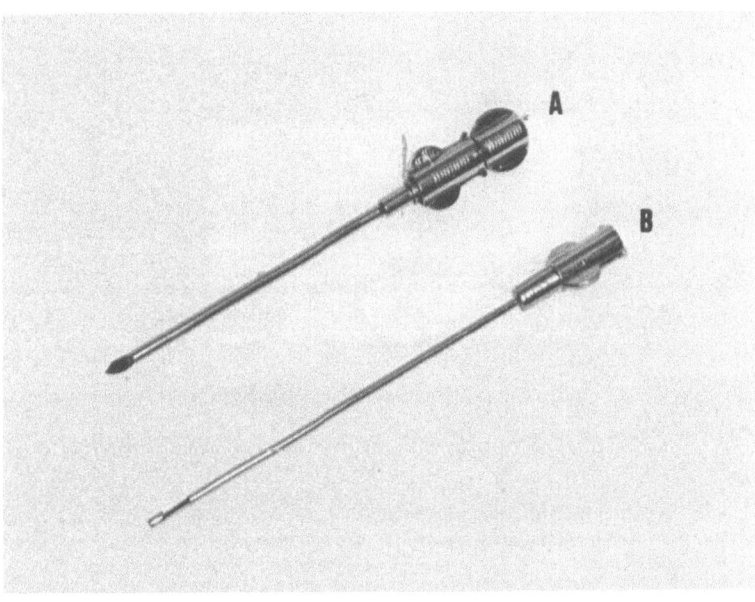

Fig. 1. Parker-Pearson needle: *A* 14-gauge needle with matching stylet; *B* 15-gauge notched aspirating needle

membrane from the knee and other superficial joints such as the ankle, elbow and wrist, while causing a minimum of trauma [9] (Fig. 1). In 1970 Aignan reported a technique for PBS of the hip under fluoroscopic control utilizing a large bore trocar and a Watanabe forceps originally designed for knee arthroscopy [1] (Fig. 2).

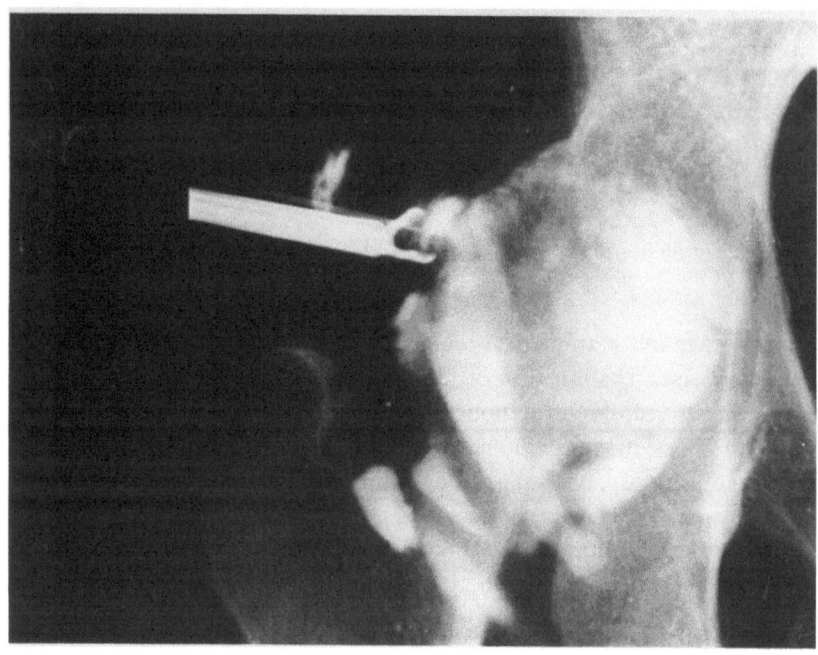

Fig. 2. PBS of the hip. Technique of Aignan [1]

Technique

In this paragraph we will discuss our technique of PBS under fluoroscopic guidance using tru-cut needles. Disposable tru-cut needles are constructed on the same basic principle as Parker-Pearson needles. They allow easy sampling of the synovial lining of the ankle, shoulder, elbow and wrist. PBS

Fig. 3. Instruments for PBS of the hip: 14-gauge 15,2 cm Tru-Cut needle introduced through a 9 cm long Jamshidi trephine needle

of the hip joint is performed with the tru-cut needle introduced through a Jamshidi or Tanzer needle (Fig. 3). Specifications and instructions for use of these different needles have been extensively discussed in the preceding chapter on percutaneous biopsy of the skeleton.

Preparation

A decision is made after complete radiological evaluation of the articular pathologic process. CT scan and CT scan combined with arthrography are always useful in localizing synovial proliferation prior to biopsy. Patient hemostasis must be checked in the days before the procedure. 100 mg Hydroxyzine orally is administered one hour prior to the examination.

Biopsy technique of the synovial membrane: the example of the hip joint

PBS is performed under local anesthesia using single plane fluoroscopy guidance. The biopsy instrument consists of a 15.4 cm long 14-gauge tru-cut needle introduced into the lumen of a 9 cm long Jamshidi trephine needle. Use of the trephine needle for approach to the joint allows sampling of several specimens during the same procedure without additional manipulation. The patient is placed supine on the table with the leg stabilized in internal rotation by sandbags. The hip joint is approached through an anterolateral route (Figs. 4 and 5). The point of skin puncture is determined with the aid of palpation and fluoroscopy. It is located 3–4 cm medial to the anterior aspect of the greater trochanter (line A) (Fig. 6). The appropriate level of puncture is determined under fluoroscopic control by placing a metallic ruler over the patient's hip in order to simulate the approaching needle on the fluorosopic screen. The ruler is appropriately tilted so that it projects over the junction of the femoral head and neck. This determines line B. The exact point of skin puncture is at the intersection of lines A and B (Fig. 6). After skin preparation, the joint is draped and superficial and deeper planes are anesthetized with 1% lidocaine. The joint is first approached with a 20-gauge needle using the usual route for arthrocentesis.

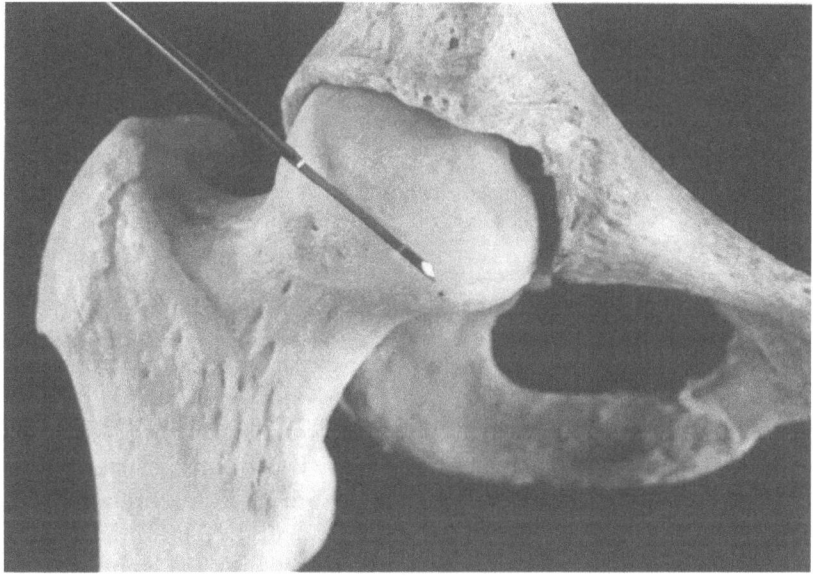

Fig. 4. Approach for PBS of the hip

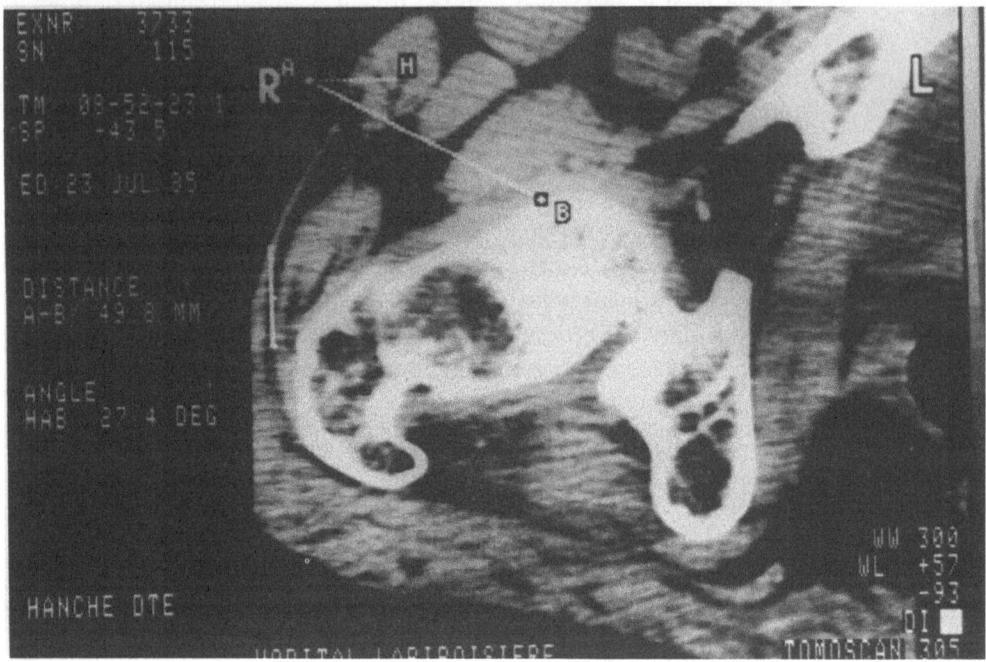

Fig. 5. Simulation of needle approach to the hip joint on a CT scan slice

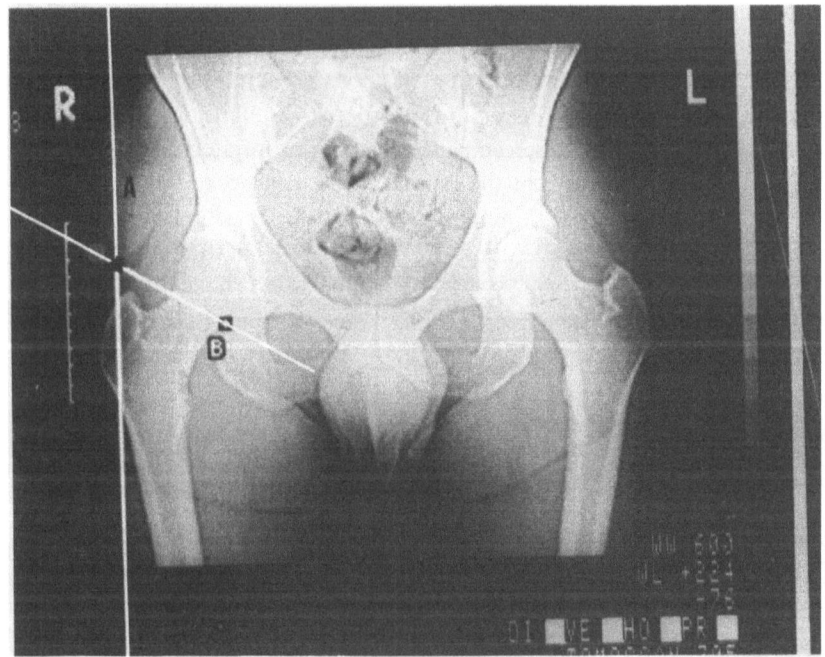

Fig. 6. Determination of point of skin puncture for PBS of the hip. *A* Vertical line 3 cm anterior and medial to the anterior aspect of the greater trochanter. *B* Parallel to the femoral head and neck junction. Puncture point is at the intersection of *A* and *B* (*)

If present, synovial fluid is aspirated for bacteriologic and cytologic examination. If no fluid is aspirated, sterile saline solution is injected and then reaspirated to be sent for analysis. One to three millimeters of Xylocaine are injected into the joint and an arthrography is performed as part of the PBS procedure. Arthrographic features may give diagnostic information and help determine the optimal biopsy site. Distension of the joint with contrast media in saline solution will also facilitate penetration of the biopsy instruments into the joint space [8]. The trephine (Tanzer or Jamshidi) is then inserted through a skin stab and advanced toward the lateral aspect of the junction of the femoral head and neck, following the caudal inclination of line **B** in Fig. 6. The needle is simultaneously directed downward at an angle of 20–30° to the horizontal plane as shown in Fig. 7. When the needle contacts the bone at the correct point, it is withdrawn 2 cm and advanced again 2 cm in a direction more horizontal to the anterior aspect of the joint. The stylet of the trephine needle is then withdrawn and replaced by the 15.4 cm long tru-cut needle, which is introduced into the joint. A gentle alteration in position and inclination of the trephine needle is usually needed to allow

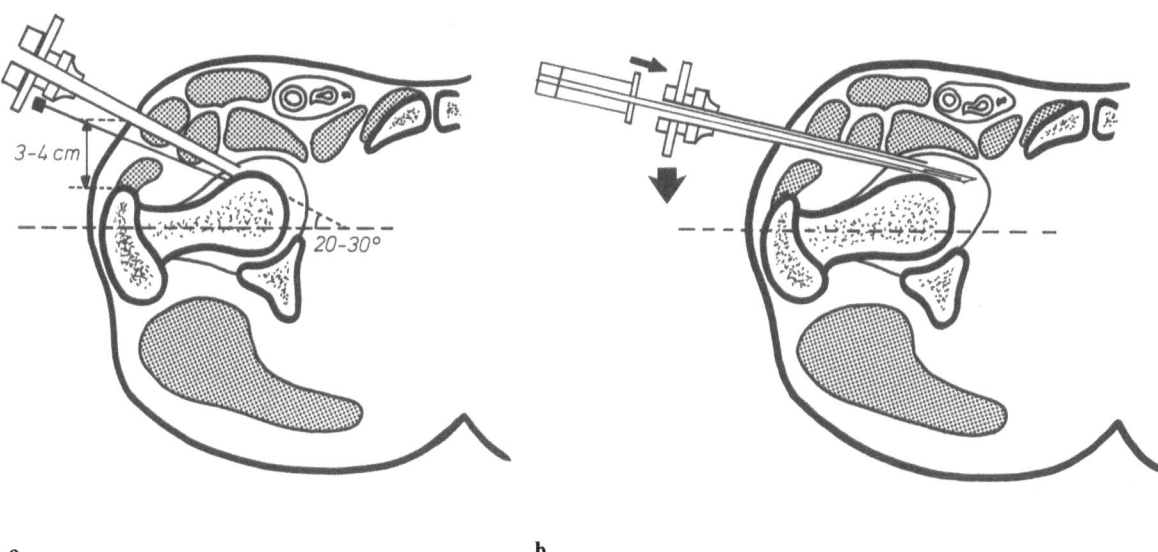

a   **b**

Fig. 7. Needle approach for PBS of the hip. The Jamshidi needle is first advanced to the lateral aspect of the junction of the femoral head and neck (**a**). The Tru-Cut needle is then introduced into the joint through the trephine (small arrow) (**b**). Gentle alteration in trephine inclination (large arrow) facilitates advancement of the Tru-Cut needle

the tru-cut to slide along the anterior aspect of the bone (Fig. 7 b). After correct placement of the tru-cut has been obtained (Fig. 8), the biopsy is performed by holding the inner needle of the tru-cut motionless in one hand, while the outer cutting needle is firmly advanced with the other hand. The external cannula of the trephine is then left in place while the tru-cut needle is removed with the specimen. Return of fluid through the lumen of the trephine after removal of the tru-cut confirms a correct biopsy site. The same process is repeated in various areas of the joint space using slight alterations in inclination and position of the trephine cannula. The inferior recess of the joint is usually the best site for tissue sampling (Fig. 9). Refilling the joint with contrast media or saline solution aids in obtaining multiple specimens. Radiographs are performed during the procedure to document the exact site of tissue sampling (Figs. 8 and 9). After the procedure, patients are advised to rest the biopsied joint for 24 hours. The need for rapid consultation in case of fever, abnormal increase in pain, swelling or any signs and symptoms which may indicate complications is carefully explained.

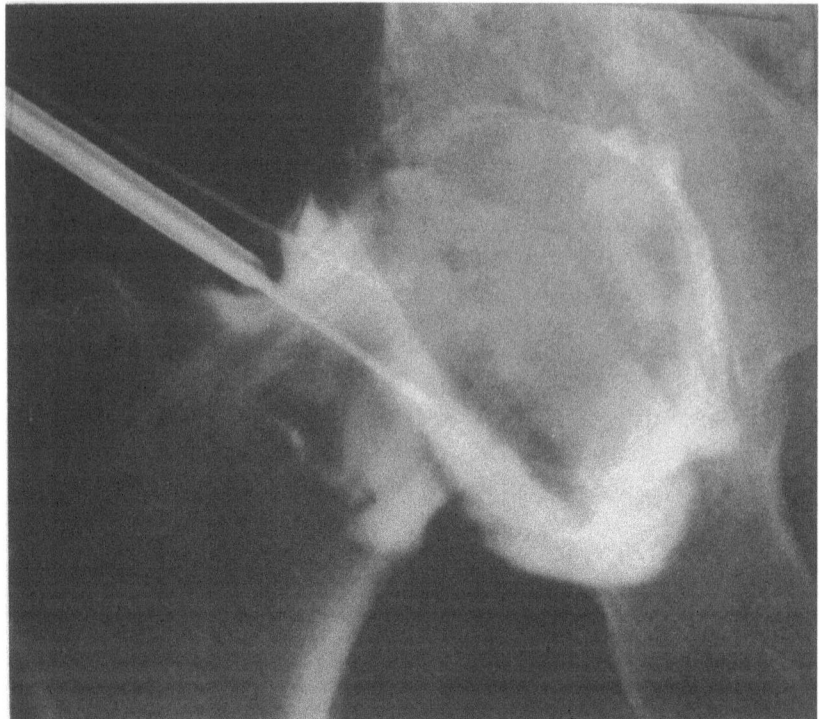

Fig. 8. PBS in hip infection due to Escherichia coli

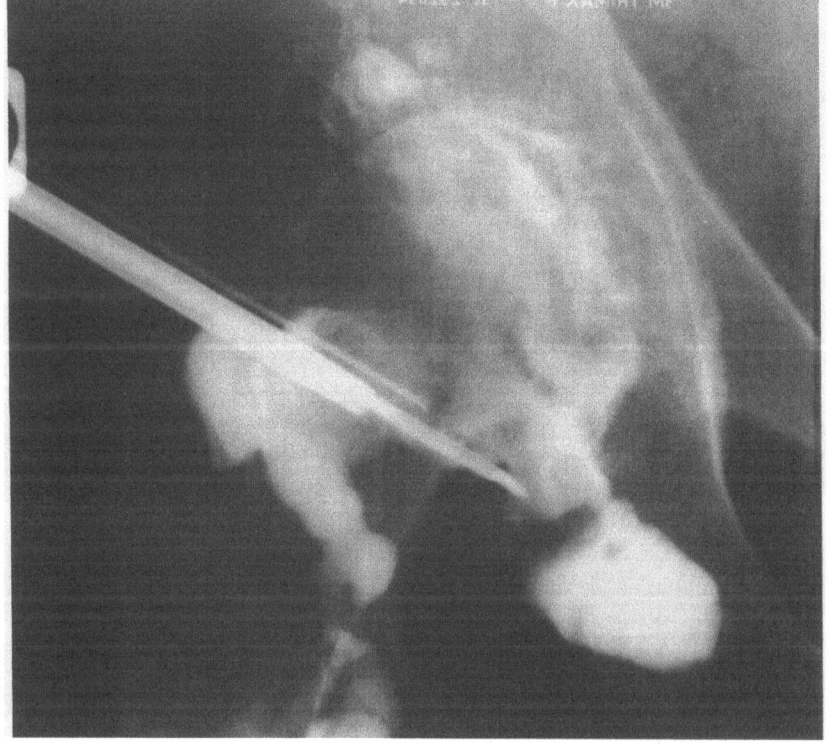

Fig. 9. Destructive arthritis of the hip in a patient with a long-standing history of Gout. PBS performed in the inferior recess of the joint demonstrated urate deposits and septic arthritis due to streptococcus beta hemolytic

Technical modification for PBS of other joints

PBS of the ankle, shoulder, elbow and wrist can be performed with either simple or aspiration tru-cut needles. Owing to the superficial location of these joints, tru-cut needles can be inserted directly without previous introduction of a trephine needle. Simple 14-gauge tru-cut needles permit sampling of large specimens. However, the entire needle is removed with each specimen. The 14-gauge tru-cut can also be introduced through a trephine needle (6 cm-long Tanzer needle) as performed in PBS of the hip, which allows sampling of several specimens using a single approach. Aspiration 17-gauge tru-cut needles obtain specimens of smaller diameter but allow multiple sampling through a single approach without the need for a trephine needle. Aspiration tru-cut needles are especially well-suited for PBS of the shoulder and wrist. Whichever needle is used, the procedure is initiated by an arthrogram performed through the usual routes of arthrocentesis. During the procedure, repeated filling of the joint cavity with contrast media or saline solution facilitates the biopsy, as mentioned in the technique for PBS of the hip joint. Only modifications in the technique from PBS of the hip joint will be discussed in the following text.

Ankle

The patient is placed supine on the table with the legs extended and stabilized in slight internal rotation with sandbags. The biopsy is performed in the anterior recess of the ankle joint (Fig. 10). The needle is inserted anterior

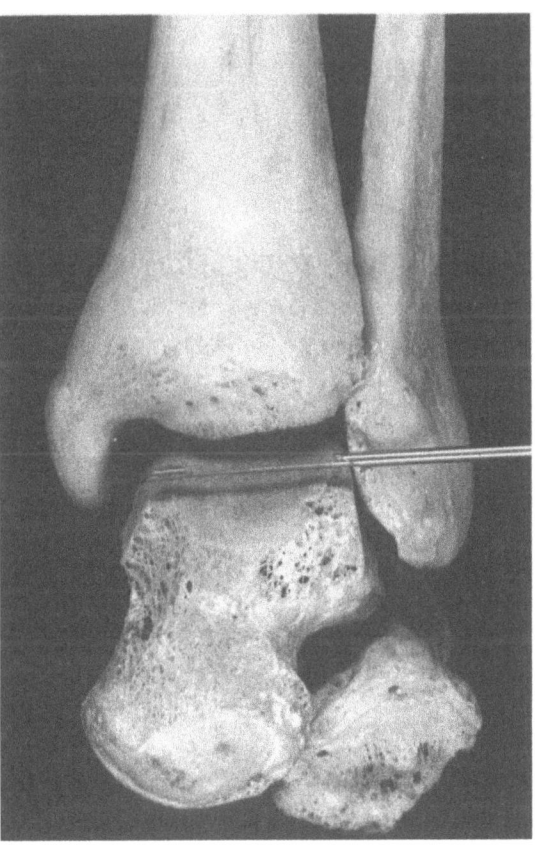

Fig. 10. Approach for PBS of the ankle

to the lateral malleolus at the level of the tibio-talar joint space and advanced horizontally toward the medial aspect of the joint (Fig. 11).

Shoulder

The anterior aspect of the gleno-humeral joint is biopsied using a descending approach (Fig. 12). The patient is positioned supine under the fluoroscope

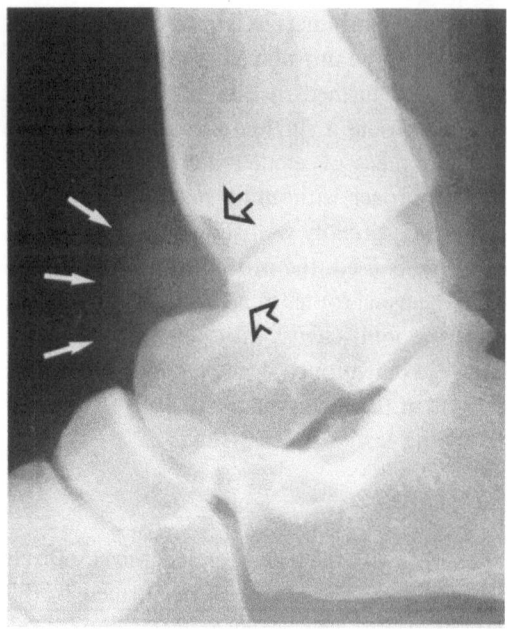

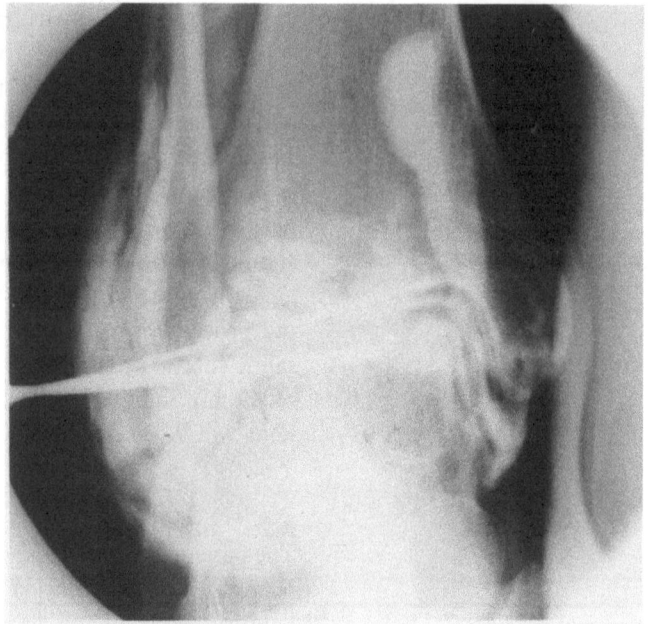

a   c

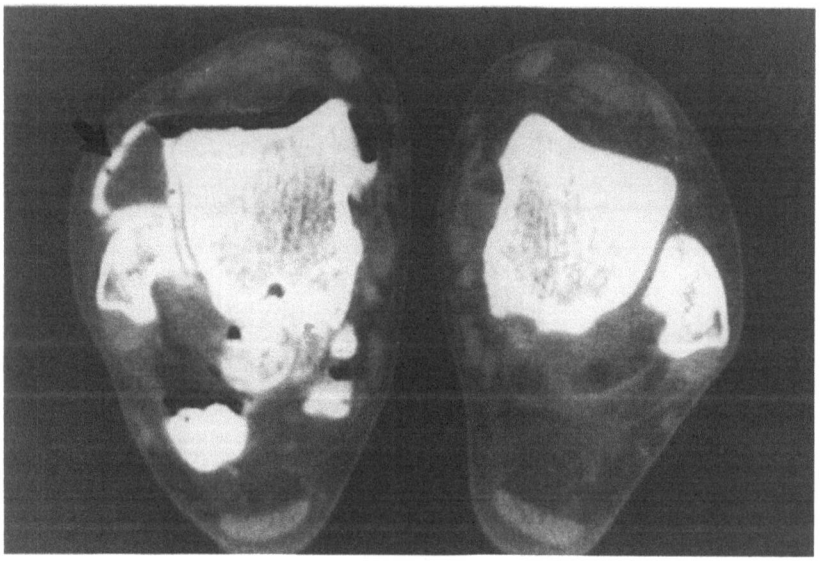

b

Fig. 11. a Soft tissue mass (solid arrows) and bone erosions (open arrows) at the anterior aspect of the ankle. **b** CT scan combined with arthrography demonstrated a synovial mass (arrow) at the anterior aspect of the lateral malleolus. **c** PBS through an anterolateral approach showed typical pigmented villo-nodular synovitis

with the hand in external rotation anchored with a sandbag. Use of a 17-gauge aspiration tru-cut needle is recommended to minimize damage to the rotator cuff tendons. The needle is inserted at a point equidistant from the acromo-clavicular joint and the coracoid process and one centimeter lateral to the vertical line passing through the glenohumeral interspace. The needle is advanced downward with a 20–30° posterior and slightly medial inclination. The needle must reach the bone and then slide over the anterior and medial aspects of the humeral head (Fig. 13).

Elbow

The biopsy is performed at the postero-lateral aspect of the elbow, between the olecranon process medially and the lateral epicondyle anteriorly (Fig. 14).

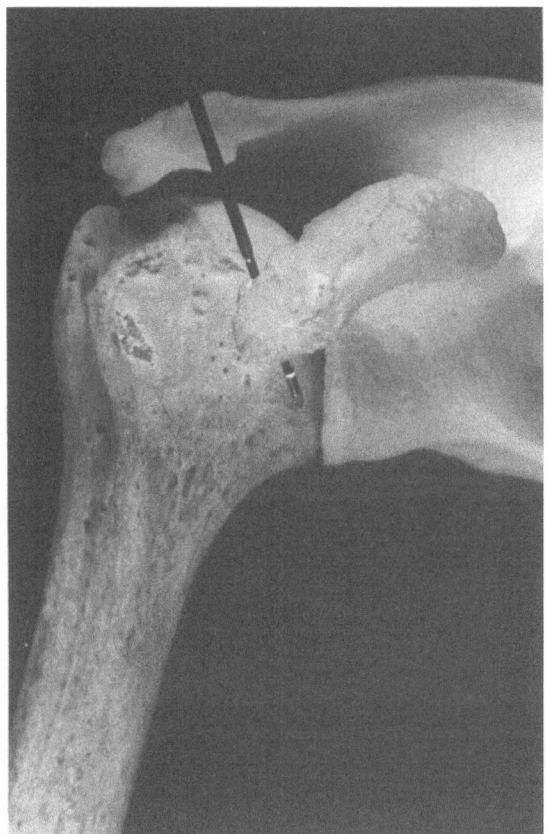

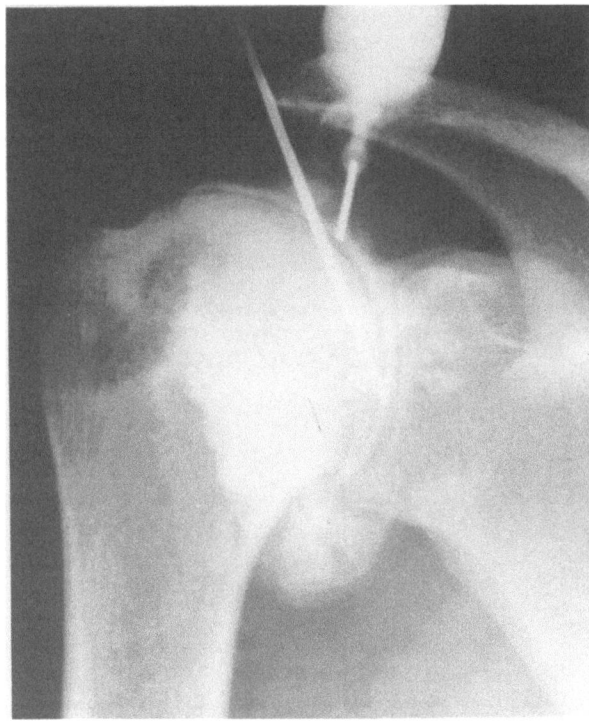

Fig. 13

Fig. 12

Fig. 12. Approach for PBS of the shoulder

Fig. 13. PBS of the shoulder

The patient is placed supine with the elbow semipronated and flexed approximately 60 degrees beneath the fluoroscope. The needle is inserted one centimeter lateral and one centimeter proximal to the upper extremity of the olecranon process and advanced downward with slight (20–30 degrees) anterior inclination to cross the posterolateral aspect of the joint (Figs. 15 and 16).

Wrist

The posterior aspect of the radiocarpal joint is biopsied through a horizontal approach (Fig. 17). The patient is placed supine on the table with his arm under the fluoroscope and the hand pronated. The needle can be inserted at the posterolateral aspect of the wrist, just below the inferior margin of the distal radius, between the extensor pollicis longus laterally and the extensor tendons medially and then advanced horizontally to the medial aspect of the radiocarpal joint (Fig. 17). A posteromedial approach between extensor carpi ulnaris and extensor tendons is also possible (Fig. 18).

Specimen handling

Specimens must be carefully examined during the procedure. Synovial tissue has a pale pink color which can be distinguished from muscle and yellow-white fibrinous exsudate or necrotic material [11]. To minimize sampling

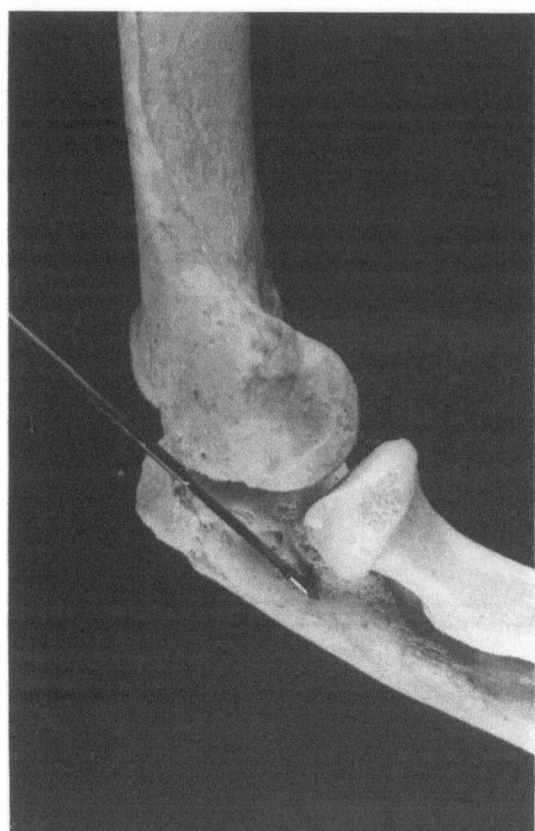

Fig. 14. Approach for PBS of the elbow

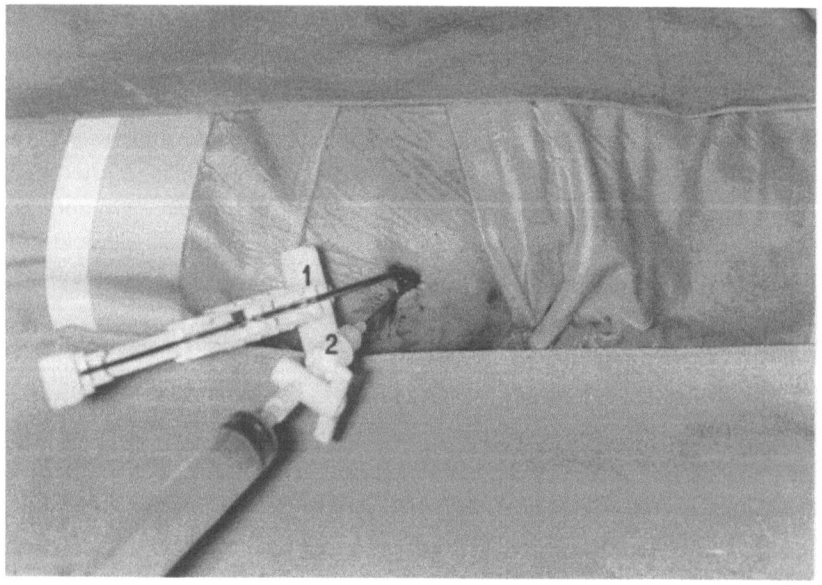

Fig. 15. PBS of the elbow: Tru-Cut needle (*1*) and thin needle for arthrography (*2*)

error, many specimens are taken from various parts of the joint lining without removal of the outer needle [11]. Several specimens are fixed in neutral formalin. When gout, calcium pyrophosphate dehydrate or hydroxyapatite crystalis-related arthropathies are suspected, absolute alcohol is also used. Other specimens and synovial fluid are taken for bacteriologic examination.

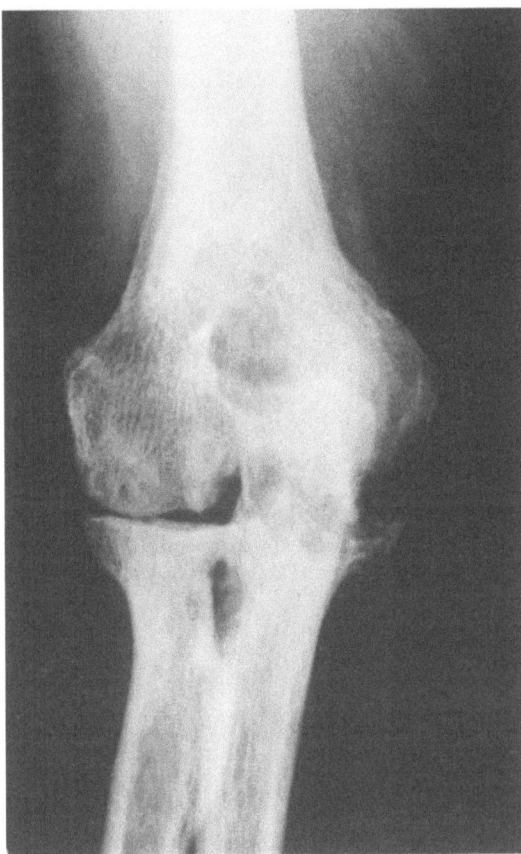

a

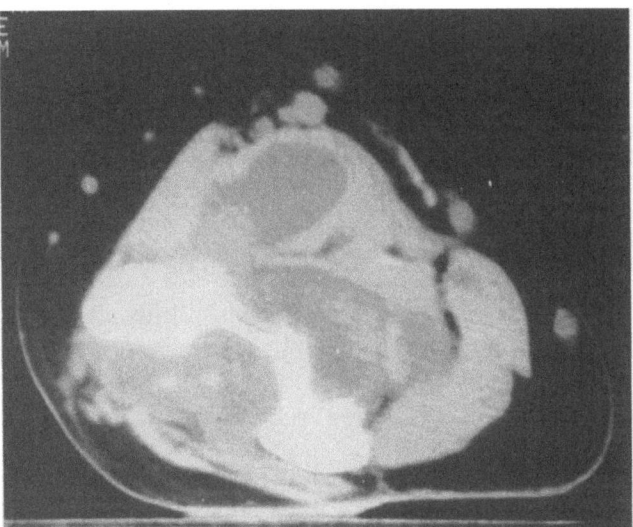

b

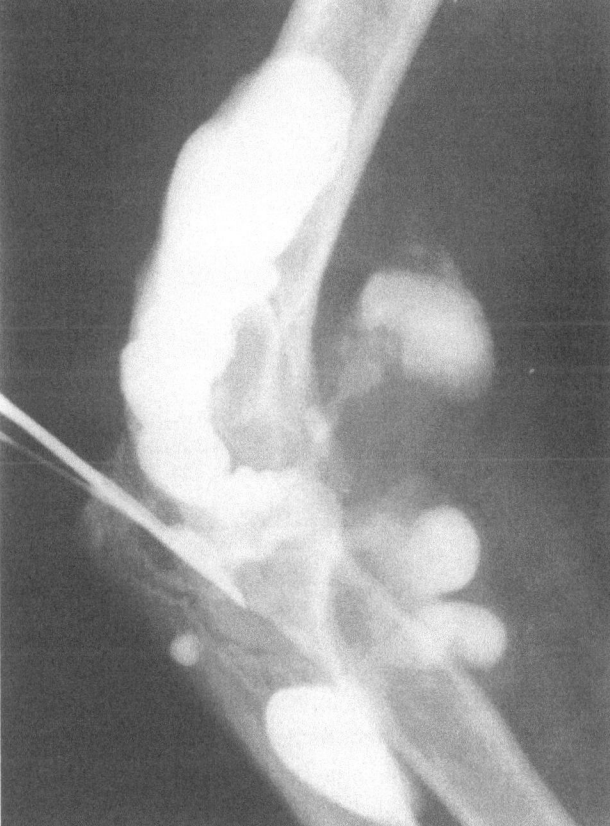

c

Fig. 16. Arthritis of the elbow with with severe destructive changes (**a**). CT scan showed large amount of fluid into the joint (**b**). PBS demonstrated that fluid was composed of cholesterol crystals (**c**). Nonspecific synovitis was found on pathological examination of synovial membrane samplings. A definite diagnosis was not obtained

Complications

Complications of PBS are rare. Some patients experience mild pain and tenderness in the days following the procedure. Joint effusion and hemarthrosis [8] are very rare. Strict antiseptic technique makes the development of a joint infection unlikely. However, this risk is much higher in patients

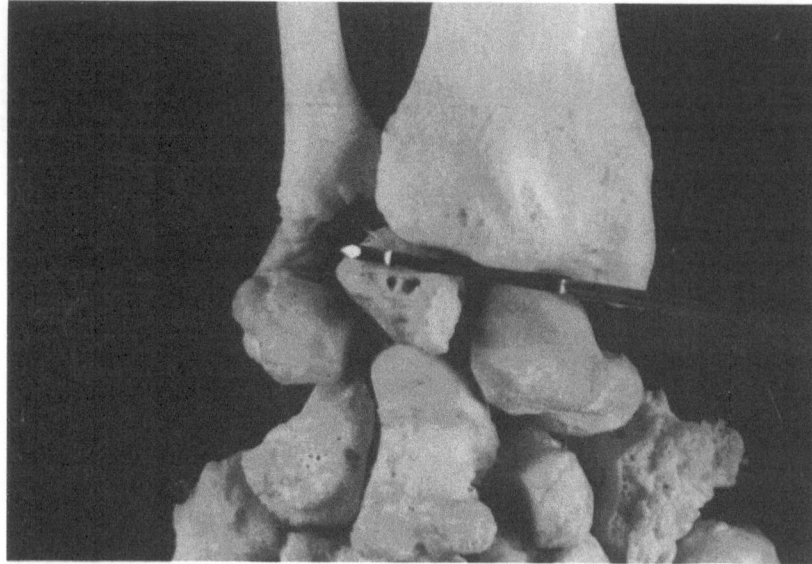

Fig. 17. PBS of the wrist: lateral approach to the posterior aspect of the joint

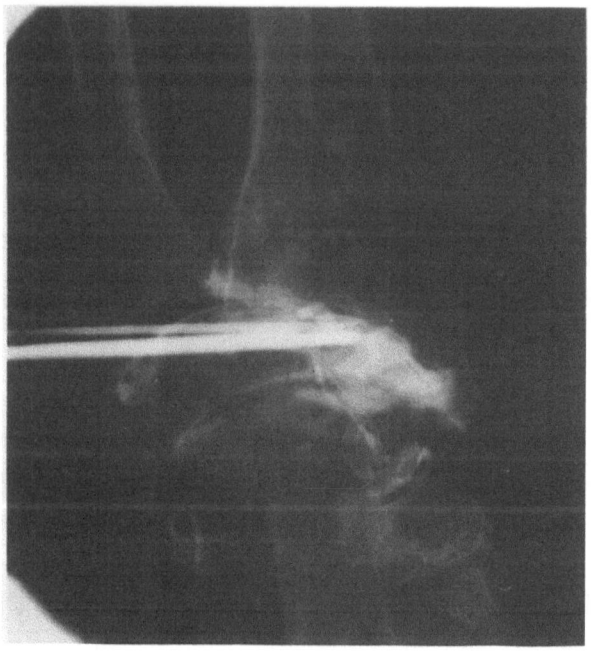

Fig. 18. PBS of the wrist through a posteromedial approach

with immune deficiency, especially in those with chronic renal failure on hemodialysis. Several reports describe intra-articular fragmentation of a Parker-Pearson needle [2, 3, 6, 7]. However, the current Parker-Pearson needle has a modified design which makes it stronger. To our knowledge, this complication has not been reported with tru-cut needles. However, this may occur with more extensive use of these needles and warrants caution when advancing the needle tip into the joint.

Results

The reported accuracy of PBS in obtaining adequate synovial specimens has varied according to the specific joint being evaluated. In most series of PBS of the knee, the accuracy rate is high, with adequate specimens obtained in

86.2% [5] to 96.2% [8] of cases. Reports concerning PBS of other joints are rare. The absence of synovial tissue upon laboratory examination, despite apparently appropriate technique and samples, is more frequent in PBS in areas other than the knee, especially in biopsies of the hip and shoulder. The articular cavity of these joints is relatively small and access to the synovium more difficult than in the knee joint [8]. Moon et al. obtained synovial tissue in 15 of 22 (68.2%) PBS of the wrist, elbow and ankle using the Franklin-Silverman needle [8]. The failure to obtain synovial tissue is frequent in PBS performed in degenerative joint disease, a condition commonly associated with fibrous atrophy of the synovium.

PBS results have been found to be of major diagnostic value in 38 [11] to 53 percent [10] of diseases undiagnosed prior to the biopsy. Other studies have described a higher percentage of biopsies helpful in diagnosis, up to 70.7 percent [8]. However, all series reported are not comparable and, as noted by Schumacher and Kulka, some of them include many patients with a previously established diagnosis [11].

Indications

Results of laboratory examination of the synovium may provide a definitive diagnosis in a limited range of articular diseases, which are summarized in

Table 1. Characteristic synovial membrane findings in rheumatic diseases*

Diagnosis	Characteristics
Tuberculosis	caseating granuloma; mycobacterium
Sarcoidosis	noncaseating granuloma
Gout	monosodium urate crystals
Pseudogout	calcium pyrophosphate crystals
Ochronosis	fragments of pigmented cartilage ("chards")
Hemochromotosis	iron predominantly in the synovial lining cells
Bacterial and fungal arthritis	organisms; intense polymorphonuclear cell infiltrate
Amyloidosis	amyloid deposit (Congo redstain)
Leukemia	malignant cells
Metastatic cancer	malignant cells
Multicentric reticulohistiocytosis	histiocytes and multinucleated giant cells
Whipple's disease	PAS positive macrophages
Pigmented villonodular synovitis	villous hypertrophy with hemosiderin and giant cells
Synovial chondromatosis	islands of metaplastic cartilage
Primary malignancies (synovioma, etc.)	malignant cells

* From [5]

Table 1 from Goldenberg and Cohen [5]. Of these conditions, suspicion of septic arthritis, either pyogenic or tuberculous, is by far the most important indication for PBS.

The non-specificity of many other synovial reactions including those of rheumatoid arthritis, seronegative spondylarthropathies and connective tissue diseases, has already been emphasized by several authors [11, 12, 13]. Even the most typical picture of rheumatoid arthritis synovitis cannot lead to a definitive diagnosis, and may be seen in other rheumatic diseases including seronegative spondylarthropathies and systemic lupus erythematosus [5]. In these conditions, synovial biopsy only contributes to a diagnosis which depends on the addition of several criteria. Finally, PBS in joints other than the knee is rarely indicated in these diseases unless intercurrent secondary joint infection is suspected.

References

1. Aignan M (1979) La ponction biopsie synoviale de la hanche. Med Hyg 37: 1153–1154
2. Bocanegra TS, McClelland JJ, Germain BF, Espinoza LR (1980) Intra-articular fragmentation of a new Parker-Pearson synovial biopsy needle. J Rheumatol 7: 248–250
3. Bonq DA, Noall D, Bennett RM (1977) Intraarticular fragmentation of synovial biopsy needle. Arthritis Rheum 20: 905
4. Forestier J (1932) Instrumentation pour médical. CR Soc Biol (Paris) 110: 186
5. Goldenberg DL, Cohen AS (1978) Synovial membrane histopathology in the differential diagnosis of rheumatoid arthritis, gout, pseudogout, systemic lupus erythematosus, infectious arthritis and degenerative joint disease. Medicine 57: 239–249
6. Guzman L, Arinoviche R (1978) Intraarticular fracture of synovial biopsy needle. Arthritis Rheum 21: 742
7. Kaklamanis PH (1978) Intraarticular fragmentation of synovial biopsy needle. Arthritis Rheum 21: 279
8. Moon MS, Kim I, Kim JM, Lee HS, Ahn YP (1980) Synovial biopsy by Franklin-Silverman needle. Clin Orthrop Rel Res 150: 224–228
9. Parker RH, Pearson CM (1963) A simplified synovial biopsy needle. Arthritis Rheum 6: 172–175
10. Polley HF, Bickel WH (1951) Punch biopsy of synovial membrane. Ann Rheum Dis 10: 277–287
11. Schumacher HR, Kulka JP (1972) Needle biopsy of the synovial membrane – experience with the Parker-Pearson technique. N Engl J Med 286: 416–419
12. Sherman MS (1951) Non-specificity of synovial reactions. Bull Hosp Joint Dis 12: 110–125
13. Sokoloff I (1961) Biopsy in rheumatic diseases. Med Clin North Am 45: 1171–1180

Chemonucleolysis in the treatment
of herniated intervertebral discs

Results and indications of lumbar chemonucleolysis

J. Roucoules[1], J.-D. Laredo[2], M. Bard[2], and D. Kuntz[1]

Departments of [1] Rheumatology (Centre Viggo-Petersen) and
of [2] Bone and Joint Radiology, Hôpital Lariboisiere, Paris, France

Chemonucleolysis (CN) consists of dissolution of the nucleus pulposus of the intervertebral disc and of its herniated fragments by a locally injected enzyme. Among feasible chemical agents, chymopapain is the only substance which has been extensively studied because of its great affinity for the nucleus pulposus. CN is used to treat nerve root pain due to a herniated disc resistant to prolonged medical management. CN has two advantages over surgery: CN obviates the need for general anesthesia and surgical exposure of the spinal canal which may be complicated by epidural fibrosis. However, because chymopapain is inactive against nerve root compression of bony origin, careful radiologic assessment is necessary for proper patient selection.

Historical review

Chymopapain is a proteolytic enzyme first isolated in 1941. It is the principal component of the papaya latex. In 1956, Lewis Thomas [80] showed that intravenous injection of chymopapain in rabbits caused their ears to sag and that this action appeared to result from a particular affinity of the enzyme for cartilage. In the light of these findings, L. Smith [74] explored the idea of using chymopapain to destroy the intervertebral disc. In France, Levernieux and De Sèze [50] conducted similar experiments but stopped their study when they noted that intradiscal injection of the enzyme was capable of causing erosions of the vertebral endplates. However, L. Smith continued animal experiments and came to the conclusion that chymopapain had a powerful nucleolytic action and that the administration of this substance was not dangerous provided an extradural approach to the disc space was used. Initial clinical trials in man, in which chymopapain was used to treat herniated discs which would normally have required surgical discectomy, were published in 1964 [74]. Their encouraging results led to extension of clinical evaluation of chymopapain CN. However in 1975, following the double blind trial of Schewtschenau et al. [72] which failed to demonstrate the superiority of the enzyme over a placebo the FDA withdrew approval of chymopapain for clinical evaluation [59]. The use of chymopapain was nevertheless continued in Canada and Europe. The trial of Schewtschenau et al. has been extensively criticized for several reasons: the double-blind code was broken prematurely, too low a dose of enzyme was used, too few patients were included, their choice of placebo was inappropriate. New studies designed to avoid these shortcomings have yielded three double-blind trials with similar results, demonstrating the efficacy of chymopapain. This led to the reintroduction of chymopapain in the United States in 1982. At present, more than 40,000 CN procedures using chymopapain have been carried out throughout the world. Extensive use of this compound has confirmed its efficacy and relative safety.

The intervertebral disc and its deterioration

The intervertebral disc consists of a central portion, the nucleus pulposus, surrounded by a lamella fibrous ring, the anulus fibrosus. The intervertebral disc is a fibro-cartilage consisting of collagen fibers, cells, and a basic substance particularly rich in proteoglycans. The latter plays an essential role in the hydration of the disc. The proteoglycan monomer consists of a protein chain with attached aminoglycans, chondroitin sulfate and keratan sulfate which are negatively charged and strongly hydrophilic (Fig. 1). Some pro-

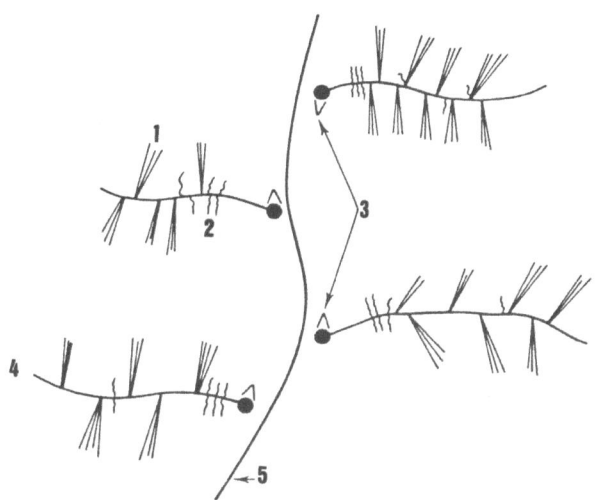

Fig. 1. Proteoglycans aggregate. *1* Chondroitine-sulfate; *2* keratansulfate; *3* linkage protein; *4* proteoglycan monomer; *5* hyaluronic acid

teoglycans are joined by a binding protein to a molecule of hyaluronic acid leading to the formation of aggregates, the volume of which is greater when the intervertebral disc is young (Fig. 1). This configuration is responsible for a very high osmotic pressure and a strongly hydrophilic nature. However, the distribution of proteoglycans is not homogeneous in the intervertebral disc. The nucleus pulposus is especially rich in proteoglycans and water. In contrast the anulus contains more collagen fibers. Collagen of types I and II are present within the intervertebral disc, but only type I is found in the anulus [29]. Few cells are present within the disc tissue. They are mainly found in the outer portion of the anulus. They consist of chondrocytes capable of producing both proteoglycans and collagen. The intervertebral disc also contains some intrinsic enzymes, the physiological role of which is not fully understood.

During the aging process, the intervertebral disc becomes dehydrated. Concurrently, the proteoglycan content falls but it is difficult to know whether this decrease is a cause or effect of disc deterioration. Dessication facilitates the protrusion of nuclear material through fissures in the anulus.

The degenerated nucleus pulposus is then able to migrate backwards to a variable extent. The first stage is simple anular bulging. When the damage is greater the strong external lamellae of the anulus (Sharpey's fibers) may rupture, resulting in disc herniation (Figs. 2 and 3). Subligamentous and extruded disc herniations must be distinguished according to whether the disc material has passed through the posterior longitudinal ligament (Figs. 2 and 3). Once in the epidural space, the extruded disc may separate from the intervertebral portion of the disc resulting in free fragments (sequestered disc fragments). It is also possible for disc material to migrate cephalad or caudad (Figs. 2 and 3).

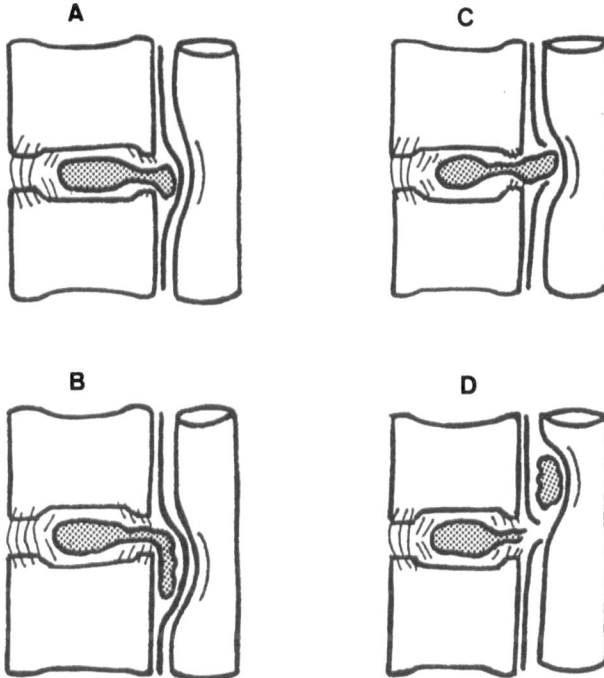

Fig. 2. Principal anatomic types of disc herniations. *A* Subligamentous disc herniation; *B* subligamentous migrated disc herniation; *C* extruded disc herniation; *D* sequestered and migrated disc herniation (free fragment)

Action of chymopapain on the intervertebral disc

Chymopapain is a proteolytic enzyme which breaks down peptide lesions between amino acids. Several animal studies have revealed the mechanism of chymopapain action on the intervertebral disc [36, 48, 49]. Although the enzyme can act on various proteins, it preferentially hydrolyzes proteoglycans and does not affect collagen. Furthermore, because of its positive charge, chymopapain can combine with negatively-charged chrondroitin sulfate and keratan sulfate. However the physiological role of this chemical bindings is doubtful since destruction of the disc by chymopapain is not affected when a negatively-charged form of the enzyme is used [12]. Regardless of the precise mechanism of chymopapain, depolymerization of proteoglycans has several consequences. It causes reduction in osmotic pressure and progressive dehydration of the disc. With the doses of chymopapain current used in CN, only the nucleus pulposus is affected since it contains most of e proteoglycans. The nucleus may even regenerate if the amount of enzyme administered is not too great [7]. By contrast, damage may extend to the anulus, cartilage and even the subchondral bone if massive amounts of chymopapain are injected. However, the action of chymopapain on the

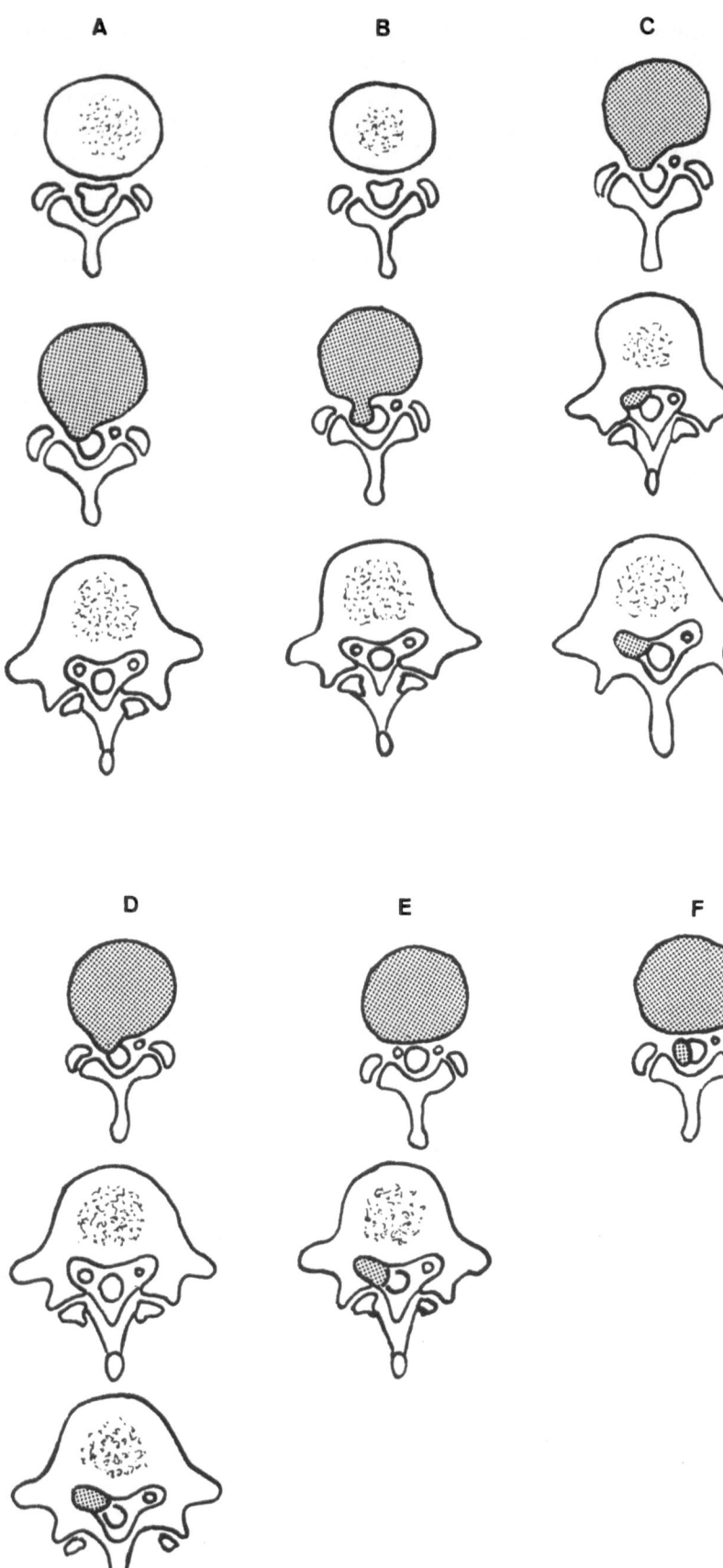

Fig. 3. Diagram of the CT appearances of disc herniation. *A* Common disc herniation; *B* large and pediculated disc herniation — in such cases, probability of extruded disc is high; *C* caudally migrated disc herniation; *D, E, F* different types of sequestered disc fragment

disc depends not only on its dosage, but on disc composition as well. The greater the proteoglycan and water content of the disc, the more effective is the action of the enzyme on the nucleus. Chemical treatment is thus less likely to be effective in the treatment of severe disc degeneration which is associated with advanced dehydration.

Several data suggest that these experimental findings in animals are applicable in man. Jenner et al. [46] found an increase in urinary glycosaminoglycan levels following CN. Enzyme analysis suggested that this increase was largely due to an increase in amounts of chondroitin sulfate, probably resulting from proteoglycan breakdown in the intervertebral disc. Furthermore, histochemical examination of discs removed surgically in patients previously treated with chymopapain confirmed the marked reduction in proteoglycan content within the nucleus and the absence of damage to the anulus [75]. Recent studies using magnetic resonance imaging (MRI) have shown progressive dehydration of the intervertebral disc following enzyme injection [39].

The action of chymopapain on disc tissue has been well delineated. However, the mechanism by which its relieves nerve root pain remains controversial. Sciatic pain may decrease very rapidly following CN. This rapid improvement may be related to a fall in intradiscal pressure and to a decrease in tension within the herniation [78]. The respective roles of chemical discolysis and of direct puncture of the nucleus have not been clearly established. A local anti-inflammatory action of chymopapain has also been suggested but has not been demonstrated. It is more difficult to interpret cases where the disappearance of the nerve root pain is progressive extending over several weeks. The action of chymopapain is very shortlived because of its rapid neutralization by a plasmatic macroglobulin. However, this is only an apparent contradiction. MRI studies have shown that disc dessication is progressive and that it continues for several weeks after enzyme injection [39]. The fall in water content in the nucleus is probably essential in decreasing hernia tension. However, the role of the decrease in size of the disc herniation in leg pain relief appears to be limited. Several months are often necessary before a significant decrease in hernia size is observed on CT scan, a period longer than that required for clinical improvement [47]. Other mechanisms may be involved in the release of disc pressure on the nerve root. Gentry et al. have suggested the importance of the disc space narrowing after injection of chymopapain [37]. According to these authors, the nerve root, tethered proximally and distally, is tightly stretched by the intervening disc herniation. The decrease in disc height which usually occurs rapidly after CN, may reduce nerve root compression by bringing together the two points of tethering. The nerve root can then pass more easily around the protruded disc.

Experimental toxicity of chymopapain

Before being used in man, chymopapain was tested in various animals [36, 48]. Diffusion of the enzyme following its intranuclear injection appears to be minimal. An epidural leak has no untoward consequences. The direct administration of chymopapain into the epidural space is well tolerated unless the amount injected is too large. Protection of neural structures by the dura mater seal may explain good tolerance of epidural leak of chymopapain. In contrast, the intradural injection leads to a subarachnoid hemmorhage due to damage of the endothelial cells of the arachnoid capillary vessels, even if the amount of substance is very small. Contamination of the cerebrospinal fluid is thus the principal risk of disc puncture at the time of CN. The intravenous administration of the enzyme, which in the animal causes damage to the microcirculation, visceral hemorrhages and sometimes coagulation disturbances, must be avoided by checking for the absence of venous opacification when performing discography prior to CN. Furthermore, intravascular toxicity of chymopapain appears to be slight for small doses.

Complications of CN

CN has three main risks: anaphylactic shock, neurological complications and spinal infection. The present discussion will be limited to the last two complications since allergic manifestations are treated in a special chapter.

Neurological complications

Neurological status must be carefully evaluated during CN. Retention of urine is seen in approximately one percent of cases [5]. This complication, which occurs shortly after treatment, requires catheterization of the bladder. However, as a general rule, control of micturition is restored by the day following CN. The mechanism of such urinary retention is not clear. The responsibility of chymopapain is unlikely since there is usually no other sign indicative of a lesion of the cauda equina. It may rather be induced by anesthetic drugs and bedrest. Bouillet [5] has also reported the beneficial influence on bladder function of getting the patient up immediately after treatment. Another neurological complication may be the onset of worsening of a nerve root deficit. In the European study involving 2,136 cases, this possibility was seen 39 times (1.83%) [5]. When signs of neurological involvement develop, they are often minimal and limited to hypoesthesia, mild motor weakness or reflex loss. However sciatic pain associated with paralysis also occurs in two cases in one thousand. It is not known whether these nerve root deficits are due to a toxic action of chymopapain (possibly favored by inflammation of nerve tissue), to damage by the puncture needle or to sudden mechanical changes. The possibility of a neurological deficit on the side opposite disc puncture and initial nerve root pain supports the last hypothesis [1].

Some severe neurological complications have also been described following CN [1, 5, 13, 17, 24, 26, 65, 87]. They occur shortly after treatment. Six cases of cerebral subarachnoid hemorrhage (three of which fatal), over fifteen cases of paraplegia and several cases of cauda equina syndromes have been reported. Although a vascular malformation may sometimes be responsible, in the majority of cases the intrathecal diffusion of chymopapain is a more likely explanation. Various causes may explain subarachnoid leakage of the substance. The dural sac may be damaged in difficult disc approach. Another explanation is that CN is performed too soon after myelography, before the thecal sac has had time to seal. In some cases a past history of discectomy at the same level may lead to intrathecal diffusion of the enzyme. It appears that most of these severe neurological complications can be avoided providing certain precautions are observed. Thus about a one-week interval should be allowed between myelography and CN.

Also, CN should be discontinued if needle approach is too difficult or requires too many maneuvers. Finally, CN should not be carried out at a vertebral level previously operated on.

Post CN discitis

Disc space narrowing is normal after CN. Erosions of the vertebral endplates are much rarer. Post CN discitis is defined by the association of these two findings. This occurs in approximately one percent of cases [5]. Incapacitating low back pain and severe lumbar spasm are constant features of post CN discitis. Neurological signs may also be noted. In some cases there is clear evidence of bacterial infection [1]. The most frequently identified microorganism is staphylococcus, either aureus or albus. However, in many cases bacteriologic studies remain negative. Since, in such patients, stabilization or healing may occur without any antibiotic treatment, the real nature of these post CN discitis remains questionable. It has been suggested that chymopapain could be responsible for chemical discitis. Identical lesions have been obtained experimentally by injecting large doses of the enzyme in animals. However, such large doses are not used in clinical practice. Experimental studies in animals by Fraser et al. [34] have also demonstrated that when chymopapain is used at usual concentration, only preparations contaminated with bacterial agents can cause comparable radiologic changes. In these cases, the microorganism may disappear spontaneously after a few weeks. These experimental findings strongly support the hypothesis that most if not all cases of post CN discitis correspond to a low grade spinal infection.

Cases of severe low back pain and lumbar spasm but without any radiologic evidence of erosions of the vertebral endplates must be differentiated from post CN discitis. This occurs in up to 25% of patients who undergo CN. In these cases Dermott and Agre [20] found no pathological evidence of any involvement of the anulus. Furthermore, pathological examination of discs removed at surgery and previously treated with chymopapain almost invariably showed that damage was confined to the nucleus [70].

Other complications

Other complications may occur following CN. Transient fever, thromboembolic accidents or a late allergic skin rash have all been reported. However follow-up over a twenty-year period of patients treated with chymopapain has failed to reveal any delayed complications.

Finally, all these complications are only very rarely fatal. In the study of Agre et al. [1], only six deaths in 30,000 patients could be imputed to chymopapain CN.

Indications of CN

CN is used to treat nerve root pain due to nerve root compression by a herniated disc. Optimal results are obtained with careful patient selection. Clinical signs of nerve root compression by a herniated disc and radiologic evidence of disc herniation are the principal factors taken into account when selecting patients suitable for CN.

Can patient selection criteria be suggested?

Some prognostic factors are common to most reported studies concerning CN. On this basis, it is possible to suggest the ideal indications for CN:

1. Nerve root pain must be more severe than low back pain.

2. Nerve root pain must be persistant after two months of conservative treatment including a minimum of two weeks of strict bed rest.

3. Leg pain must be limited to a single nerve root dermatoma. When present, objective clinical signs of nerve root involvement such as localized hypoesthesia, slight motor weakness or loss of a deep tendon reflex are valuable in confirming the involvement of a single nerve root.

4. Clinical signs of disc pressure on the nerve root, i.e. an antalgic position of the spine or a positive homolateral or contralateral straight-leg raising are mandatory.

5. Radiologic evidence of disc herniation compressing a nerve root must be obtained. The level of this herniated disc must correspond to the level of the nerve root clinically involved.

6. Some local contraindications to CN such as free disc fragments, lateral recess stenosis and central stenosis of the spinal canal must be eliminated. They are discussed in detail below.

In everyday practice it is sometimes difficult to conform to a list of criteria in deciding to opt for CN. In fact, an infinite range of clinical situations exists. However one must be aware of the increasing risk of failure of treatment whenever the patient fails to meet the above criteria. It is also essential that the patient be informed when the best chances of success cannot be assured. Some dissuasive factors such as psychological problems and worker's compensation claims must also be taken into account. However, these factors do not constitute a contraindication to enzyme injection.

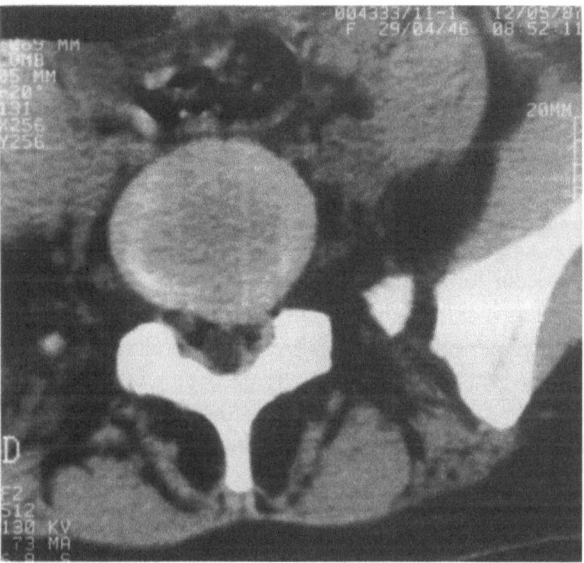 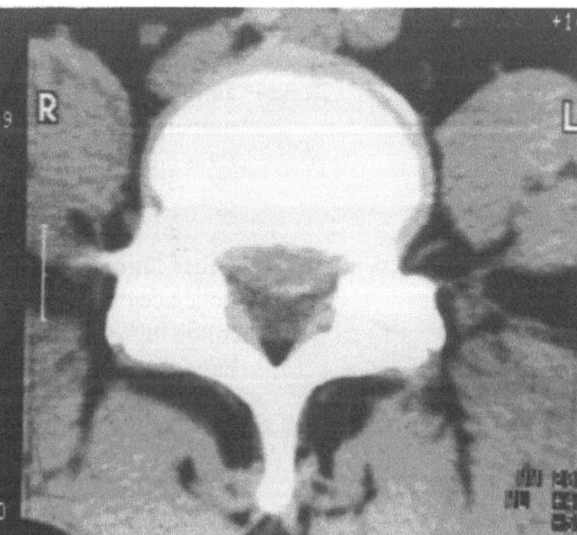

Fig. 4   Fig. 5

Fig. 4. Common disc herniation at L 5-S 1 level with compression of the left S 1 nerve root

Fig. 5. Midline disc herniation at L 5-S 1 level. Impingement on the nerve root is less apparent than in Fig. 4

How to carry out neuroradiological investigation

Radiologic demonstration of a disc herniation corresponding to the clinical level and the elimination of local contraindications are indispensable prior to CN. CT scan is at present the most efficient radiological tool for patient selection [41] (Figs. 4 and 5). It has no side effects as compared to myelography and can be performed as an outpatient procedure. CT scan is also the best method for detection of foraminal herniated disc and disc herniation at L 5-S 1 interspace. CT is also more efficient than myelography in excluding causes of nerve root compression unresponsive to CN, such as a sequestrated disc or nerve root compression of bony origin in a narrowed lateral recess. For these reasons CT scan should always be performed prior to CN. If a herniated disc with nerve root compression corresponding to the clinical level is demonstrated, additional radiologic investigations are unnecessary. In contrast, myelography is required when no definite nerve root compression corresponding to the clinical level is demonstrated (Fig. 6). Myelography

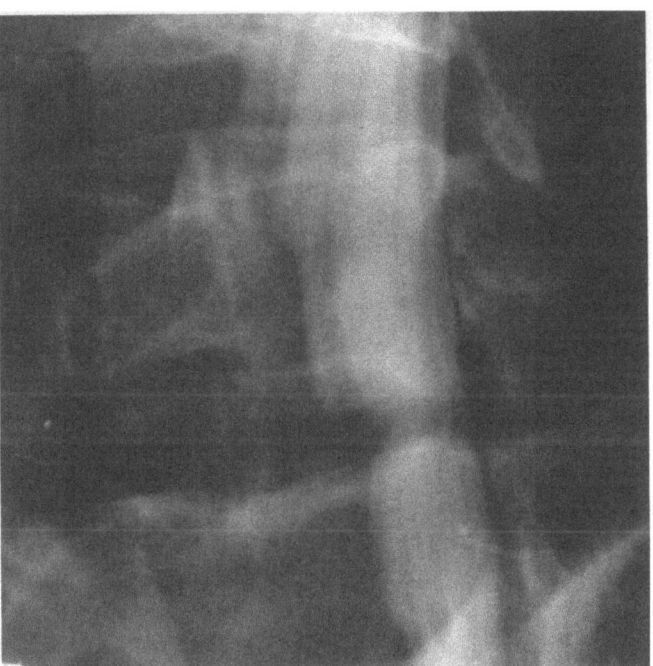

Fig. 6. Myelogram: oblique view demonstrating encroachment on the right S 1 nerve root at the level of the L 5-S 1 disc space

offers the possibility of films in upright position which sometimes reveal a dynamic disc herniation that is not apparent when the patient is lying flat. In our experience, CT scan and myelography are sufficient to decide whether CN is appropriate. Troisier et al. [82] also performed discography as an examination independent from CN procedure. According to these authors, the size of the disc protrusion, the density of contrast media filling the herniated disc and the reproduction of nerve root pain at the time of injection are of great prognostic value. However, since discography is necessary at the time of CN and carries its own complications, we do not perform it as a separate pretherapeutic investigation. The true potential of MRI imaging in the selection of patients has yet to be determined. This procedure might be extremely valuable in detecting a free disc fragment, but the inaccuracy of bony investigation and the present high cost of the technique are factors which limit its use.

Two final questions may be raised concerning indications of CN. What should be thought of enzyme injections at several spinal levels? What is the role of CN in the treatment of low back pain?

CN at several levels

Fulfillment of the criteria mentioned above imply that chymopapain is injected at a single intervertebral level. The association of two disc herniations with nerve root involvement at two different levels is a rare event. Some authors performed injection of chymopapain at several levels and treated both the level of the disc herniation and adjacent interspaces when simple disc bulging was present (Fig. 7). We do not agree with this practice for

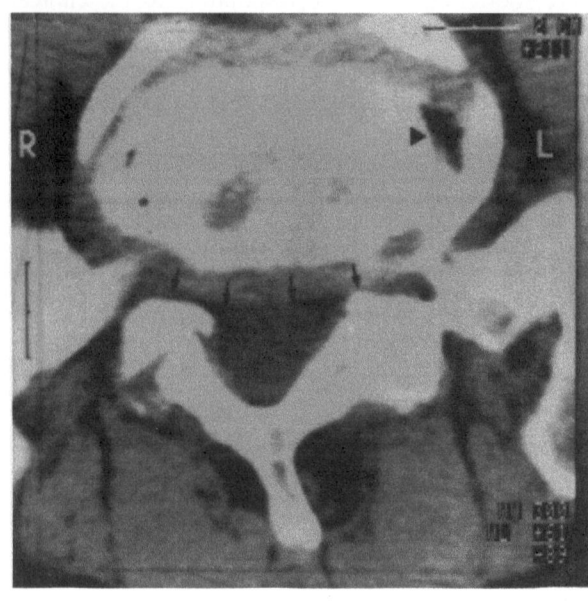

Fig. 7. Degenerative disc disease with symmetrical disc bulge (arrows) and vacuum phenomenon (arrowhead)

several reasons. There is no demonstrated advantage to performing preventive CN on disc bulging which is not responsible for nerve root compression. In addition, performance of CN at several levels may have long-term deleterious effects on the lumbar spine. Finally, a number of teams which frequently performed enzyme injections at two or three levels have abandoned this practice in favor of a more severe patient selection. However an exception to the rule of treating only one level may be legitimate when radiologic investigations reveal the chance finding of a second image of nerve compression by a disc herniation not associated with any nerve root pain. CN at one level only would carry a strong risk of causing decompensation of the second disc lesion as a result of modifications in the dynamic of the spine. Under such circumstances, it may be appropriate to treat both herniations especially since the injection of chymopapain cannot be repeated.

Role of CN in the treatment of low back pain

CN is contraindicated in cases of isolated low back pain without signs of nerve root involvement for several reasons. CN is of no value in the treatment of non-discogenic low back pain such as facet arthropathy with or without referred pain. It is always difficult to determine the precise cause of low back pain and to confirm the sole responsibility of an intravertebral disc.

Even if this was possible, it is rather difficult to conceive of the usefulness of chymopapain injection to reduce pressure exerted by the disc on the posterior vertebral ligament, when CT scan follow-up of patients after CN usually shows the persistance of a disc protrusion at levels injected. Furthermore residual low back pain is frequent after chemical treatment. CN induces accelerated disc degeneration and disc bulge which may create or increase low back pain syndrome [40]. In addition, disc narrowing caused by CN may also induce degenerative changes in the facet joints. Attempts involving CN in low back pain have generally given disappointing results [52]. Psychological and social factors, the importance of which is recognized in low back pain, are such that aggressive management techniques should be limited as far as possible.

Are there exceptions to this rule of not using chymopapain in the treatment of low back pain? There is no definite answer to this question. However it may be reasonable and legitimate to comply with the request of some patients with chronic incapacitating low back pain and focal disc herniation, who satisfy the very strict criteria listed in Table 1. In such cases CN could be attempted if the patient, fully informed of the risks of failure, is insistant that CN should be first attempted prior to lumbar fusion.

Table 1. Selection criteria in discogenic low back pain (all criteria are required)

Positive criteria	Negative criteria
Young adult (under 40)	no psychologic component
Chronic incapacitating low back pain for more than 1 year unresponsive to conservative treatment	no workmen's compensation claim
Clinical symptoms typical of discogenic pain	no other obvious cause of low back pain such as facet joints arthropathy
Radiologic evidence of focal disc herniation	no evidence of severe degenerative disc disease on conventional X-rays with symmetrical disc bulge on CT
Single level involved	
Reproduction of usual pain at discography performed as a preliminary separate procedure	

Contraindications of CN

The use of chymopapain is not always possible (Table 2). Under some circumstances chymopapain CN may be dangerous. This applies when there is sensitivity to papaya, if the patient's cardiovascular state is too precarious, or if sciatic pain occurs during pregnancy. It would also be unwise to repeat CN. Even though Sutton [76] feels that the risk is overestimated, prior CN with chymopapain increases the frequency and severity of anaphylactic shock. Despite progress in this area, there is still no absolute method for the detection of patients allergic to chymopapain and the prevention of allergic accidents. There are other contraindications to CN. When discec-

Table 2. Contraindications of CN

General contraindications	Local contraindications
Sensitivity to papaya	sciatic with complete motor loss
Prior CN	hernia sequestration
Precarious vascular state	lateral recess stenosis (width < 3 mm)
Pregnancy	spinal stenosis (central canal less than 12 mm in the area of herniated disc)
	large disc herniation (measuring more than 50% of the midsagittal dimension of the spinal canal)
	calcified disc herniation
	prior surgical discectomy at the same level

tomy has already been performed at the level which must be injected, the risk of intrathecal diffusion and the frequency of poor results are such that chemical treatment should be avoided. The same applies to patients suffering from sciatic pain with paralysis. The only treatment ensuring the rapid and regular relief of root compression is surgery. Slight motor weakness which frequently accompanies nerve root pain is not a contraindication of CN. Finally, it is fairly easy to comply with all of these precautions. In contrast, it is more difficult to detect specific local anatomical conditions which are often responsible for failure of CN. This may include hernia sequestration, narrow lumbar canal or stenosis of a lateral recess. These three potential sources of failure are briefly reviewed below.

Lateral recess stenosis

Disc bulging, osteophytic formations or subluxation of a facet joint can encroach upon the limited space available to the nerve root in the lateral recess, causing back and unilateral leg pain [52]. Lateral recess stenosis is a comprehensive contraindication to CN since chymopapain cannot have any effect when a nerve root compression is of bony origin. This is supported by findings at surgery following the failure of chemical treatment [18]. However, clinical diagnosis of nerve root compression in the lateral recess is difficult to carry out [52]. Low back pain usually predominates over leg pain. Leg pain is often claudicant in nature. In typical cases, straight leg-raising, owing to the pain in the leg, is not significantly reduced. Objective neurologic signs such as wasting, motor weakness or diminution of reflex activity are usually absent. Finally, it is often difficult to distinguish lateral recess syndrome from disc herniation on clinical grounds and in fact, the two conditions may coexist. Myelography is usually negative or shows a flattening of the nerve root sheath at the pedicle level. CT scan is a more efficient technique. CT measurements of the lateral recess must be taken at its upper part, i.e. the superior border of the pedicle, with CT window setting

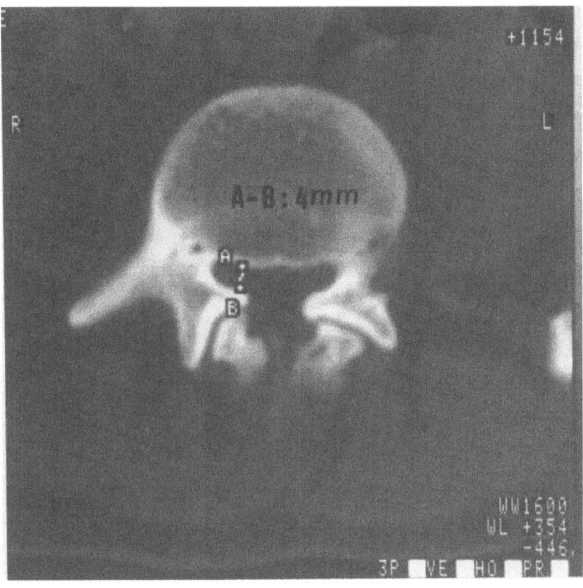

a

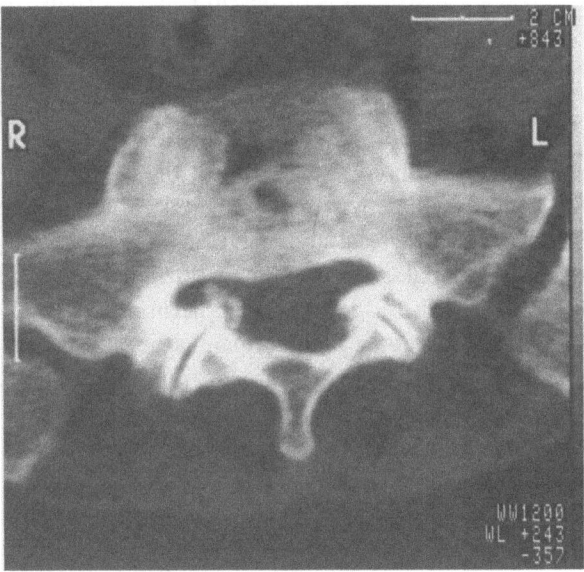

b

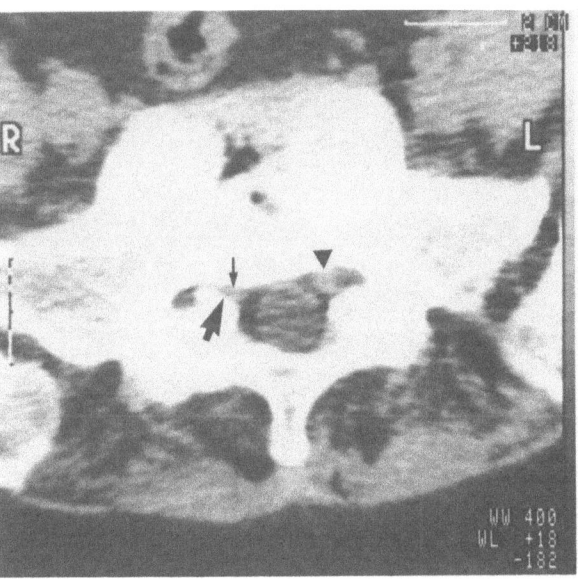

c

Fig. 8. CT evaluation of the lateral recess. **a** Measurement of the lateral recess is performed at the upper part of the pedicle, with bone window setting. **b** Acquired stenosis of the right lateral recess of S 1 caused by calcifications arising from the facet joint (bone window setting). **c** Same slice as **b**, with soft tissue window setting, demonstrating compression of the right S 1 nerve root (small arrow) by the calcification (large arrow), as compared with normal left lateral recess and S 1 nerve root (arrowhead). This is a contraindication to CN

adapted to bone distance measurement (Fig. 8 a). Absolute narrowness, which contraindicates CN, is defined by a sagittal width of less than 3 mm. However, to have any clinical significance, a narrowed lateral recess must be associated with CT evidence of nerve root encroachment (Fig. 8 b, c).

Spinal stenosis

Spinal stenosis may be of developmental origin or result from a degenerative process. In true segmental degenerative stenosis, CN will increase disc bulging and dural compression. However when stenosis is moderate and when there is a true disc herniation, the injection of chymopapain may be attempted but the patient must be warned that the chances of success are reduced [3].

When the stenosis is of developmental origin, and if a disc herniation is responsible for nerve root compression, it is theoretically possible that enzymatic digestion of the disc may restore the canal to its former condition (Fig. 9). However the anular bulging and intervertebral space narrowing resulting from CN may create a central degenerative stenosis with dural compression or bony entrapment of the nerve root in the lateral recess or foramen. For these reasons enzyme injection is contraindicated in absolute narrowness of the central canal, i.e., when the midline sagittal diameter of the central canal is less then 12 mm (absolute developmental stenosis of the central canal in the region of the herniated disc) [25, 67]. Developmental stenosis of the upper lumbar spine is not a contraindication if the region interested by the herniated disc has normal or subnormal dimensions.

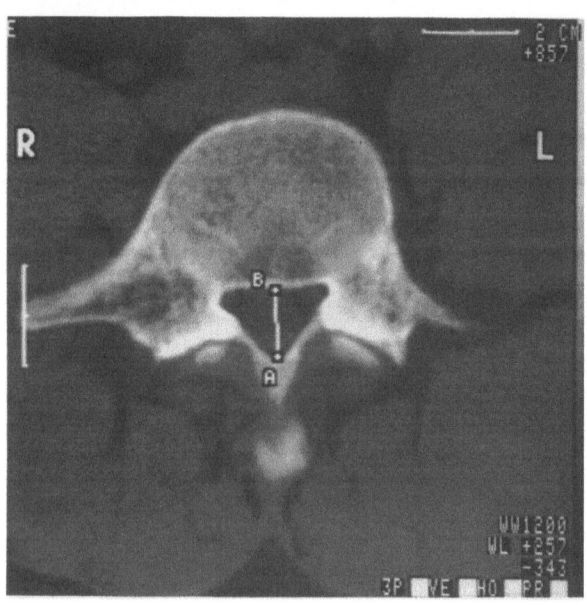

Fig. 9. Midsagittal diameter of 13 mm measured at pedicle level with bone window setting. This represents relative narrowness of the central spinal canal

Sequestered herniation (free disc fragments)

Since chymopapain injected in the intervertebral disc should not affect epidural disc fragments separated from the parent disc, sequestered discs constitute a comprehensive contraindication to CN. Post CN operative findings have confirmed that free fragments are actually one of the main causes of CN failure [18]. Disc sequestration is clinically suspected when low back pain disappears suddenly, while at the same time, nerve root pain is exacerbated. However, this event is rather rare. Deburge et al. found no definite correlation between clinical symptoms and disc sequestration [19]. The presence of a free fragment is often a surprise radiological discovery. Nevertheless radiologic investigation remains difficult and often disappointing. In fact, CT demonstration of an epidural disc fragment definitely separated from the intervertebral space (Fig. 10) is the only specific radiologic criterion preventing the use of chymopapain, but, according to Dillon et al. [22], this sign was identified in only 42.8% of cases. Other CT features and myelographic findings are not pathognomonic of disc sequestration and have only indicative value. Particular emphasis has been placed upon the value of an irregular appearance of the edges of the hernia (polypoid margins) and the degree of migration of disc material from the parent disc [19]. However the extent to which these two findings must be considered as contraindications to CN is a matter of controversy. No clear comparison of treatment results in the presence or absence of these two features exists in the literature. Polypoid irregular margins are as frequent in extruded as in sequestered disc herniations [19]. It is likely that the probability of disc sequestration increases with the degree of migration of the herniation. However, the degree

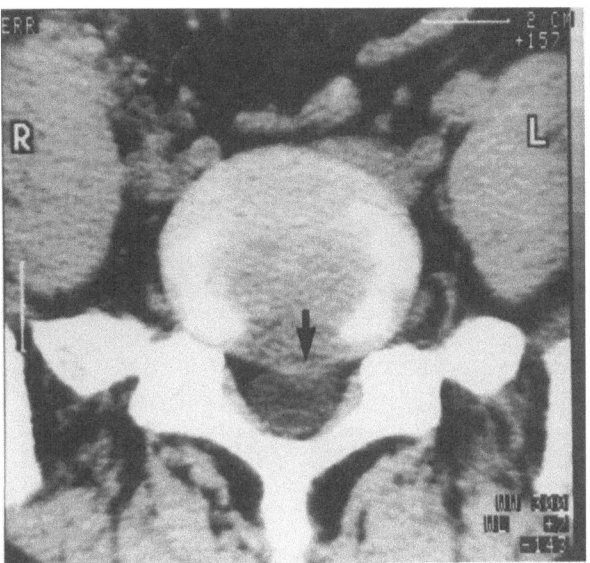
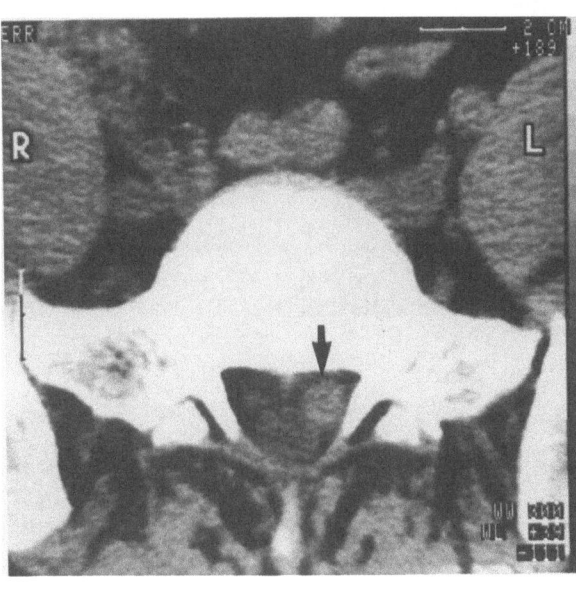

a   b

Fig. 10. Free disc fragment: at disc level (**a**), only a small protrusion (arrow) is visible at the left posterolateral aspect of the disc space (L 5-S 1). One centimeter caudally (**b**) a free disc fragment (arrow) surrounded by fat is found in the extradural space and obliterates the left S 1 nerve root

of the disc migration which would make disc sequestration highly probable and CN unwise, is not known. Finally moderate disc migration is present in a majority of herniated discs and is not a contraindication to CN. The maximum distance of migration which the authors accept is 1 cm above or below the intervertebral space, i.e. two contiguous slices of 5 mm thickness (Fig. 11). Eschard et al. [28] believe that a sequestered herniation is highly probable when, during discography, all of the contrast medium is distributed behind the disc without the disc itself being opacified. In any event, this disc incontinence must be distinguished from an ordinary epidural leak which does not alter the indication for CN.

Large disc herniations

Large herniated discs are likely to be extruded through the posterior longitudinal ligament. Fries et al. [35] showed the ratio of the size of the herniated disc to the antero-posterior diameter of the thecal sac to be correlated with disc extrusion; they found that 90 percent of herniated discs larger than half the sagittal diameter of the thecal sac were extruded through the posterior longitudinal ligament at surgery (Fig. 12). However it has not been estab-

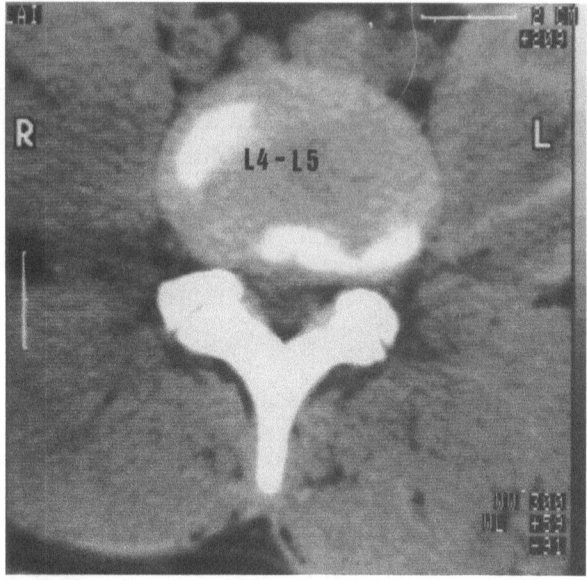

a

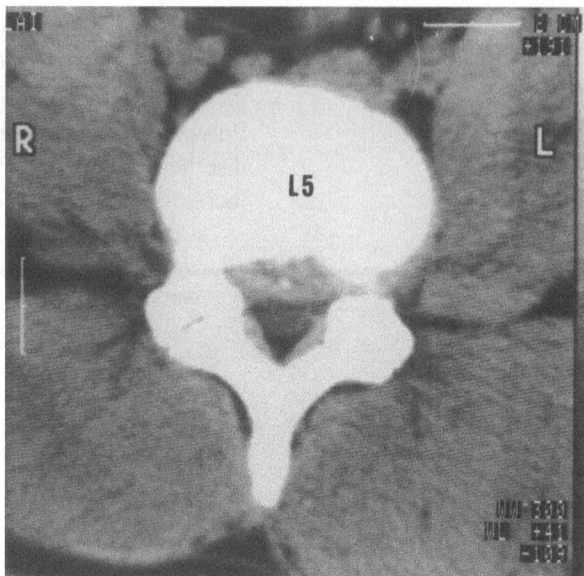

b

c

Fig. 11. Caudad disc migration: small right hernia at L 4-5 disc level (**a**). Disc material is still present in the upper (**b**) and lower parts (**c**) (arrow) of the right lateral recess of L 5

lished that suspicion of an extruded disc is a contraindication to CN. Many authors do not perform CN in very large herniated discs but give no reason for their decision. However, recent studies have reported that success rate of CN decreases in cases of large disc herniations, i.e. those measuring more than 50 percent of the midsagittal dimension of the spinal canal [60, 66]. Therefore, the authors are used to refer patients with very large herniations to the surgeon.

Calcified disc herniations

CN is contraindicated in highly calcified herniated discs since chymopapain is likely to be ineffective. However, true calcification of herniated disc ma-

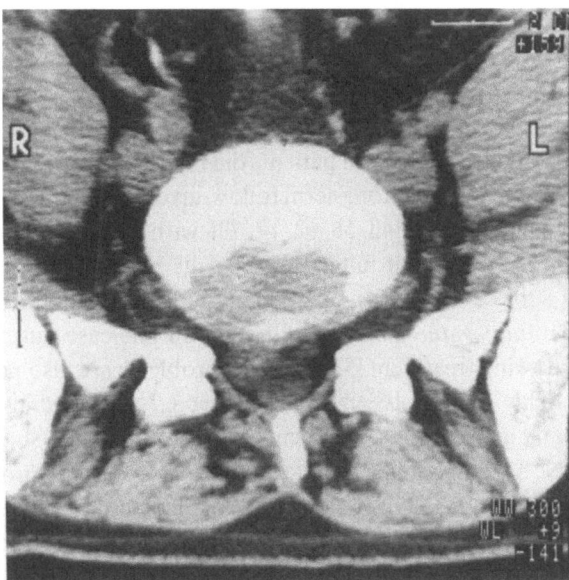

Fig. 12. Large and pediculated disc herniation with an anteroposterior dimension equivalent to that of the thecal sac. Extruded disc is highly probable. Disc sequestration is possible

terial must be differentiated from partial bony avulsion of the vertebral endplates which is rather frequently associated with large disc herniation and does not contraindicate CN.

Results

The effectiveness of CN in the treatment of nerve root pain due to disc herniation is no longer contested. Since the initial study of Schewtschenau et al. [72] which showed no significant difference between the action of the enzyme and of a placebo, three double blind trials have demonstrated the effectiveness of CN. According to Fraser [33], 80% of patients were markedly improved by chymopapain and only 50% by the placebo. In the study of Javid et al. [45] involving 108 CN procedures, the success rate obtained with chymopapain injection was almost twice as great as with the placebo injection, with 82% of good results six months after injection; furthermore, following failure of the placebo, a successful result was almost always obtained with the use of chymopapain. In France, Feldman et al. [30] also confirmed the efficacy of CN but their figure for good results did not exceed 65% after three months follow-up. The high level of placebo effects in all of these series is noteworthy, and this raises two questions. Was the decision that medical treatment had failed taken too hastily? Does disc puncture in itself reduce disc pressure within the disc herniation?

In addition to double blind trials, a large number of uncontrolled series have been published. The methodology differs from one trial to another, often making interpretation of the results difficult [21]. Criteria required to perform CN, the number of discs injected, and the scoring of results are not always identical [44, 81, 88]. However an increasing number of authors used MacNab's criteria [54]. The dose of chymopapain used also varies but the dosage most often employed is 4,000 U per disc. Despite these discrepancies, the overall results of these studies which involve a very large number of patients provide extensive information. Thus on the basis of 3,854 published cases, Simmons et al. were able to calculate that the mean percentage success rate, defined as sufficient improvement to return to normal activity, was 76%. This figure is similar to that of European publications. Return to work of patients following CN is satisfactory. Three patients out of four returned to work before the end of the fourth month after chymopapain injection. Only 13% of patients did not return to work and only 17% changed their job [43, 56]. Long-term follow-up of patients after CN has been achieved by several teams [43, 56, 62, 77, 79], with a time interval frequently exceeding ten years. Overall initial good results persist with time but the percentage of subjects free of any pain slightly decreases [79]. Recurrence of disc herniation is rare. When it occurs it is in one case out of two during the year following treatment [53, 77]. Late problems are also possible, especially after a delay of four to eight years after CN. According to Mansfield [56], as many as 5 percent of patients undergo low back surgery in the years after CN. Several studies have attempted to find some prognostic factors of CN results [15, 53, 83]. It would appear that the sex of the patient, the duration of symptoms before CN, the duration of a severe degenerative disc disease and the number of vertebral levels treated (if not more than two) do not modify the success rate. Moreover, the age of the subject has no influence on results and even adolescents may be treated with chymopapain [51]. However, results are often unsatisfactory in patients claiming workers compensation.

Following CN, sudden disappearance of pain after treatment is possible though rare. More often, nerve root pain lessens gradually. Improvement starts by the third day in one case out of two and continues over the subsequent weeks [30]. If a moderate motor weakness is present before treatment, it improves in one case out of two. Finally, results concerning nerve root pain are satisfactory in 60% of cases by the end of the third week [20] but progress may continue for several months. However if no improvement is seen by the end of six weeks, failure may be considered as definitive. Evolution of low back pain often differs from that of leg pain. Low back pain may decrease rapidly after enzyme injection, but worsening is reported in two patients out of three in the post operative days [32]. In 25–30% of cases, low back pain may be particularly severe and accompanied by major spasm of the paravertebral muscles [69, 20]. The possibility of such severe painful reactions must be well known since they are alarming and may disappear only gradually. However such acute attacks of low back pain are of no prognostic significance. Length of bed rest after CN seems to have no influence on their onset. A lumbar corset may shorten these painful episodes but does not prevent them [57]. After a few months or years, low back pain usually improves but disappears in only one third of cases while one half of treated patients has no leg pain [43].

A number of studies has dealt with radiologic changes following the injection of chymopapain. Decrease in the height of the treated disc is almost constant (average 25%) [47]. Disc space narrowing usually occurs early within one week and is maximal after four to eight weeks, subsequently showing no change [31, 38]. However late reexpansion of the disc is possible and may be seen in 15% of cases if surveillance is sufficiently long [77, 79]. It is not known whether notable disc space narrowing is more common when treatment is successful since data from the literature are in disagreement. CT scan studies of patients following CN have provided very valuable information [10, 23, 38, 47, 55]. It is presently known that disc bulge is a frequent consequence of chemical treatment and that the disappearance of the herniation is slow and inconstant. In fact a focal disc protrusion is still visible in one case out of two at the end of three months, but this proportion decreases with time [47]. Early disappearance of the herniation on CT scan would appear to be correlated with the success of treatment. Finally, when CT scans are repeated, it is not uncommon to note the appearance of a vacuum phenomenon within the treated disc [38]. Recently magnetic resonance imaging has also been used to evaluate changes in injected discs [39]. After treatment their signal decreases notably, in particular in T 2 weighted images, reflecting a fall in water content [39]. This loss of signal does not appear immediately after CN but develops progressively over a few weeks, at the same time that the disc narrows. From the fourth month onward the signal loss is complete at all treated levels. These changes would appear to be independent of the therapeutic result.

Comparison of CN and surgical discectomy

Since CN is an alternative to disc surgery, these two types of treatment must be compared. The frequency of serious complications following the injection of chymopapain is no greater than that of discectomy and minor complications would seem to be rarer. The relative efficacy of the two approaches is more difficult to assess. In many series of CN a good or excellent result is reported in approximately 80% of cases. However failing rates vary greatly from one study to another and close comparison of disc surgery and CN is difficult. Some studies have been published in which patients are divided into two groups according to the mode of treatment. In open trials, results are fairly homogeneous and show no significant difference in efficacy between CN and discectomy [16, 61, 63, 85, 86]. However return to activity would seem to be more rapid after enzyme treatment. Leavitt et al. [49] reached a similar conclusion in a randomized series. In contrast, two other randomized trials give the advantage to surgery, but in these studies the failure rate of CN was as high as 50% [14, 27]. In the end, it is difficult to reach any conclusion concerning the respective merits of the two types of treatment. New prospective trials would be desirable particularly since use of the microscope in disc surgery appears to improve the results of discectomy [58]. The costs of each method have also been compared. CN is the less expensive technique, except in the case of worker's compensation claimants who more often undergo secondary surgery because of the frequent failure of enzyme injection [64, 68]. However the cost of surgery seems to decrease when microdiscectomy is involved, in which case it would differ little from CN [58].

Surgery after CN

Following failure of CN, one patient in two undergoes surgery. A minimum interval of two months must be allowed to elapse between the injection of chymopapain and surgery. The surgical procedure does not seem to be particularly difficult. The surgeon almost never finds evidence of epidural fibrosis and most often discovers that the disc residue is minimal. However persistance of a disc herniation at the level treated, reflecting true inactivity of the enzyme, is possible. The result of this surgery is satisfactory in two thirds of cases. It is all the better when compression on the nerve root is found at surgery.

The causes for failure of CN are more easily understood by analysis of operative findings. Broadly speaking, one patient out of two has a persistant disc herniation. Disc herniation may be sequestered but subligamentous migration of the disc may also be associated with failure of CN. In one case out of three, persistence of nerve root pain is explained by the existence of a narrow lateral recess. In other cases the surgeon finds a disc protrusion or more rarely a narrowing of the foramen by subluxated facet joints due to disc narrowing. Finally, no cause of nerve root compression is found in 10% of patients [84, 89]. It should be noted that, according to some authors, spinal segmental instability may sometimes be responsible for the failure of CN. Under such circumstances a posterior lumbar interbody fusion would give satisfactory results [73].

Other enzymes

When the nucleus pulposus is severely degenerated, its proteoglycan and water content decreases notably. Chymopapain is then probably less active. Furthermore the immunogenic power of this substance is such that repeated injection is not possible. This has led some authors to study other enzymes, such as collagenase or iniprol, which have a less selective hydrolytic site of action. Collagenase has been the most extensively studied. When administered in the animal, the toxicity of this substance would seem to differ little from that of chymopapain, and in particular epidural injection is not dangerous while introduction of the compound into the subarachnoid space causes a subarachnoid hemorrhage [8, 71]. However some studies indicate the need for caution. Histological results obtained from discs removed surgically after failure of collagenase CN are contradictory. According to Artigas et al. [2], the anulus, vertebral endplates and epidural fat all show changes. However Brown found no extranuclear abnormalities. A few clinical trials have been carried out with this enzyme and the initial results are encouraging [42]. In a double blind trial, collagenase was even found to be markedly more active than a placebo, the percentage of good and very good results reaching 80% [9].

Recently a French team carried out disc injections of triamcinolone (hexatrione*) [6]. It would seem that this compound, which is a fluorinated corticosteroid with a highly atrophic action, is capable or relieving patients suffering from sciatic pain of disc origin. However there is no evidence currently available to show that triamcinolone produces disc nucleolysis.

Conclusion

CN is undoubtedly effective. However in the absence of a strictly managed trial comparing CN and surgery, and especially microscopic discectomy, their relative roles in the treatment of disc herniation cannot be completely assessed. A careful radiologic evaluation is necessary in order to detect the anatomical features responsible for most of the inefficacy of chemical treatment. Although CN is associated with little danger, it is not reasonable to expose the patient to the cumulative risks of two successive treatments.

---

* Triamcinolone hexacetonide

References

1. Agre K, Wilson RR, McDermott DJ (1984) Chymiodactin post-marketing surveillance. Demographic and adverse experience data in 29,075 patients. Spine 9: 479–485

2. Artigas J, Brock M, Mayer HM (1984) Complications following chemonucleolysis with collagenase. J Neurosurg 61: 679–685

3. Benoist M, Bouillet R, Mulholland R (1983) Chimionucléolyse. Résultats d'une enquête Européenne. Acta Orthop Belg 49 [Suppl 1]: 33–48

4. Bitz DM, Ford LT (1977) An evaluation of narrowing following intradiskal injection of chymopapain. Clin Orthop 129: 191–195

5. Bouillet R (1983) Complications du traitement de la hernie discale. Etude comparée des complications du traitement chirurgical et de la nucléolyse par la chymopapaine. Acta Orthop Belg 49 [Suppl 1]: 49–75

6. Bourgeois R, Frot B, Folinais D, David M, Benacerraf R, Palazzo E, Vigneron AM, Kahn MF (1986) Traitement de la lombosciatique par hernie discale par nucléorthèse à l'hexacetonide de triamcinolone. Presse Med 15: 41

7. Bradford DS, Cooper KM, Oegama TR (1983) Chymopapain, chemonucleolysis and nucleus pulposus regeneration. J Bone Joint Surg [Am] 65: 1220–1231

8. Bromley J, Hirst J, Osman M (1980) Collagenase. An experimental study of intervertebral disc dissolution. Spine 5: 126–132

9. Bromley JW, Varma AO, Santoro AJ, Cohen P, Jacobs R, Berger L (1984) Double-blind evaluation of collagenase injections for herniated lumbar discs. Spine 9: 486–488

10. Brown B, Stark E, Dion G, Ono H (1985) Computed tomography and chymopapain chemonucleolysis: preliminary findings. AJR 6: 51–54

11. Brown M, Tompkins J (1986) Chemonucleolysis (discolysis) with collagenase. Spine 11: 123–130

12. Buttle D, Abrahamson M, Barret A (1986) The biochemistry of the action of chymopapain in relief of sciatica. Spine 11: 688–694

13. Cormier C (1986) La nucléolyse chimique à la chymopapaine. Etude du résultat à long terme des 67 premiers malades traités à l'Hôpital Cochin. Mémoire pour les prix de l'internat en Médecine

14. Crawshaw C, Frazer A, Merriam W, Mulholland R, Webb J (1984) A comparison of surgery and chemonucleolysis in treatment of sciatica. A prospective randomized trial. Spine 9: 195–198

15. Dabezies E, Beck C, Shoji H (1986) Chymopapain in perspective. Clin Orthop 206: 10–14

16. Dabezies E, Brunet M (1978) Chemonucleolysis versus laminectomy. Orthopedics 1: 26

17. Davis RJ, North RB, Campbell JM, Suss RA (1984) Multiple cerebral hemorrhages following chymopapain chemonucleolysis. J Neurosurg 61: 169–171

18. Deburge A, Rocolle J, Benoist M (1985) Surgical findings and results of surgery after failure of chemonucleolysis. Spine 10: 812–815

19. Deburge A, Benoist M, Boyer D (1984) The diagnosis of disc sequestration. Spine 9: 496–499

20. Dermott D, Agre K, Brim M, Demma F, Nelson J, Wilson R, Thisted R (1985) Chimiodactin in patients with herniated lumbar intervertebral disc(s). An open label multicenter study. Spine 10: 242–249

21. Deyo RA (1984) Chymopapain for herniated intervertebral disc: a methodologic analysis and an agenda for future research. Spine 9: 474–478

22. Dillon WP, Kaseff LG, Knackstedt VE, Osborn AG (1983) Computed tomography and differential diagnosis of the extruded lumbar disc. J Comput Assist Tomogr 7: 969–975

23. Drouillard J, Lavignolle B, Philippe JC, Eresue J, Drouillard N, Senegas J, Tavernier J (1982) Scanographie et chimionucléolyse des hernies discales. Intérêts et limites. A propos de 20 cas. J Radiol 63: 276–272

24. Dyck P (1985) Paraplegia following chemonucleolysis. A case report and discussion of neurotoxicity. Spine 10: 359–362

25. Eisenstein S (1976) Measurements of the lumbar spinal canal in 2 racial groups. Clin Orthop 115: 42–46

26. Eguro H, Joliet PD (1983) Transverse myelitis following chemonucleolysis. J Bone Joint Surg [Am] 65: 1328–1330

27. Ejeskar A, Nachemson A, Herberts P, Lysell E, Andersson G, Irstam L, Peterson L (1985) Surgery versus chemonucleolysis for herniated lubar disc. A prospective study with random assignment. Clin Orthop 174: 236–242

28. Eschard JP, Seignon B, Gatfosse M, Brochot P, Baudrillard JC, Etienne JC, Gougeon J (1986) L'incontinence discale: une contre-indication à la nucléolyse. Nouv Press Med 15: 2166

29. Eyre D, Muir H (1976) Types I and II collagens in intervertebral disc. Biochem J 157: 267–270

30. Feldman J, Menkes C, Pallardy G, Chevrot A, Horreard P, Zenny J, Godefroy D, Amor B (1986) Etude en double-aveugle du traitement de la lombo-sciatique discale par chimionucléolyse. Rev Rhum 53: 147–152

31. Flanagan M, Chung B (1985) Roentgenographic changes in 188 patients 10–20 years after discography and chemonucleolysis. Spine 11: 444–448

32. Flanagan N, Smith L (1986) Clinical studies of chemonucleolysis patients with ten to twenty-year follow-up evaluation. Clin Orthop 206: 15–17

33. Fraser RD (1982) Chymopapain for the treatment of intervertebral disc herniation. A preliminary report of a double-blind study. Spine 7: 608–612

34. Fraser RP, Fracs, Osto O, Vernon-Roberts B, Path F (1986) Disctis following chemonucleolysis. An experimental study. Spine 11: 679–687

35. Fries JW, Abodeely DA, Vijungco JG, Yeager VL, Gaffey WR (1982) Computed tomography of herniated and extruded nucleus pulposus. J Comp Assist Tomogr 6: 874–887

36. Garvin P, Jennings R, Smith L (1965) Chymopapain: a pharmcologic and toxicologic evaluation in experimental animals. Clin Orthop 42: 204–223

37. Gentry L, Strother G, Turski P, Javid M, Sackett J (1985) Chymopapain chemonucleolysis: correlation of diagnostic radiographic factors and clinical outcome. AJR 6: 311–320

38. Gentry L, Turski P, Strother C, Javid M, Sackett J (1986) Chymopapain chemonucleolysis: CT changes after treatment. AJR 6: 231–329

39. Gibson M, Buckley J, Mulholland R, Worthington B (1986) The changes in the intervertebral disc after chemonucleolysis demonstrated by magentic resonance imaging. J Bone Joint Surg [Br] 68 B: 719–723

40. Gotfried Y, Bradford DS, Oegema TR (1986) Facet joint changes after chemonucleolysis induced space narrowing. Spine 11: 944–950

41. Greenough CG, Dimmock S, Edwards D, Ransford AO, Bentley G (1986) The role of computered tomography in intervertebral disc prolapse. J Bone Joint Surg [Br] 68 B: 729–733

42. Hedtmann A, Steffen R, Kramer J (1987) Prospective comparative study of intra-discal high-dose and low-dose collagenase versus chymopapain. Spine 12: 388–392

43. Jaabay G (1986) Chemonucleolysis. Eight to ten-year follow-up evaluation. Clin Orthop 206: 24–31

44. Javid MJ (1980) Treatment of herniated lumbar disk syndrome with chymopapain. JAMA 243: 2043–2048

45. Javid MJ, Nordby EJ, Ford LT, Hejna WJ, Whisler WW, Burton C, Millet K, Wiltse LL, Widell EH, Boyd RJ, Newton SE, Thisted R (1983) Safety and efficacy of chymopapain (chymiodactin) in herniated nucleus pulposus with sciatica. Results of a randomized, double-blind study. JAMA 249: 2489–2494

46. Jenner J, Buttle D, Dixion A (1986) Mechanism of action of intradiscal chymopapain in the treatment of sciatica: a clinical, biochemical and radiological study. Ann Rheum Dis 45: 441–449

47. Konings J, Williams F, Deutman R (1986) Computed tomography analysis of the effects of chemonucleolysis. Clin Orthop 206: 32–36

48. Krempen JF, Minnig DI, Smith BS (1975) Experimental studies on the effect of chymopapain on nerve root compression caused by intervertebral disk material. Clin Orthop 106: 336–349

49. Leavitt F, Garron D, Whisler W, D'Angelo C (1980) A comparison of patients treated by chymopapain and laminectomy for low back pain using a multidimensional pain scale. Clin Orthop 146: 136–143

50. Levernieux J, Seze S de (1968) Le renouveau de la discographie lombaire. In: Sèze S de, Ryckewaert A, Kahn MF, Dreyfus P (eds) L'actualité rhumatologique. Expansion Scientifique Française, Paris, pp 207–214

51. Lorenz M, Mc Culloch J (1985) Chemonucleolysis for herniated nucleus pulposus in adolescents. J Bone Joint Surg [Am] 67: 1402–1404

52. McCulloch JA (1977) Chemonucleolysis. J Bone Joint Surg [Br] 59: 45–52

53. Maciunas R, Onofrio B (1986) The long term results of chymopapain chemonucleolysis for lumbar disc disease. Ten-year follow-up results in 268 patients injected at the Mayo-clinic. J Neurosurg 65: 1–8

54. MacNab I (1971) Negative disc exploration. J Bone Joint Surg [Am] 53: 891–895

55. Mall JC, Kaiser JC (1984) Post-chymopapain (chemonucleolysis) clinical and computed tomography correlation; preliminary results. Skeletal Radiol 12: 270–275

56. Mansfield F, Polivy K, Boyd R, Huddleston J (1986) Long-term results of chymopapain injections. Clin Orthop 206: 67–69

57. Margue A (1986) Etude randomisée de l'influence du lever précoce sur le résultat de la chimionucléolyse dans le traitement des sciatiques d'origine discale. Mémoire de CES de rhumatologie 1986, Université de Paris V

58. Maroon JC, Abal A (1985) Microdiscectomy versus chemonucleolysis. Neurosurgery 16: 644–649

59. Martins AN, Ramirez A, Johnston J, Schwetschenau PR (1978) Double-blind evaluation of chemonucleolysis for herniated lumbar disc. J Neurosurg 49: 816–827

60. Mulawka S, Weslowski D, Herkowitz N (1986) Chemonucleolysis. The relationship of the physical findings, discography and myelography to the clinical result. Spine 11: 391–396

61. Nordby E, Lucas G (1973) A comparative analysis of lumbar disk disease treated by laminectomy of chemonucleolysis. Clin Orthop 90: 110–115

62. Nordby E (1986) Eight-to 13-year follow-up evaluation of chemonucleolysis patients. Clin Orthop 206: 18–23

63. Nordby EJ (1985) A comparison of discectomy and chemonucleolysis. Clin Orthop 200: 279–283

64. Norton W (1986) Chemonucleolysis versus surgical discectomy. Comparison of costs and results in worker's compensation claimants. Spine 11: 440–443

65. Parkinson D (1983) Late results of treatment of intervertebral disc disease with chymopapain. J Neurosurg 59: 990–993

66. Postacchini F, Lami R, Massobrio M (1987) Chemonucleolysis versus surgery in lumbar disc herniations: correlation of the results to pre-operative pattern and size of herniation. Spine 12: 87–96

67. Postacchini F, Ripani M, Carpano S (1983) Morphometry of the lumbar vertebral. An anatomic study in two caucasoid ethnic groups. Clin Orthop 172: 296–303

68. Ramirez L, Javid M (1985) Cost effectiveness of chemonucleolysis versus laminectomy in the treatment of herniated nucleus pulposus. Spine 10: 363–367

69. Revel M, Boumier P, Chevrot A, Feldmann J, Amor B, Menkes C (1985) Prise en charge immédiate et tardive des patients traités par nucléolyse. Rhumatologie 27: 25–28

70. Roggendorf W, Brock M, Gorge H, Curio G (1984) Morphological alterations of the degenerated lumbar disc following chemonucleolysis with chymopapain. J Neurosurg 60: 518–522

71. Rydevik B, Brown M, Ehira T, Nordborg C (1985) Effects of collagenase on nerve tissue. An experimental study on acute and long-term effects in rabbits. Spine 10: 562–566

72. Schwetschenau P, Ramirez A, Johnston J, Barnes E, Wiggs C, Martins A (1976) Double-blind evaluation of intra-discal chymopapain for herniated lumbar discs. Early results. J Neurosurg 45: 622–627

73. Sepulveda R, Kant A (1985) Chemonucleolysis failures treated by Plif. Clin Orthop 193: 68–74

74. Smith L (1964) Enzyme dissolution of the nucleus pulposus in humans. JAMA 187: 137–140

75. Suguro T, Oegema R, Bradford D (1986) The effects of chymopapain on pro-lapsed human intervertebral disc. A clinical and correlative histochemical study. Clin Orthop 213: 223–231

76. Sutton JC (1986) Repeat chemonucleolysis. Clin Orthop 206: 45–49

77. Sutton C (1986) Chemonucleolysis in the management of lumbar disc disease. A minimum six year follow-up evaluation. Clin Orthop 206: 56–60

78. Takahashi K, Inoue S, Takada S, Nishiyama H, Mimura M, Wada Y (1986) Experimental study of chemonucleolysis with special reference to the change of intradiscal pressure. Spine 11: 617–620

79. Thomas J, Wiltse L, Widell E, Spencer C, Zindrick M, Ted Field B (1986) Chemonucleolysis. A ten year retrospective study. Clin Orthop 206: 61–66

80. Thomas L (1956) Reversible collapse of rabbit ears after intravenous papain and prevention of recovery by cortisone. J Exp Med 104: 245–259

81. Troisier O, Dewerpe P, Pelleray B (1982) Bilan de 5 années de traitement par nucléolyse de 150 radiculalgies et 10 lombalgies discales. Rev Rhum Mal Os-teoartic 49: 377–383

82. Troisier O, Cypel D (1986) Discography an element of decision surgery versus chemonucleolysis. Clin Orthop 206: 70–78

83. Valat JP, Eveleigh M, Fouquet B, Martin P, Le Goff P, Alcalay M, Bontoux D, Chales G, Busson F, Pawlotsky Y, Passuti J, Lanoiselee J, Prost A, Moneger M, Vialle M, Bureau T, Bregeon C (1986) La chymionucléolyse dans le traitement des lombo-radiculalgies d'origine discale. Etude coopérative de 333 cas. Rev Rhum Mal Osteoartic 53: 467–471

84. Vignon E, Bochu M, Vignon G, Vial B, Charhon S, Delmas P (1985) Etude des échecs de la nucléolyse. Rhumatologie 37: 7–12

85. Watts C, Hutchison G, Stern J, Clark K (1975) Comparison of intervertebral disc disease treatment by chymopapain injection and open surgery. J Neurosurg 42: 397–400

86. Weinstein J, Lehmann T, Hejna W, Neill Th, Spratt K (1986) Chemonucleolysis versus open discectomie. A ten-year follow-up study. Clin Orthop 206: 50–55

87. Weitz EM (1984) Paraplegia following chymopapain injection. A case report. J Bone Joint Surg [Am] 66: 1131–1135

88. Wilste LL, Widell EH, Yuan HA (1975) Chymopapain chemonucleolysis in the lumbar disc disease. JAMA 231: 474–479

89. Zaleske D, Ehrlich M, Huddleston J (1977) Combined biochemical and clinical investigation of chemonucleolysis failures. Clin Orthop 126: 121–126

Anesthesia and complications in chymopapain chemonucleolysis

Marie-Christine des Essarts

Department of Anesthesia, Hôpital Lariboisière, Paris, France

Chemonucleolysis is a painful procedure which requires anesthesia. In addition, chymopapain, an enzymatic substance extracted from the latex of the unripe papaya, is antigenic. In approximately 0.5% of cases, chymopapain chemonucleolysis (CN) produces anaphylactic shock. For these reasons, CN requires the presence of an anesthesiologist experienced in radiologic procedures and capable of intervening in case of anaphylactic shock [1, 4, 12].

Methods of anesthesia

Each step in the chemonucleolysis procedure, needle approach, discography and enzyme injection, requires effective anesthesia. Several methods of anesthesia are possible.

General anesthesia

General anesthesia was recommended during the early years of chemonucleolysis [9, 12, 23].

Advantages

With general anesthesia, fluoroscopic guidance during the approach is more accurate because the patient is immobile, and difficult punctures are accomplished more easily.

Disadvantages

General anesthesia is a complicated procedure necessitating intubation and controlled respiration. The drugs used (sedatives, curares) have histamine release and antigenic properties, which add to those of chymopapain [13]. The risk of hypotension is increased, because sympathethic reactions are reduced under general anesthesia [13]. With the use of halogen anesthetics, epinephrine can cause cardiac arrhythmia. Several reports indicate that CN accomplished under general anesthesia is associated with an increased incidence of anaphylactic shock as well as neurologic complications. Of 58,333 cases in a phase IV study by Smith-Armour, 0.6% of anaphylactic reactions occurred under general anesthesia, while 0.4% of anaphylactic reactions occurred under local anesthesia [4].

Local anesthesia with neuroleptanalgesia

This is the most widely used method at present [1, 8, 9, 12, 14, 15, 21, 23].

Advantages

With local anesthesia, positioning of the patient is facilitated. In case of a nerve root injury, the patient can indicate pain in the lower leg. He can also report early warning signs which may preceed anaphylactic shock and should be recognized immediately. In this case, the procedure should be discontinued at once, and the anesthesiologist must begin the required treatment without delay.

Disadvantages

Since the patient is not asleep, he must be willing to cooperate during the procedure.

Management

Effective local anesthesia is ensured with 1% lidocaine. Approximately 10 minutes before the procedure, drugs are administered intravenously to produce analgesia, relaxation and relief of anxiety. Drugs which do not produce histamine release are selected [12, 13]. The analgesic drugs, fentanyl or phenoperidine, are used in association with either a benzodiazepine, such as diazepam, flunitrazipam or midazolam, which produce a "diazanalgesia", or with a neuroleptic such as droperidol, which produces a neuroleptanalgesia. These drugs can be associated with hydroxyzine, which combines anti-anxiety and antihistamine properties. An example of a detailed protocol is figured in Table 1. During discography, the analgesic is readministered, so that chymopapain will be well tolerated when it is injected approximately 10 minutes later. Respiration must be carefully monitored at the time of the second injection of analgesic, since there is a risk of respiratory failure. The patient spontaneously breathes oxygen-enriched air or a mixture of 50% $N_2O - 50\%$ O_2.

Other techniques

Epidural anesthesia, and electric anesthesia in conjunction with drugs are practiced in some hospitals. These methods can produce a prolonged analgesic effect without risk of respiratory failure [24].

Pre-operative assessment

Pre-anesthesia consultation

This should be done about one week before chemonucleolysis. The anesthesiologist fills out an anesthesia chart after a careful history and clinical examination. Special attention should be paid to cardiovascular and anaphylactic risks.

Cardiovascular risk

Coronary patients, or patients with unstable cardiovascular conditions who may not tolerate the abrupt hemodynamic variations of severe anaphylactic shock, must not undergo chemonucleolysis [1]. Effective beta-blocking therapy, making the patient resistant to epinephrine, is also a contraindication [5].

Table 1. Example of detailed protocol. Chemonucleolysis L4–5, 5-20-87

GAR ..... M 38   100 Kilo./1 m 68

Time	9	10	11	12

Blood
Pressure

Cardiac
Rate

G 5% 500 ml   RINGER

FENTANYL
100 δ IV

HALOPERIDOL
5 mg IV

DIAZEPAM
5 mg IV

HYDROXYZINE
100 mg IV

X. Ray

LIDOCAINE 1%   PUNCTURE   DISCOGRAPHY   PAPAIN 4000 U   RECOVERY ROOM

Anaphylactic risk

The following elements are carefully investigated:
- prior allergy to food;
- prior allergy to drugs;
- atopy, including asthma in the young patient, nasal allergy (hay fever), and eczema;
- prior contact with papain.

Sensitivity to chymopapain has not been established in such patients. However, in these cases, it is wise to perform immunoallergic studies in order to eliminate allergic patients prior to the procedure [2, 6, 15, 18, 19]. It appears that patients with food or drug allergies present a greater risk than those with a history of papaya intake or atopy [6, 19]. The presence of three of these factors suggests latent sensitivity to chymopapain [6, 19]. Derivatives of papain are currently found in many very common products such as fruit juices, some beers, coca-cola, meat tenderizer, toothpaste, cosmetics, drugs prescribed for pharyngitis and dyspepsia, and contact lens cleaners. Thus, all patients are exposed and may be immunized. According to American and French statistics, 1% of the population has hidden anti-papain "IgE" [6, 10, 26]. In a survey of 60,255 patients, women were found to be three times as likely to experience anaphylactic shock (0.9%) as men (0.3%) [1, 4].

Immunoallergic assessment

The aim of these tests is to determine the presence of specific anti-chymopapain antibodies. This involves a number of tests, the following of which are currently the most common:

In vivo:

the Prick test consists of a minute quantity of chymopapain introduced by epidermic puncture.

In vitro:

the radio-allergo sorbent test (RAST) detects the presence of the specific IgE antibodies;

the Human Basophil Degranulation Test (HBDT) determines the percentage of degranulated basophilic cells after chymopapain has been introduced. Even when these tests are positive, the prediction of clinical risk is uncertain [14]. At present, there is no ideal analysis that can accurately predict who is a candidate for CN without risk [19]. However, the Prick test is a valuable tool because it reveals the release of vasoactive substances from cutaneous mast cells. The advantage is that the results of this test are known immediately. It is also safe [6–15] and inexpensive, and allows accurate detection of highly sensitive patients, as shown by McCulloch's study (Table 2) [15]. In the state of current knowledge, the Prick test is restricted to patients at risk. A positive result contraindicates chemonucleolysis.

Table 2. McCulloch's study [15]

	Patient group	Skin test result	Effect of therapeutic chymopapain injection
Group A	35 normal control subjects	all negative	not done
Group B	3 previous anaphylactic reactions	2 positive 1 negative	not done not done
	1 allergic to ingestion of papaya	negative	no reaction
Group C	238 chemonucleo-lysis candidates	6 positive	5 immediate allergic reactions 1 no reaction
		232 negative	3 reactions
Group D	6 repeat CC	5 negative 1 positive	5 no reactions not done

When local contraindications have been eliminated, the anesthesiologist looks for contraindications of a general nature (Table 3) [22].

Table 3. Contraindications to chymopapain chemonucleolysis [22]

Absolute contraindications	Relative contraindications
Pregnancy	unstable cardiovascular condition
Sensitivity to papain, iodine	under 15 years old
Prior chemonucleolysis (13–47% risk of sensitivity one month later) [6, 15, 19, 25]	over 70 years old
Positive Prick test	infectious skin condition (procedure delayed)

Preventive treatment

In theory it must:
- inhibit the formation of histamine (ex., Tritoqualine);
- inhibit the formation of other mediators of anaphylaxis (experimental drugs);
- block cardio-vascular and bronchial histamine receptors using anti-H 1, such as hydroxyzine, dexchlorpheniramine and anti-H 2, such as cimetidine, ranitidine;
- reduce activation of complement at the time of injection of contrast material (ex., tranexamic acid).

In practice, there is no way to guarantee total protection against anaphylactic shock, except by avoiding the suspected drug [1, 7, 13]. With preventive medication, reactions seem to be less serious but not less frequent [7, 20]. A study of 59,042 cases revealed that 0.55% of premedicated patients experienced anaphylactic shock. In non-premedicated patients, 0.32% suffered anaphylactic shock [4]. Premedication relaxes the patient and eliminates his anxiety, which in itself can decrease the risk of anaphylactic shock [12, 13]. Moss recommends the combination of an anti-H 1 (diphenhydramine 4 × 50 mg orally) and an anti-H 2 (cimetidine 4 × 300 mg orally) administered 24 hours before chemonucleolysis [20].

Laxenaire et al. suggest 100 mg of an anti-H 1 such as hydroxyzine at bedtime for three nights prior to chemonucleolysis, and one gram of tranexamic acid IV slowly (one to two minutes) fifteen minutes before the procedure, to inhibit complement activation [12]. To these authors, the addition of an anti-H 2 does not seem to offer a significant advantage. Of 299 subjects with a history of anaphylactic reaction, 293 (98%) had received anti-H 1 and anti-H 2 treatment [12]. Furthermore, Cimetidine is not devoid of unwanted secondary effects, as it blocks the negative feedback effect of histamine release and modifies hepatic metabolism of drugs. Cortisone drugs do not seem to play a preventive role [4, 12, 13, 17, 18]. The preventive treatment used in our experience is figured in Table 4.

Chemonucleolysis monitoring

This is usually done in the radiology department, where the X-ray equipment is more effective than in the operating room. This room must be equipped

Table 4. Suggested protocol

8 days before the procedure	— pre-anesthesia consultation — prescription of preventive treatment to all patients — pre-operative *blood analysis*: blood group, CBC, ESR, PT, PTT, RH, search for abnormal antibodies, serum electrolytes — electrocardiogram, chest X-ray
3 days before	— hydroxyzine*, 100 mg at bedtime
2 days before	— hydroxyzine*, 100 mg at bedtime
1 day before	— hydroxyzine*, 100 mg at bedtime, hospitalization
2 hours before the procedure	hydroxyzine 100 mg orally
1 hour before	diazepam 10 mg i.m.; atropine 0.50 mg i.m.
During the procedure	sedation by the association of: — fentanyl, 50 to 100 μg + 100 to 150 μg i.v. at the time of discography — hydroxyzine**, 100 mg i.v. — diazepam, 5 to 10 mg i.v. — droperidol, 10 mg i.v. (uncooperative patient)

  * Sometimes hydroxyzine is combined with:
— predrisolone, 10 mg × 3/24 hours × 3 days;
— tritoqualine, 200 mg × 3/24 hours × 3 days
  ** Hydroxyzine must be carefully diluted and injected into a large vein

with material for anesthesia and reliable equipment for resuscitation, should a problem arise. Required material is as follows:
- oxygen;
- suction;
- cardiovascular monitoring equipment, including EKG scope and Dinamap (blood pressure cuff);
- material for respiratory assistance;
- defibrillator;
- tray of drugs for anesthesia and resuscitation;
- saline solution (Ringer's Lactate).

Sterile conditions and minimal risk of sepsis must be respected just as in the operating room. The anesthesiologist must supervise placement of the patient in the lateral decubitus position. The patient is made as comfortable as possible, with protection of all compression points. Monitoring is particularly strict during discography (use of contrast iodine) and at the time of chymopapain injection. Injection of a test dose of chymopapain (0.2 ml) 15 minutes before giving the entire dose has been suggested in order to limit anaphylactoid reactions. In practice, however, this does not provide any additional protection, as indicated by the study of Agre et al. [1] (Table 5). 52% of events occur within four minutes, 83% within ten minutes and 96% within twenty minutes [1]. Pulse and blood pressure must be checked frequently and any changes in respiration, hemodynamics, or skin should be

Table 5. Time of onset of anaphylaxis in relation to dose*

Dosing schedule	Number of patients with anaphylaxis	Number of patients at time of onset of anaphylaxis in minutes			
		0–4	5–10	11–20	>20
Test dose only (0.2–0.3 ml)	52	34	15	3	0
Therapeutic dose only	35	17	12	6	0
Test dose followed** by therapeutic dose					
Interval less than 5 min	12	6	4	0	2
Interval 5–10 min	28	12	11	5	0
Interval over 10 min	42	20	10	7	5
Interval not reported	2	0	1	1	0
Totals	171***	89	53	22	7
		52.0%	31.0%	12.9%	4.1%

   * From [1]
  ** Time of onset is from last injection of chymodiactin
 *** Time of onset was not reported for 23 patients not included in this table

carefully noted. The appearance of any abnormal sign, especially if associated with a major decrease in blood pressure, with or without broncho-constriction, requires immediate cessation of the procedure, and warrants appropriate treatment by the anesthesiologist without delay as detailed belows.

Monitoring after chemonucleolysis

The patient should be taken to the recovery room for 2 to 3 hours. Close monitoring of the following is necessary:
– respiration (risk of secondary respiratory failure after analgesic drugs);
– allergic reaction (delayed reactions can occur in 4% of cases);
– neurological status (urinary retention, paralysis).

At least one large vein catheter must be kept in place for several hours in case complementary analgesia or resuscitation is required.

Anaphylactic reactions

Anaphylactic shock, e.g., severe hemodynamic problems sometimes associated with bronchospasm, occurs in 0.5% of patients treated by CN. This is the most serious complication of CN. In a recent report of 60,255 cases, there were 317 confirmed cases of anaphylactic shock, including three deaths [4]. True anaphylactic reaction is the result of specific IgE antigen-antibody reactions. Chymopapain can also act directly on the mast cells and basophils, causing histamine release and anaphylactoid reactions. However, this mechanism probably has a limited role because plasma distribution of chymopapain is reduced, and the enzyme is rapidly inactivated by an $\alpha 2$ plasma macroglobulin [14]. Mediators, clinical signs and treatment of both anaphylactic and anaphylactoid reactions are identical.

Mediators released [11, 13, 14]

Several mediators are released:
- Histamine is the major mediator and the best known. It is stocked in the granules of circulating basophils and tissue mast cells;
- ECF-A, eosinophilic chemotactic factors of anaphylaxis;
- SRS-A, a slow-reacting substance of anaphylaxis (formed by the association of three leukotrienes);
- kinins and prostaglandins.

The role of these different mediators has not been proven since their dosage is still experimental. They are released or secreted at the time of activation of the basophils and mast cells. They bind themselves to receptors distributed throughout the cardiovascular system, bronchial tree, skin and digestive organs. The histamine receptors are of two types: H1 and H2. The vaso-active molecules lead to capillary vasodilatation with reduction in effective plasma volume, as well as to contraction of smooth muscle in the lungs and intestine, which explain the clinical signs.

Clinical signs

These are variable in both degree and severity. Any combination of clinical signs is possible, but the most severe reaction is anaphylactic shock [11, 13, 14].

Cardiovascular signs

Initially, vasoplegia with sudden hypotension, sinus tachycardia or various arrhythmias, transient increase in cardiac output with increase in systolic stroke volume, decrease in systemic arterial resistance and transitory increase in pulmonary arterial resistance can occur. If treatment is delayed or inefficient, a decrease in cardiac output, coronary constriction and myocardial ischemia can rapidly occur. In other cases, these complications may also begin suddenly.

Bronchial-pulmonary signs

Bronchospasm with decrease in thoracic compliance and elevated bronchial resistance, alveolar hypoventilation, and frequently edema of the upper airway can occur. Pulmonary edema may result when anaphylactic shock is prolonged.

Skin signs

These reflect cutaneous vasodilatation. In severe shock, cutaneous signs are absent.

Digestive system signs

There are irregular intense abdominal pains accompanied by diarrhea and vomiting.

Warning signs

In addition, *warning signs* can appear before blood pressure falls. The following signs are cause for alarm [8–14]: generalized or localized burning sensations, discomfort, itching, nausea, respiratory difficulties (gasping for breath). A tingling sensation is sometimes felt at the moment of chymopapain injection, but this has no further consequences.

Treatment

Treatment is handled in a rigorous and highly urgent manner. Primarily it consists of [8, 11, 12, 13, 14]:
- discontinuation of chymopapain injection;
- administration of 100% oxygen;
- rapid fluid replacement with non-histamine releasing cristalloid solution (Ringer's Lactate). Approximately 5–7 liters can be necessary;
- administration of epinephrine (adrenalin) in 1/10 mg increments until blood pressure stabilizes.

Epinephrine is the drug of choice in allergic reactions. Its α + effects inhibit vasoplegia and its β 1 effects relieve bronchospasm. In the mast cells, adrenalin increases the intracellular concentration of cyclic AMP and reduces the release of mediators [11]. However, its β 2 action can lead to serious arrhythmias and angina. Its administration is therefore performed under permanent cardioscopic control.

Secondarily, as necessary, treatment may be completed by:
- cortisone injection: one gram of hydrocortisone i.v.;
- administration of salbutamol spray and aminophylline to relieve bronchospasm;
- antihistamines;
- sodium bicarbonate.

Usually, improvement is rapid without sequelae. After the anaphylactic shock has been controlled, the patient is transferred to an intensive care unit where monitoring and resuscitation are available, until the hemodynamic situation stabilizes (minimum 24 hours).

References

1. Agre K, Wilson RR, Brim M, McDermott DJ (1984) Chymodiactin postmarketing surveillance. Demographic and adverse experience data in 29,075 patients. Spine 9: 479–485
2. Bernstein DI, Gallagher JS, Ulmer A, Bernstein IL (1985) Prospective evaluation of chymopapain sensitivity in patients undergoing chemonucleolysis. J Allergy Clin Immunol 76: 458–465
3. Bouillet R (1983) Complications du traitement de la hernie discale. Acta Orthop Belg 49: 49–76
4. Chymodiactin postmarketing surveillance data (1986) Interim report, May 1986. Smith Laboratories, Northbrook, IL, U.S.A., Laboratoire Armour-Montagu, Levallois-Perret, France
5. Cornaille G, Leynadier F, Modiano, Dry J (1985) Gravité du choc anaphylactique chez les malades traités par bêta-bloqueurs. Presse Méd 14: 790–791
6. Cornaille G, Guerin B, Leynadier F, Menkes CJ, Dry J (1987) Tests cutanés et immunoglobulines E spécifiques à la chymopapain après chimionucléolyse. Presse Méd 16: 881–884
7. Grammer LC, Ricketti AJ, Schafer MF, Patterson R (1984) Chymopapain allergy: case reports and identification of patients at risk for chymopapain anaphylaxis. CL Orthop Relat Res 188: 139–143
8. Hall BB, McCulloch JA (1983) Anaphylactic reactions following the intradiscal injection of chymopapain under local anesthesia. J Bone Joint Surg [Am] 65 g: 1215–1219
9. Kahn M, Clement PH, Charriaud A, Jorrot JC (1983) L'anesthésie au cours des chimionucléolyses pour hernies discales. Cah Anesthésiol 31: 301–302
10. Kapsalis AA, Stern IJ, Bornstein I (1978) Correlation between hypersensitivity to parenteral chymopapain and the presence of IgE antichymopapain antibody. Clin Exp Immunol 33: 150–158

11. Korinek AM, Viars P (1987) Choc anaphylactique et anaphylactoide. In: Korinek AM, Viars P (eds) Infectiologie, allergologie en anesthésie réanimation. Cours supérieur d'anesthésie et de réanimation. Editions Arnette, Paris, pp 47–64

12. Laxenaire MC, Gilet B, Moneret-Vautrin DA (1985) Anesthésie pour chimionucléolyse à la chymopapaine. Symp Int Alternatives in Spinal Surgery, June 17–18, Paris

13. Laxenaire MC, Moneret-Vautrin DA, Vervloet D, Alazia M, Francois G (1985) Accidents anaphylactoides graves peranesthésiques. Ann Fr Anesth Reanim 4: 30–46

14. Levy JH, Roizen MF, Morris JM (1986) Anaphylactic and anaphylactoid reactions. A review. Spine 11: 282–291

15. McCulloch JA, Canham WD, Dolovich J (1985) Skin tests for chymopapain allergy. Ann Allergy 55: 609–611

16. McDermott DJ, Agre K, Brim M, Demma FJ, Nelson J, Wilson RR, Thisted RA (1985) Chymodiactin in patients with herniated lumbar intervertebral disc(s). An open-label, multicenter study. Spine 10: 242–249

17. Moneret-Vautrin DA, Laxenaire MC, Mouton C, Widmer S, Pupil P (1985) Modification de la réactivité cutanée dans l'anaphylaxie aux myorelaxants et hypnotiques, après administration d'anti-H 1, d'anti-H 2 et de tritoqualine. Ann Fr Anesth Reanim 4: 225–230

18. Moneret-Vautrin DA, Benoist M, Laxenaire MC, Croizier A, Gueant JL (1985) Allergie à la chymopapaine: intérêt de tests prédictifs avant chimionucléolyse. Ann Fr Anesth Reanim 4: 313–315

19. Moneret-Vautrin DA, Mouton C, Laxenaire MC, Roland J, Aussedat P, Occelli G, Gerard H (1986) Détection d'une sensibilisation à la chymopapaine. Bilan chez 111 candidats à la chimionucléolyse. Sem Hop Paris 62: 3499–3504

20. Moss J (1985) Use of prophylactic combined H 1 and H 2 antagonists reduces mortality in chymopapain anaphylaxis. Ann Fr Anesth Reanim 4: 221–224

21. Noury D, Lavenac G, Milon D, Sauvage J, Bodin JM, Saint-Marc C (1984) Neuroleptanalgésie pour la nucléolyse chimique. Cah Anesthésiol 32: 397–401

22. Roucoules J, Thurel C, Bard M, Laredo JD (1985) Le point sur la nucléolyse discale. In: Seze S de, Ryckewaert A, Kahn MF, Guerin C (eds) L'actualité rhumatologique. Expansion Scientifique, Paris, pp 274–282

23. Schaffer J, Piepenbrock S (1984) Chemonucleolysis. General or local anesthesia. Br J Anaesth 56: 1306

24. Troisier O, Gozlan E, Durey A, Rodineau J, Gounot-Halbout MC, Pelleray B (1980) Traitement des lombo-sciatiques par injection intra-discale d'enzymes proteolytiques (nucleolyse). Nouv Press Med 9: 227–230

25. Tsay YG, Jones R, Calenoff E, Sun J, Arndt L, Crispin B, McDermott D, Stern I (1984) Chymopapain-induced hypersensitivity following chemonucleolysis. Spine 9: 769–771

26. Tsay YG, Jones R, Calenoff E, Sun J, Arndt L, Crispin B, Rock H (1984) A preoperative chymopapain sensitivity test for chemonucleolysis candidates. Spine 9: 764–768

Technique of lumbar chemonucleolysis

J.-D. Laredo[1], J. Busson[1], M. Wybier[2], and M. Bard[1]

[1] Department of Bone and Joint Radiology, Hôpital Lariboisière, and
[2] Department of Bone and Joint Radiology, Hôpital Cochin,
Paris, France

Intrathecal injection of chymopapain causes fatal hemorrhage in rabbits and dogs [12]. To prevent the possibility of chymopapain leaking into the subarachnoid space, chemonucleolysis should not be performed immediately after myelography in order to allow any dural leak to seal [12]. McCulloch recommends a minimum delay of 24 hours [12]. We agree with other practitioners and require at least seven days between myelography and enzyme injection [1].

Material

Chemonucleolysis is carried out in the X-ray room under aseptic operating room conditions. Sterile drapes, gloves, and surgical gowns are used.

The X-ray room

Good fluoroscopic visualization of the anatomic details of the low lumbar spine is necessary for performance of chemonucleolysis. Both lateral and AP views of the lumbar spine must also be available during chemonucleolysis in order to control needle position. Approach to the disc space is facilitated by a portable C-arm image intensifier unit which provides biplane fluoroscopic control and radiographs without moving the patient. However, this can also be accomplished using a single plane fluoroscopy table and an auxiliary tube allowing cross-table X-rays (Fig. 1). If neither is available, chemonucleolysis can still be carried out with a single plane fluoroscopy table alone; AP and lateral views of the lumbar spine are then obtained by rotating the patient. However, this is less comfortable.

The chemonucleolysis tray

A standard chemonucleolysis tray contains (Fig. 2):
1 per cent Xylocain;
10 ml syringes for local anesthesia;
2.5 ml syringes for contrast media and chymopapain injections;
disposable needles for local anesthetic infiltration;
disposable needles for discography and chymopapain injection (Fig. 3):
   – for the L 4-5 interspace and upper levels a set including a 9 centimeter 18-gauge needle and a 15.2 centimeter 22-gauge needle (Becton-Dickinson),

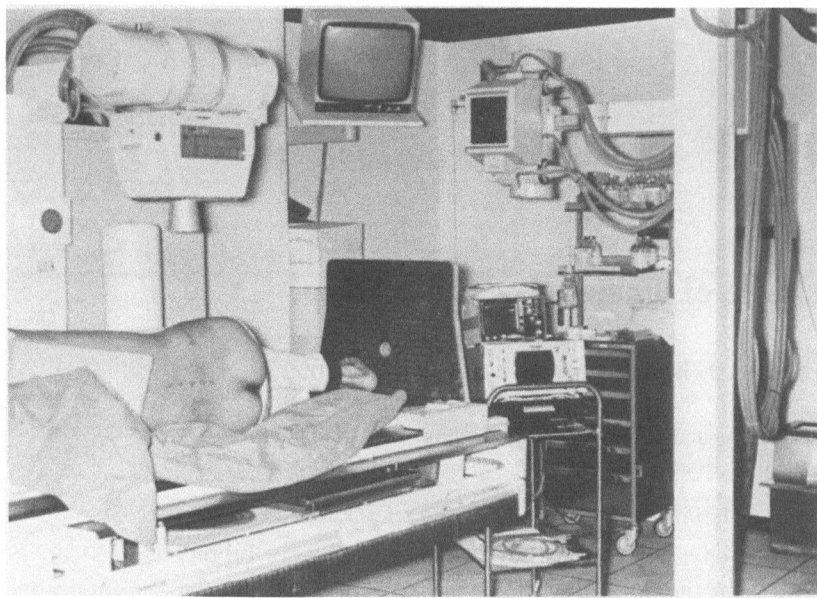

Fig. 1. Installation for chemonucleolysis with single plane fluoroscopy and auxiliary tube allowing cross-table X-rays

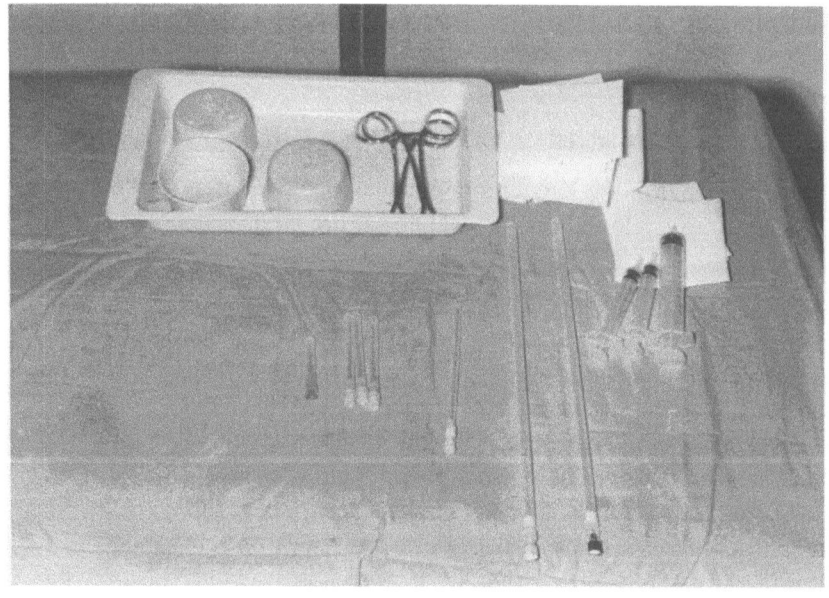

Fig. 2. Standard chemonucleolysis tray

     — for the L 5-S 1 interspace, and systematically in corpulent patients, a 15.2 centimeter 18-gauge needle and a 20.3 centimeter 22-gauge needle (Fig. 4);

sterile water;

a water-soluble intrathecally well-tolerated contrast media (Iohexol);

Chymopapain — 2 kinds are available:

     — Chymodiactin (Smith Labs), the Chymopapain used in our patients,

     — Discase (Baxter-Travenol Labs).

No substantial difference in activity or other characteristics exists between these two chymopapain preparations. 4,000 U per disc is considered an adequate dose [12]. A single dose package of 4,000 U is recommended. The enzyme must be refrigerated up to the time of use and mixed with

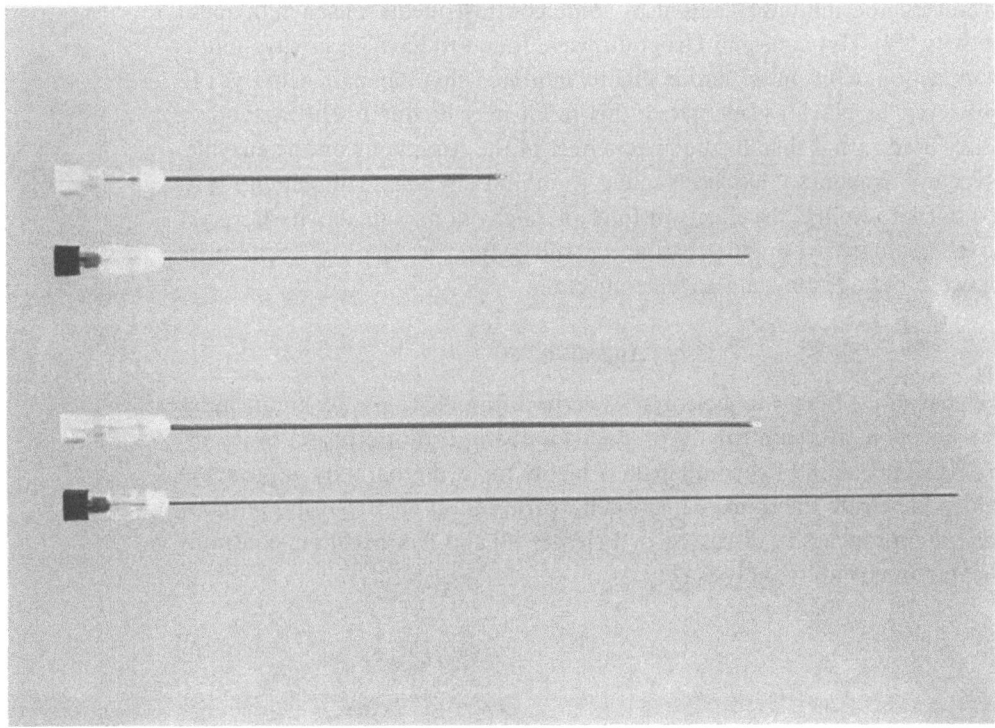

Fig. 3. Set of two disposable needles for chemonucleolysis at L 4-5 (below) and L 5-S 1 (above) interspaces

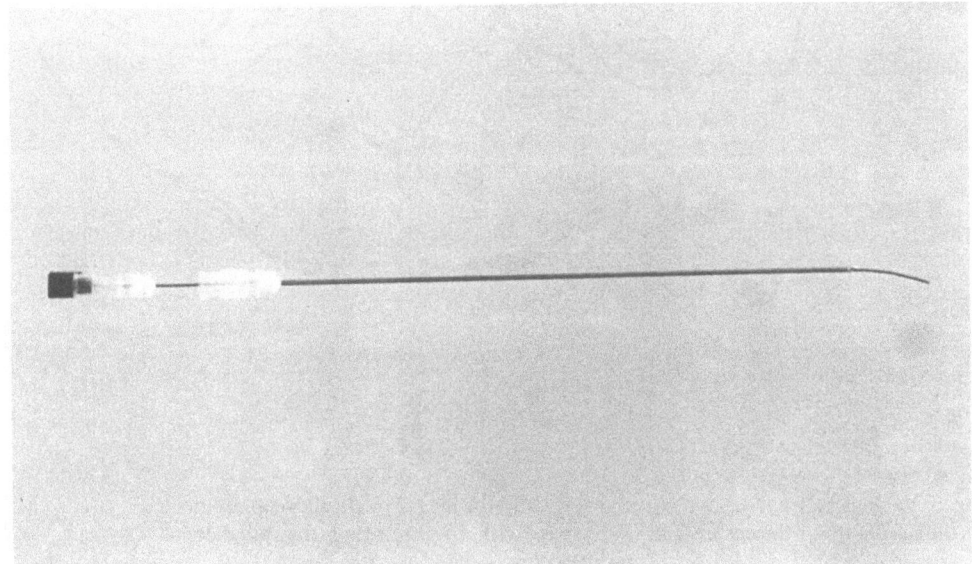

Fig. 4. 22-gauge needle with bent tip introduced through the 18-gauge needle

distilled water (2 cc for a single dose of 4,000 U) at room temperature. The solution should be used immediately and cannot be stored.

Intradiscal injection of radiopaque fluids prior to intradiscal injection of chymopapain has been believed to have some inhibitory effect on the enzyme activity [14]. This raised the question as to the amount of time that should elapse between discography and chymopapain injection. Naylor et al.

evaluated the inhibitory action of some contrast media on chymopapain activity [14]. Hypaque and Urografin were found to have no in vitro inhibitory action, while metrizamide slightly inhibited chymopapain activity [14]. However, as Naylor et al. stated, this result may be due to the method of assay used rather than to the direct effect of the Amipaque on the enzyme. Recently, Iopamidol has been found to inhibit chymopapain activity [17]. To our knowledge, the effects of Iohexol on chymopapain activity have yet to be evaluated. For this reason, we still respect a delay of 15 minutes between discography and enzyme injection.

Approach

The technique of chemonucleolysis is derived from discography but includes the injection of a proteolytic product for therapeutic purposes. Since intrathecal injection of chymopapain is highly toxic, discography approaches which penetrate the spinal canal, such as transdural *posterior* and *posterolateral* approaches as discussed by Erlacher [8] and Keck [10] are contraindicated in chemonucleolysis (Fig. 5).

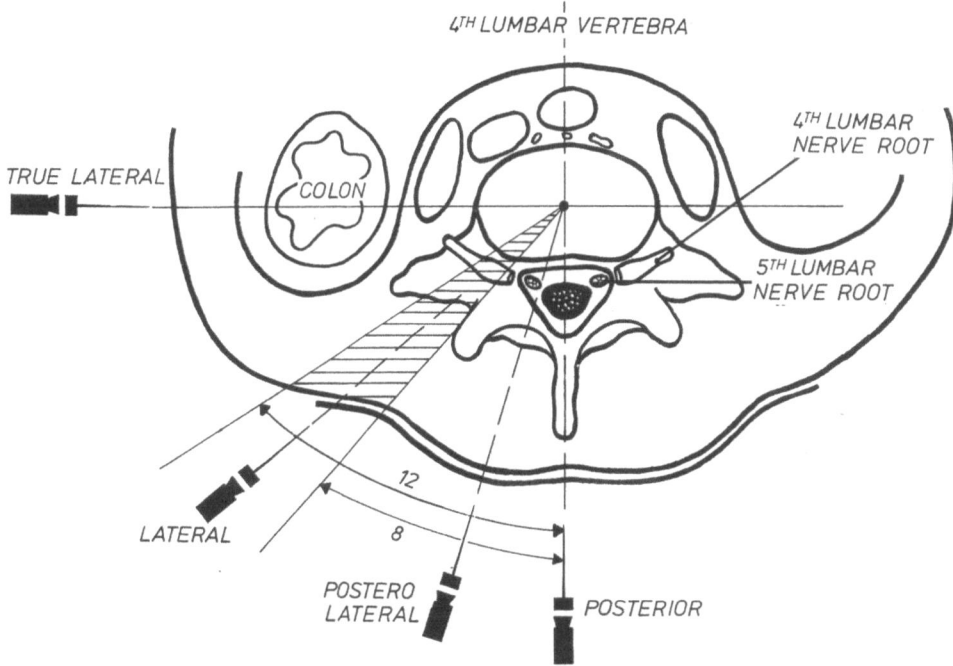

Fig. 5. Various approaches to the disc space. Only the lateral route allows placement of the needle tip in the center of the disc space without penetrating the spinal canal and the colon

A *true lateral* approach may be complicated by infectious discitis since the colon may be punctured before disc penetration (Fig. 5).

A *lateral*, extradural approach [2, 7, 9, 13, 18] must be used for chemonucleolysis. This approach is termed "lateral" as opposed to the posterolateral approach of Erlacher, but it is in fact a posterolateral extraspinal approach through the paravertebral muscles (Fig. 5).

The needle is inserted at 8 to 12 centimeters from the midline of the spinous processes and advanced at 40 to 60 degrees to the sagittal plane, depending on the disc space level treated and patient morphology (Fig. 6).

Either a right or left-sided approach can be used for chemonucleolysis. The side of approach is not related to the side of the sciatic pain. At L 4–5 and higher disc spaces, the right-sided approach is habitually used. At L 5-S 1, the iliac crest and broad transverse apophysis of L 5 are two obstacles to proper needle placement. Therefore, the preferable side of approach is determined on the basis of the L 5-S 1 AP view.

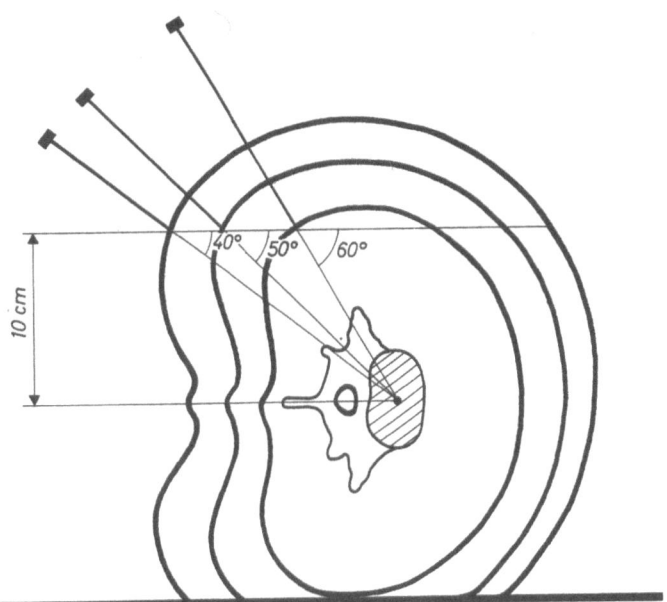

Fig. 6. Influence of patient morphology on disc space approach. With the same point of skin puncture, appropriate angle of approach may vary from to 40 to 60 degrees to the sagittal plane

Positioning the patient

L 4–5 chemonucleolysis can be performed with the patient in either a strict lateral or a prone oblique position. Each position has both advantages and disadvantages and the choice between them depends mainly on personal preferences and experience.

The principal advantages of the strict lateral position are clear visualization of bony obstacles, easy approach to the disc space in a direction parallel to the vertebral endplates, and easy control of the precise placement of the 18-gauge needle tip at the posterior border of the disc and of the 22-gauge needle at mid-disc height. The main disadvantage of the lateral position is that it requires clear, three-dimensional visualization of the anatomy. Exact positioning of the patient is crucial.

The main advantage of the prone oblique position is the possibility of fluoroscopic guidance along the long axis of the needle toward the center of the disc space [19, 20]. Disadvantages are greater difficulty in approaching the needle in a direction parallel to the vertebral endplates and in placing the needle tip at precisely mid-disc height.

Higher lumbar disc spaces can also be approached with either a lateral or a prone oblique patient position. The L 5-S 1 disc space is approached using a lateral patient position since the iliac crest precludes the prone oblique technique.

Technique of chemonucleolysis for L 4–5 and higher levels with lateral positioning of the patient

The patient is placed on his left side on a radiolucent table. Knees, ankles and the left arm are protected by adapted positioning and foam cushions. The head is elevated (Fig. 7). A folded sheet is placed beneath the flank in order to correct lateral deviation of the spine due to lateral decubitus. Lumbar lordosis is reduced by asking the patient to flex hips and knees.

During the first part of the approach, the radiographic tube must be strictly vertical, orthogonal to the X-ray table. The image intensifier is positioned over the disc interspace to be treated. Perfect lateral positioning of the patient is carefully checked so that a true lateral view of the inferior lumbar spine is obtained on the intensifier screen. Strict lateral positioning is crucial for the correct approach to the lumbar spine.

The spinous processes and iliac crest are sought by palpation. The needle insertion point is just above the iliac crest at 8 to 10 cm lateral to the midline,

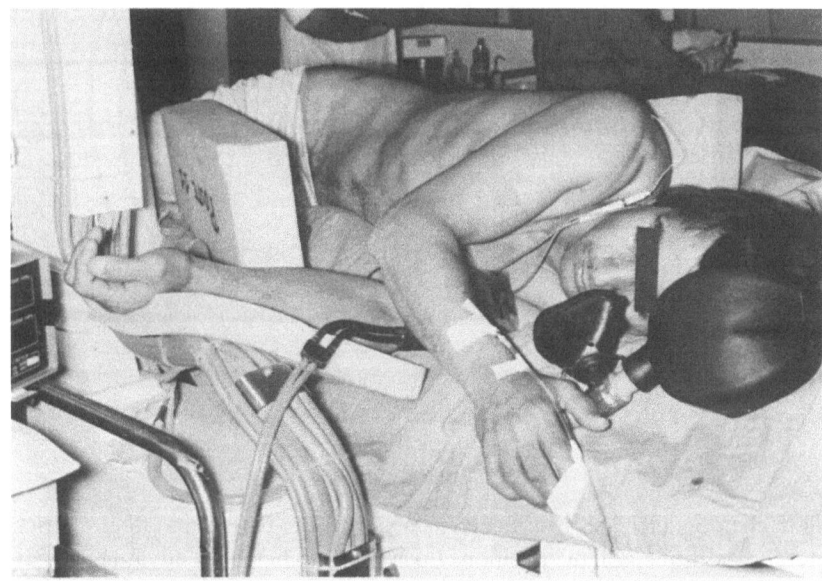

Fig. 7. Patient in lateral position

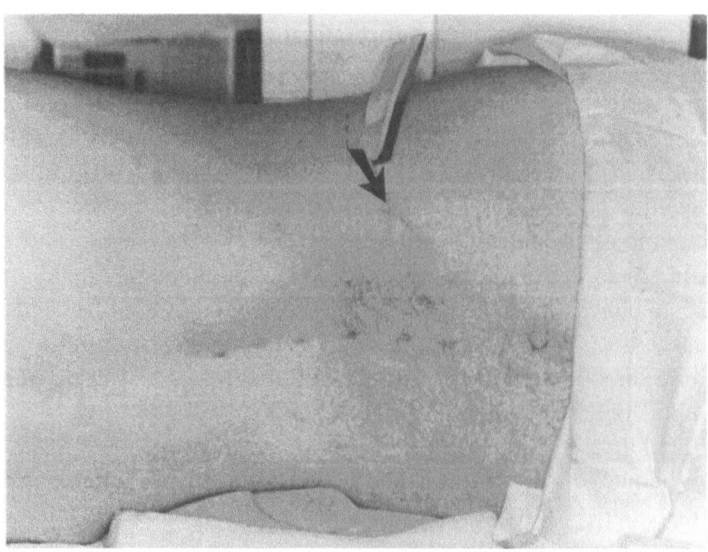

Fig. 8. Point of skin puncture immediately above the iliac crest at 8–10 cm from the midline. A metallic ruler is placed on the patient's side to simulate the approaching needle

depending on the patient's build (Fig. 8). The exact level of needle insertion and the angle of caudal approach are determined using a metallic ruler placed on the side of the patient in order to simulate the approaching needle on the fluoroscopic screen (Fig. 8). The ruler is tilted as desired so that it projects between the needle insertion point and the posterior margin of the L 4–5 disc space as visualized on the lateral view screen (Fig. 9). The projected approach must avoid the transverse processes and be as parallel as possible to the vertebral endplates. If necessary, the needle insertion point is corrected to meet these criteria. Once determined, this angle is marked on the patient's

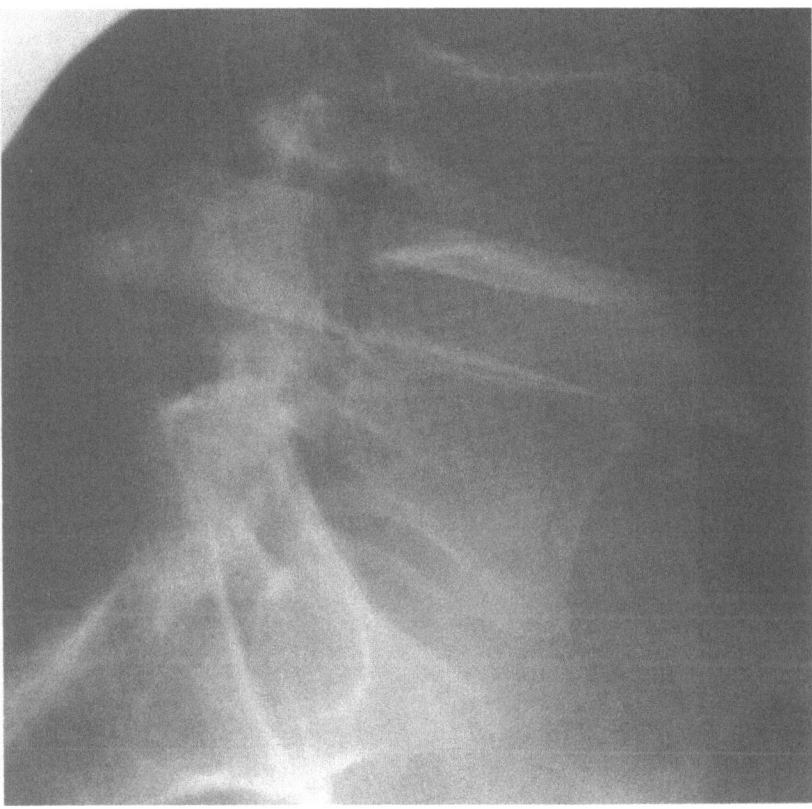

Fig. 9. The metallic ruler is tilted to determine the appropriate angle of approach

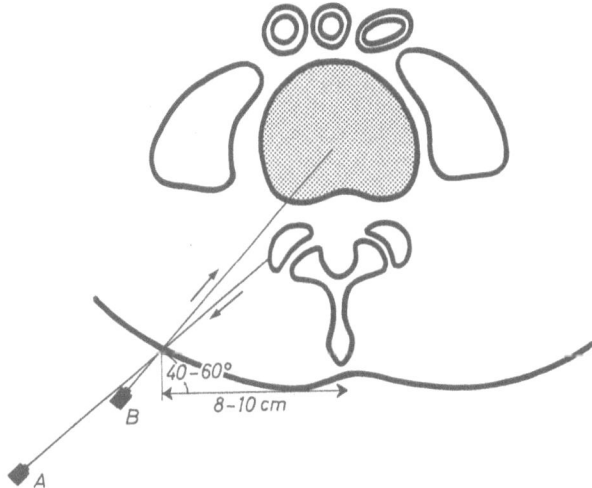

Fig. 10. The needle is first advanced to the facet joint (*A*). When contact with the facet joint is obtained, the needle is withdrawn 2–3 cm and then advanced in a more sagittal direction (*B*)

skin. The skin is disinfected, and a sterile surgical field is prepared. The skin
and superficial planes are anesthetized with an infiltration of 1% Xylocaine.

An 18-gauge needle is then inserted at an angle of 40–60 degrees to the
sagittal plane (Fig. 10) and needle advancement is controlled on the inten-

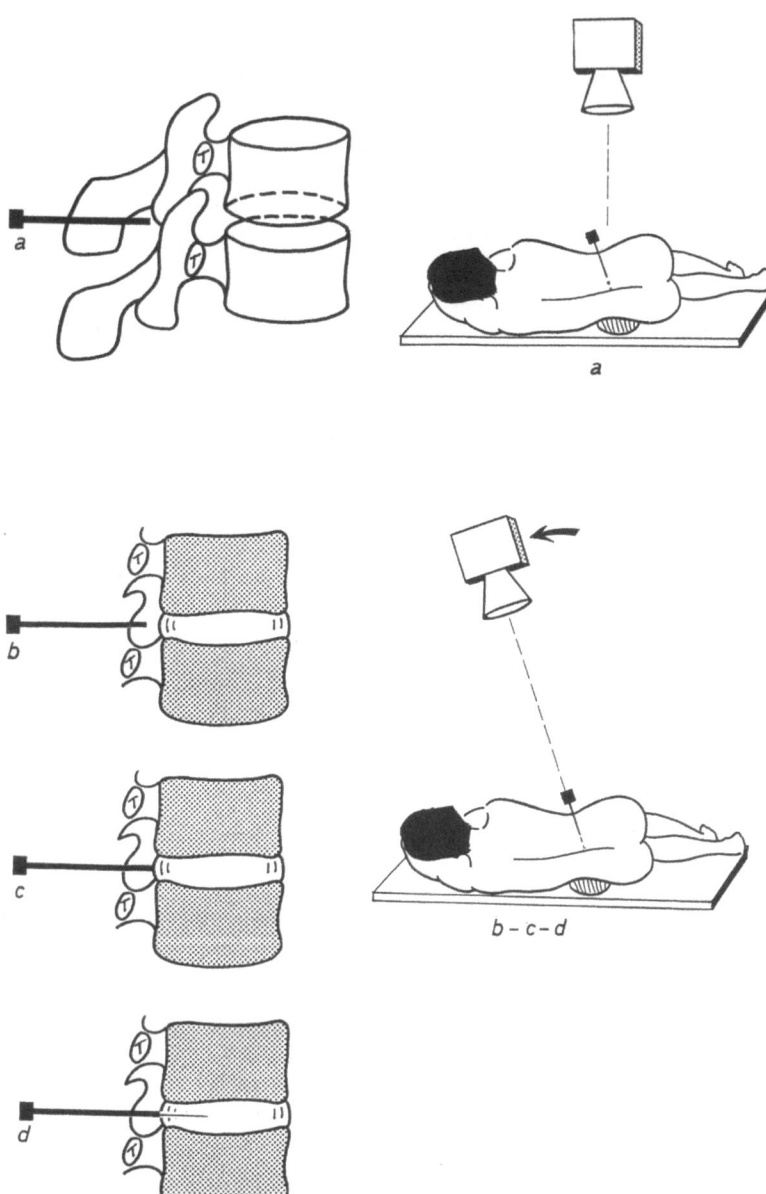

Fig. 11. During the first part of needle approach, the X-ray beam is strictly vertical, perpendicular to the radiographic table (**a**). Once the needle tip has passed the line joining the transverse processes (*T*), the X-ray is tilted to profile the disc space (**b, c, d**)

sifier screen. The tip of the needle is first advanced to the facet joint. When
contact with the facet joint is obtained, the angle of needle insertion must
be reduced. The needle is withdrawn 2–3 cm and then advanced in a more
sagittal direction (Fig. 10). The X-ray beam must not be tilted during this
part of the procedure (Fig. 11).

If the level of the puncture point has been adequately determined according to the guidelines above, and if the needle is correctly advanced
toward the disc space, the facet joint is the only bony structure which can
be encountered. However, the cause of bony obstruction can be determined

on the lateral view image intensifier [13]. Bony impingement at the level of the pedicles is probably due to the transverse process. Obstruction at the level of the lower half of the vertebral body and opposite disc space is probably due to the facet joint. If the transverse process causes obstruction, the site of needle insertion is moved in a cephalad or caudal direction. Problems may arise if too much attention is paid to the angle of needle insertion and too little to the overall size and shape of the patient. A clear three-dimensional mental image of the lower lumbar region is essential [13].

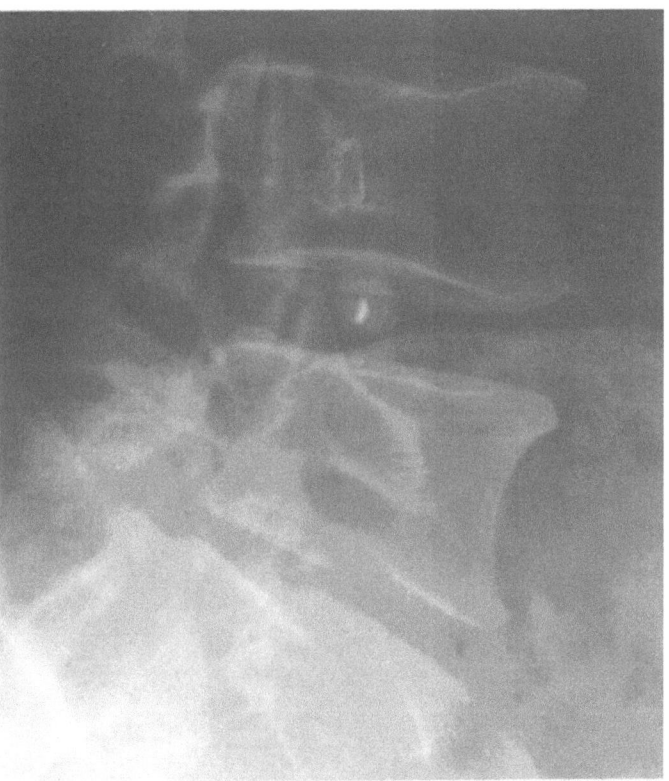

Fig. 12. On an oblique view, the ideal position of the needle tip is at the junction between the internal and middle third of the disc width

Throughout the procedure, patients tend to rotate the pelvis away from the needle, introducing an oblique dimension to the lateral radiograph, which confuses the three-dimensional image and invalidates the technical details mentioned above. If difficulties do arise, it is advisable to pause and confirm that the patient remains in a truly lateral position [13].

Once the needle has passed the line joining the transverse processes, the direction of the X-ray beam is tilted to adequately demonstrate the disc interspace and the needle is advanced toward the center of the disc (Fig. 11). A firm gritty sensation is felt when the tip of the needle encounters the anulus of the disc. This contact should be obtained when the needle reaches a line joining the posterior border of the vertebral bodies as shown on a lateral view [13] (Fig. 11). If the tip of the needle passes anterior to this line before contact with the anulus, the needle will not bisect the nucleus but will pass antero-lateral to it. If this occurs, the angle of needle insertion is too sagittal and should be changed to a more coronal direction.

Once the needle is adequately placed against the anulus, the patient is rotated to an oblique position and the X-ray beam is tilted to exactly parallel the direction of the needle. Position of the needle tip in relation to the disc

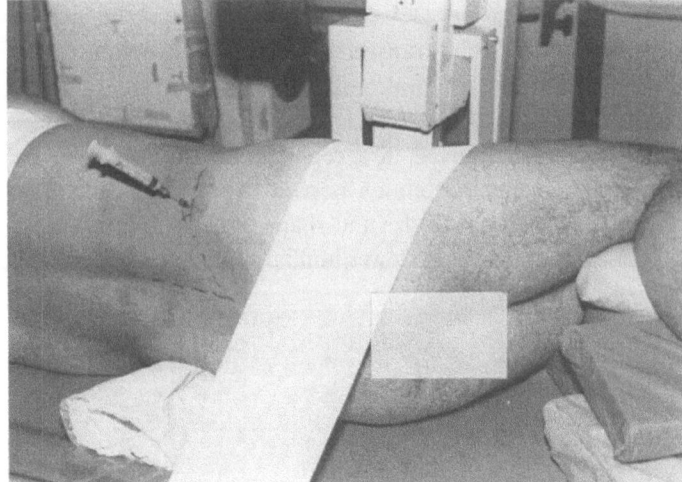

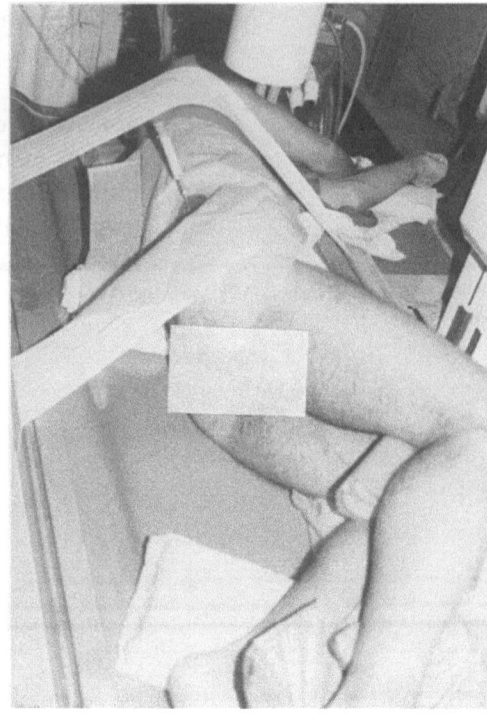

Fig. 13. Needle approach in a patient in strict lateral position

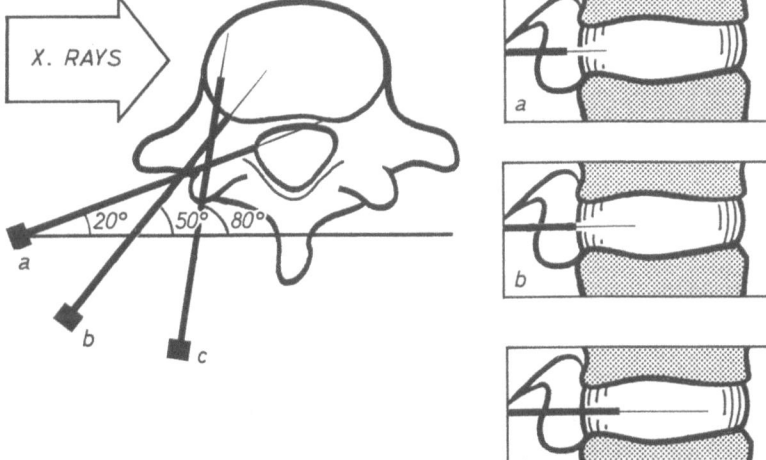

Fig. 14. Needle projection on lateral fluoroscopy. **a** Incorrect needle placement, too medial and coronal approach. **b** Correct needle placement. **c** Incorrect needle placement, too lateral and sagittal approach

width is then checked on the X-ray intensifier screen. The ideal needle position is at the junction between the internal and middle thirds of the disc width (Fig. 12). With this position, the thin needle will reach the center of the disc with no curvature. If the needle is slightly lateral to this point, near the center of the disc, the nucleus can still easily be reached by applying a mild curvature to the 22-gauge needle tip as mentioned below (chemonucleolysis technique for L 5-S 1). The patient is then returned to a strict lateral position. The stylet of the thick needle is withdrawn and the absence of CSF reflux is verified. If CSF is obtained, the procedure must be stopped and chymopapain must not be injected. If there is no reflux of CSF, the procedure can continue. The 15.2 cm 22-gauge needle is introduced into the lumen of the 18-gauge needle and advanced to the center of the disc (Fig. 13).

If the needle bisects the disc, a 1 cm advance of the needle should be accompanied by an apparent advance of approximately one-half cm on the lateral view screen (Fig. 14 b) [6]. If the needle is too lateral, advancement of the needle by 1 cm will be accompanied by an apparent advancement of nearly 1 cm (Fig. 14 c) [6]. Conversely, if the needle passes across the back of the disc space, a 1 cm advance will be accompanied by very little apparent advance on the screen (Fig. 14 a). Attention must also be paid during this part of the procedure to the proper placement of the needle at the midheight of the intervertebral disc. Curvature of the needle can be utilized to bring the needle tip to this position. Once the needle is in the center of the disc space according to the instructions given in Fig. 15, an AP view is performed using a C-arm fluoroscopy unit or a cross-table X-ray tube. If neither is available, the patient is gently rotated into a prone position. On the AP view, the tip of the 22-gauge needle must be close to the midline or, in any case, between the internal margins of the pedicles. It also must be equidistant from the two vertebral endplates (Fig. 15).

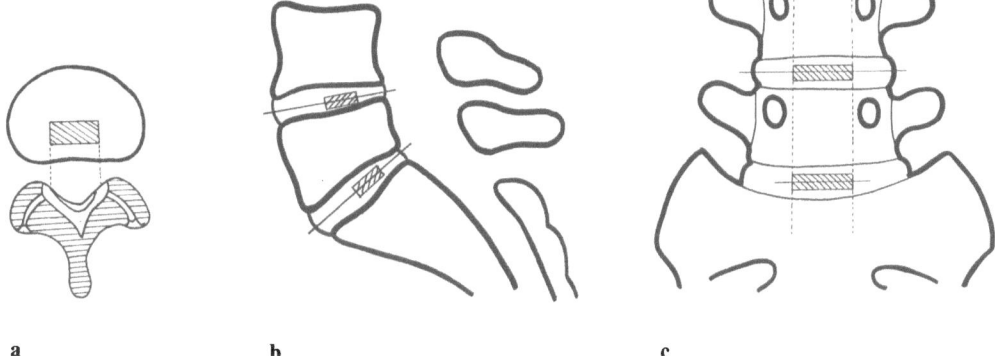

a   b   c

Fig. 15. Target volume in chemonucleolysis, according to Nazarian [15]. **a** Horizontal section, **b** sagittal section, **c** anterior view

Technique of chemonucleolysis for L 4–5 and higher levels with prone oblique positioning of the patient

The patient is placed in an oblique prone position with his right side up so that the disk space is approached through a right posterolateral route (Fig. 16) [20]. A radiolucent block is placed under the patient's left side in order to widen the space between the lower ribs and the iliac crest. The patient is asked to flex his right hip and knee to steady the oblique position while his left lower limb remains extended. For comfort, a small cushion is placed under his right knee (Fig. 16).

The X-ray beam is tilted in order to profile the disc space. Exact oblique positioning is then determined under fluoroscopic control by slowly rotating the patient until the right L 4–5 facet joint is imaged within the dorsal third of the disc space as shown in Fig. 17. When this correct position is obtained, the right superior facet of L 5, the lower endplate of L 4 and the iliac crest delimit a "triangle of puncture" on the fluoroscopic screen (Fig. 17).

After disinfection of the skin and local anesthesia of superficial and deeper planes, the distal tip of the 18-gauge needle is placed on the patient's skin so that it projects fluoroscopically in the center of the triangle. The needle is then inserted through the skin following the angle of inclination of the X-ray tube as shown in Fig. 18.

After a 2- or 3-cm advancement of the needle, its direction is controlled

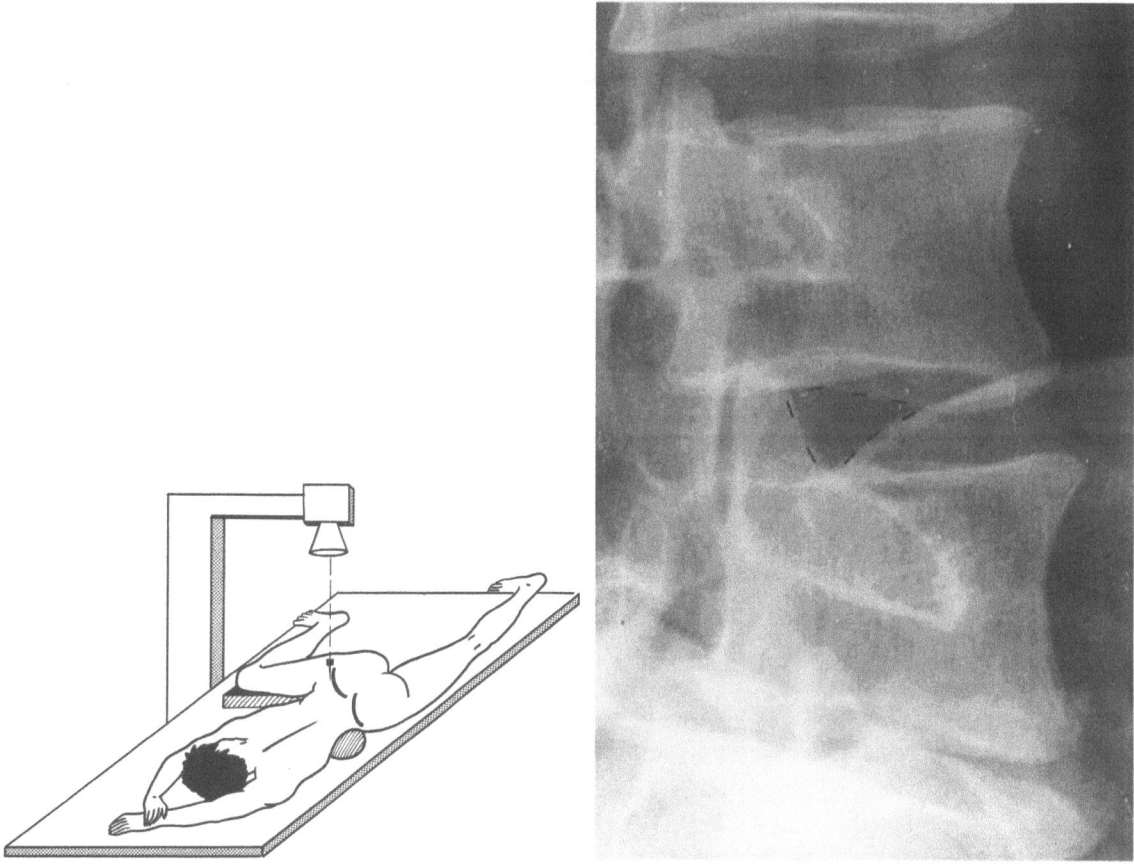

Fig. 16   Fig. 17

Fig. 16. Patient in oblique position. The X-ray beam is tilted to profile the disc space

Fig. 17. Roentgenogram in a patient in oblique position: the "triangle of puncture" is bordered by the superior facet joint, the inferior endplate and the iliac crest

on the fluoroscopic screen: a punctiform projection of the needle seen in the center of the triangle indicates a correct needle approach (Fig. 19).

The needle is further advanced under fluoroscopic control until the disc is reached. If the needle accidentally encounters the right L 4 nerve root, it must be withdrawn 2–3 cm. The patient is rotated into a 5- to 10-degree more oblique position and the same procedure is repeated.

Once the 18-gauge needle has reached the anulus, its stylet is removed and the 22-gauge needle is introduced into its lumen and advanced 2 cm to the center of the disc. The patient is then moved to a strict lateral position. The X-ray beam is tilted to adequately profile the disc and the 22-gauge needle is slightly advanced or withdrawn until its distal tip reaches the exact center of the disc space. The position of the 22-gauge needle is then controlled on an AP view and the procedure is completed as mentioned above in the lateral position technique.

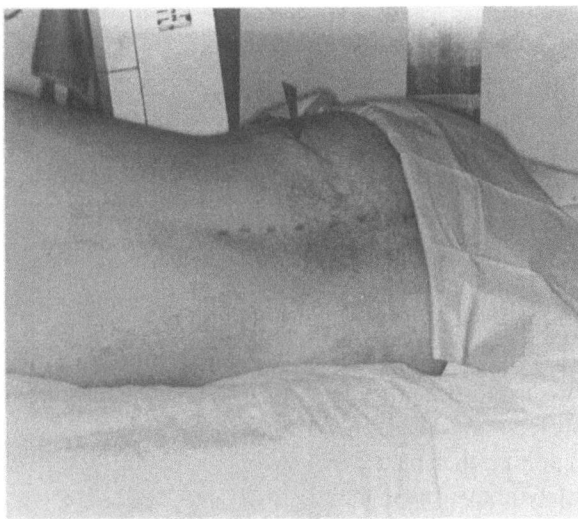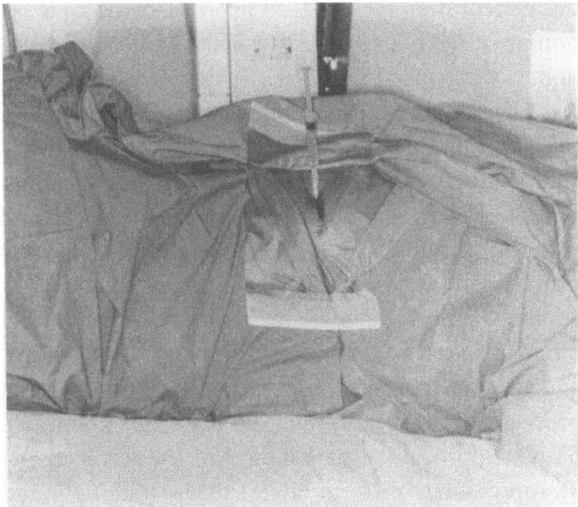

Fig. 18. Needle insertion along the direction of the X-ray beam

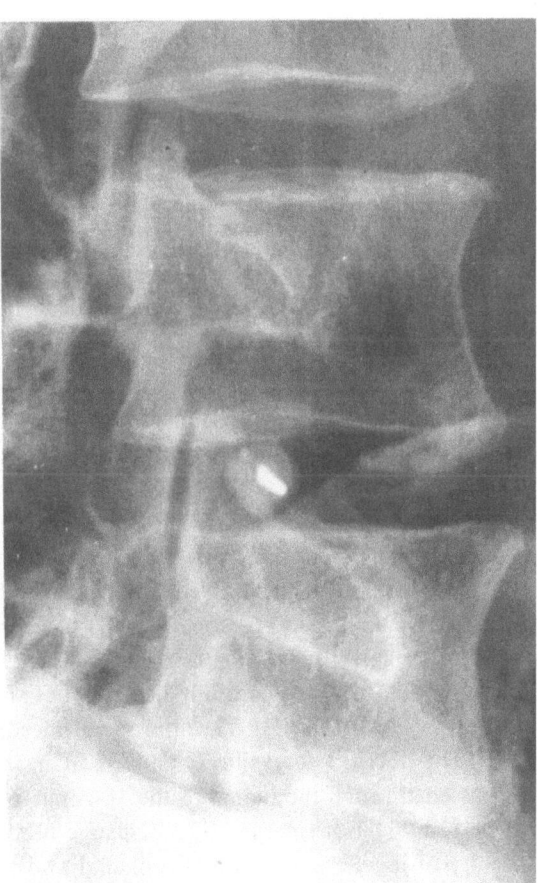

Fig. 19. Correct needle insertion: the needle appears as a dot in the center of the triangle of puncture

Technique of chemonucleolysis for the L 5-S 1 interspace

A lateral patient position is recommended for the L 5-S 1 approach, since the iliac crest precludes direct puncture in the oblique position. Furthermore, since in most cases the direction of approach cannot be strictly parallel to the vertebral endplates, true lateral fluoroscopic control with exact positioning of the patient is of particular importance in assessing bony obstruction.

The L 4–5 technique with a lateral patient position is used with the following modifications. The side of approach is chosen based on the AP view of the L 5-S 1 interspace. In the case of unilateral sacralization, the non-sacralized side will be chosen for the approach. In other cases, the side with the maximum distance between the inferior border of the L 5 transverse process and the superior border of the sacral wing is selected, and the patient is placed on his opposite side on the radiolucent table. A large radiolucent block is placed beneath the decubitus flank in order to lower the iliac crest and to open the intervertebral disc and the space between the transverse process and the sacral wing. This manoeuver considerably facilitates the L 5-S 1 disc approach. A mark indicating the needle insertion point is then placed at approximately 10-cm lateral to the midline immediately above the iliac crest.

Adequate caudal inclination for needle approach is then determined using a metallic ruler placed upon the side of the patient in order to simulate the approaching needle as mentioned in the L 4–5 technique. The ruler is tilted as required so that it projects between the insertion point and the posterior border of the L 5-S 1 disc space as visualized on the lateral screen (Fig. 20 a). Absence of bony obstruction by the L 5 transverse process is verified. The more the approach parallels the disc space, the easier it will be. Once determined, this angle is marked on the patient's skin. After disinfection of the skin and local anesthesia of superficial and deeper planes, the L 5-S 1 15.2 cm 18-gauge needle is inserted at the same angle to the sagittal plane as for the L 4–5 technique, i.e. 50 ± 10 degrees, but it is caudally oriented at the same time as discussed above.

Relation of the needle to the vertebral bodies is of particular importance in assessing bony obstruction during the L 5-S 1 approach. Obstruction by a broad transverse process may necessitate complete repositioning of the needle; a more medial and inferior approach to needle insertion may be tried.

Finally, the 18-gauge needle should reach the anulus at the posterior and superior margin of the L 5-S 1 disc space, as shown in Fig. 20 b.

The procedure is then completed in a way similar to the L 4-5 technique except that, for anatomical reasons (Fig. 21), a curved 22-gauge needle should be used to reach the center of the disc (Fig. 22). The terminal 2-cm of the 22-gauge needle must be carefully bent so that the bevel is on the convex aspect of the curve (Fig. 23). The appropriate needle curve is selected on the basis of the oblique view (Fig. 22 and 24). This curved 22-gauge needle is then passed through the lumen of the 18-gauge needle and is advanced to the point at which it emerges. The 22-gauge needle is rotated to bring the concavity of its curve medial and cephalad and advanced to the center of the disc (Fig. 25). Adequate rotation of the 22-gauge needle during its progression allows continuous correction of its position.

Discography

Discography is carried out as part of the operative procedure of chemonucleolysis to confirm proper placement of the needle within the nucleus pulposus prior to enzyme injection [1]. Some authors perform discography

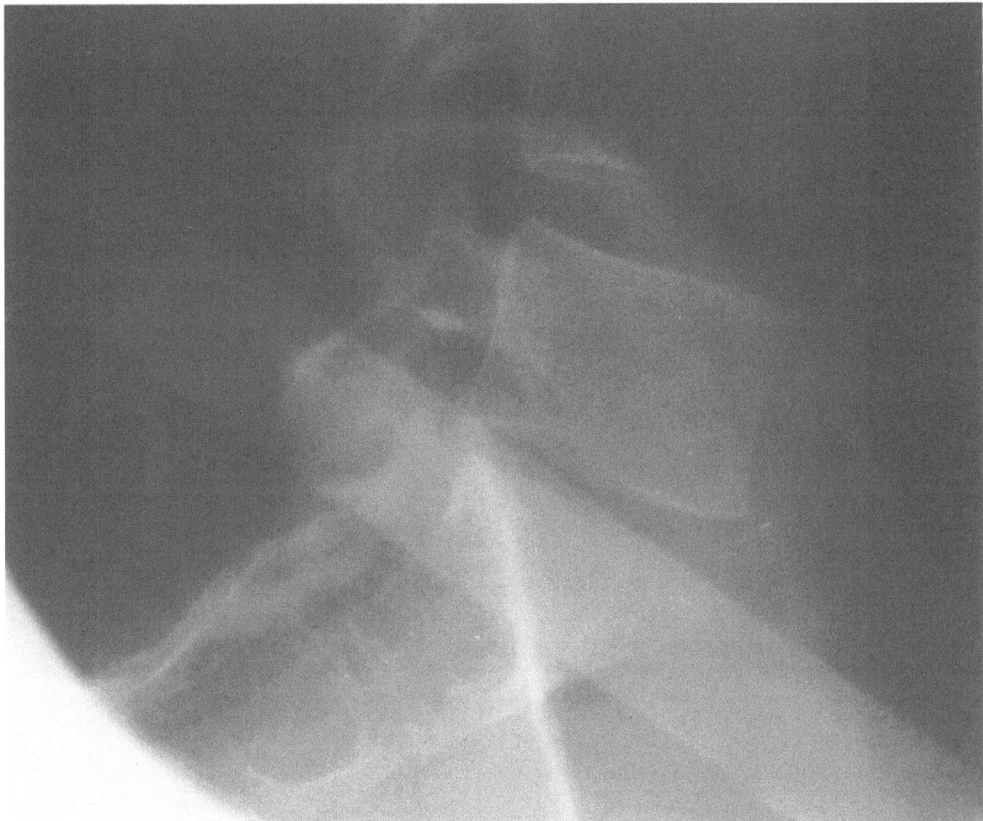

a

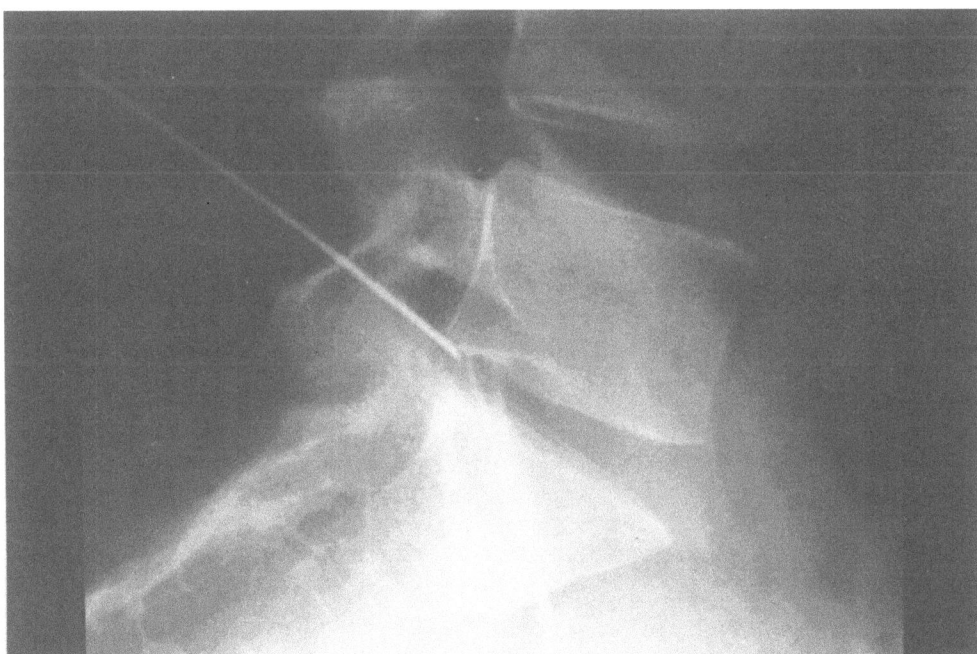

b

Fig. 20. Technique of approach to the L5-S1 interspace. **a** Metallic ruler simulating the approaching needle helps to determine appropriate caudad tilt during needle advancement. **b** Placement of the 18-gauge needle at the posterosuperior margin of the disc space

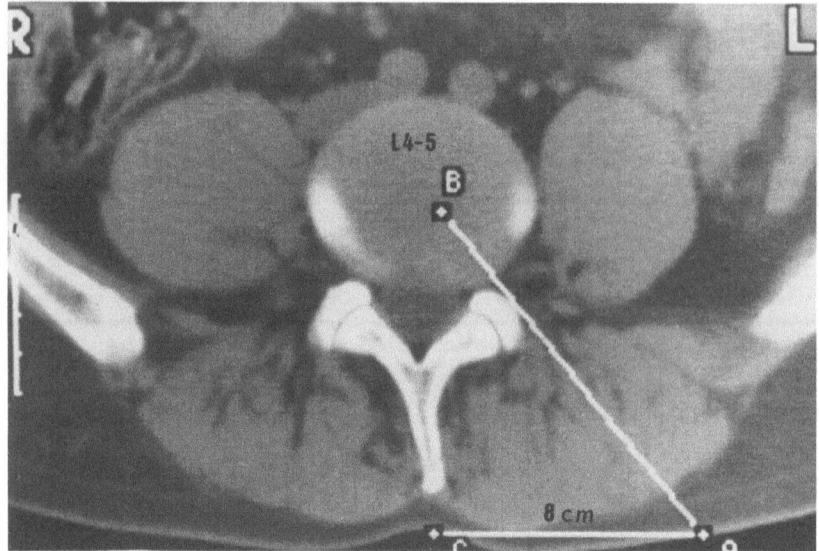

a

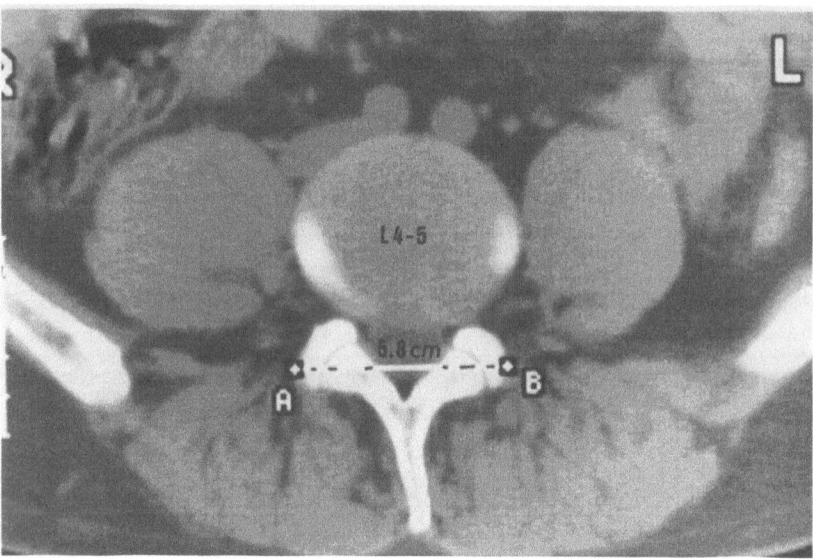

b

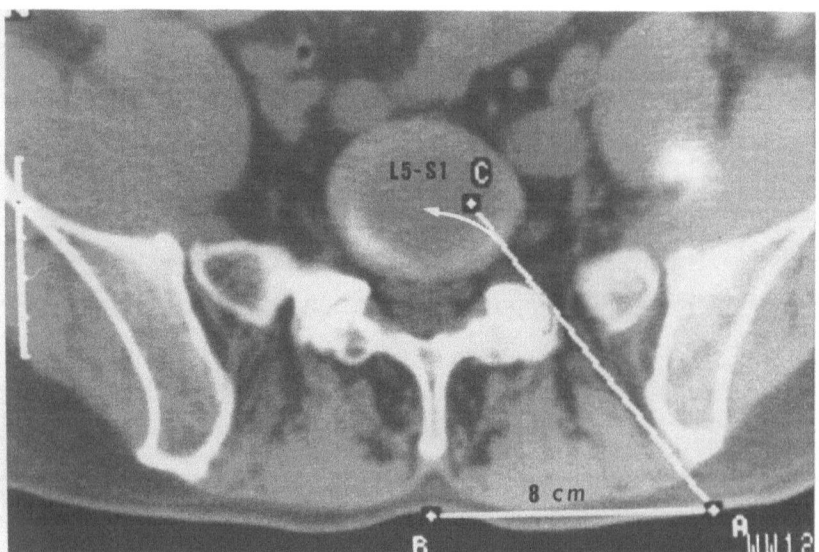

c

Fig. 21. Why a curved 22-gauge needle should be used at L 5-S 1 level: At L 4-5, needle placement in the center of the disc is usually possible using a straight needle (**a**). At L 5-S 1, the needle tip must be curved to reach the center of the disc space (**c**) because the width of the neural arch (**d**) is much greater than at L 4-5 level (**b**)

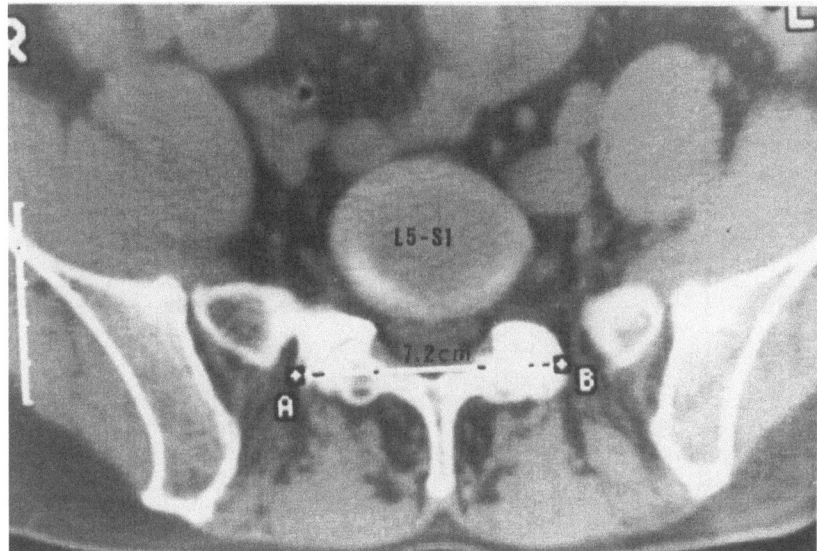

Fig. 21 **d**

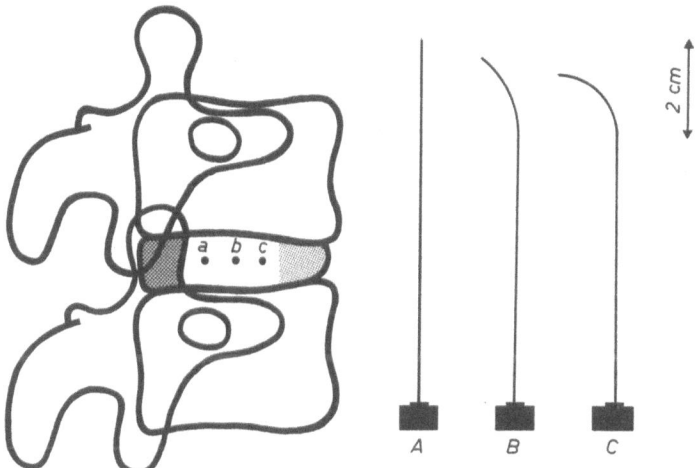

Fig. 22. Technique of approach to the L 5-S 1 interspace. When the 18-gauge needle has reached the anulus, the patient is turned to an oblique position and the X-ray beam is tilted to profile the disc. In this position, the needle will appear as a dot projecting on the disc space. Needle position in relation to the disc space indicates how much the 22-gauge needle must be curved to reach the center of the disc. Needle projection on the medial and lateral part of the disc (dotted areas) indicates needle position is incorrect. Needle projection in **a** indicates that the center of the disc can be reached without curving the needle (*A*). With needle projection in **b**, the needle curve must be moderate (*B*). With needle projection in **c**, needle curve must be maximal to reach the center of the disc space (*C*)

using an important volume of contrast media (1–2 ml). We use only a small amount (0.5 ml) of contrast material [1, 12]. With this small volume of contrast media, the disc herniation is often poorly opacified (Fig. 26). However no contraindication comes from the discographic appearance of either intervertebral disc or disc herniation. Furthermore, filling of the disc space with contrast material may increase disc pressure and cause leakage of enzyme along the needle into the paravertebral space. Another reason for using a small amount of contrast material is the alleged inhibitory action

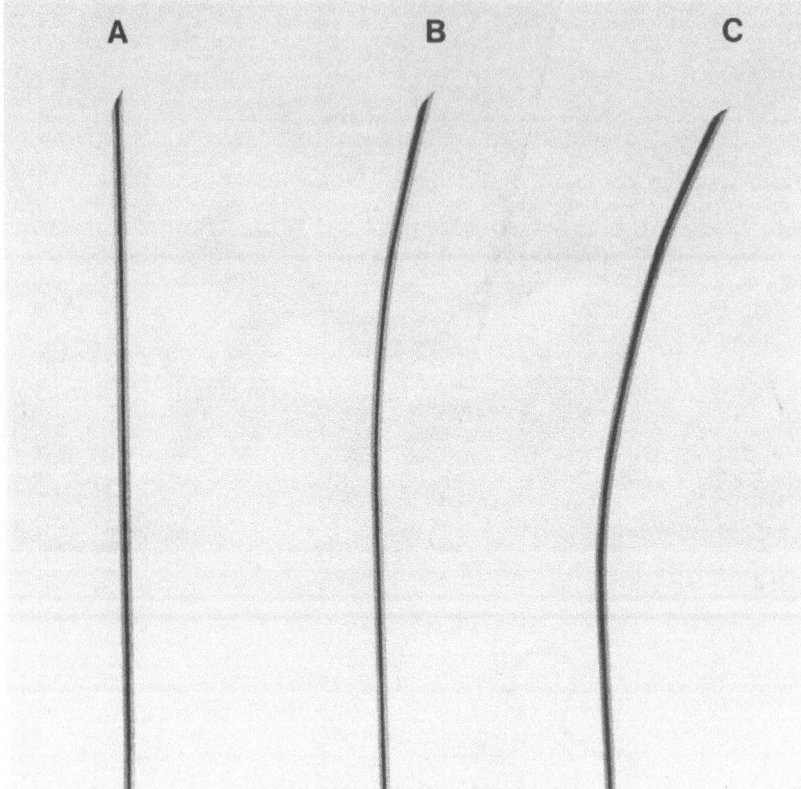

Fig. 23. Straight (*A*) and curved (*B* and *C*) 22-gauge needles used to reach the center of the disc space

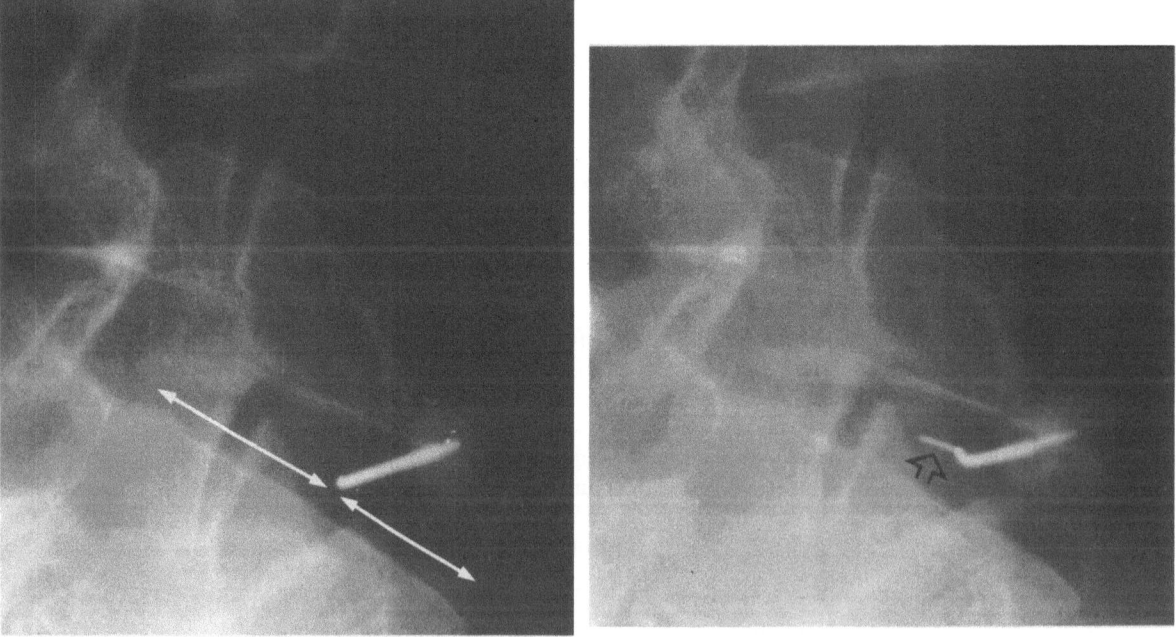

a

b

Fig. 24. Oblique view of the L5-S1 interspace. **a** The tip of the 18-gauge needle projects close to the midwith of the disc space. In that particular case, the 22-gauge must be moderately curved (as in *B* of Figs. 22 and 23) to reach the center of the disc space. **b** The curved 22-gauge needle had been introduced through the 18-gauge needle: note how its distal curve (arrow) allows easy placement of the needle tip in the center of the disc space

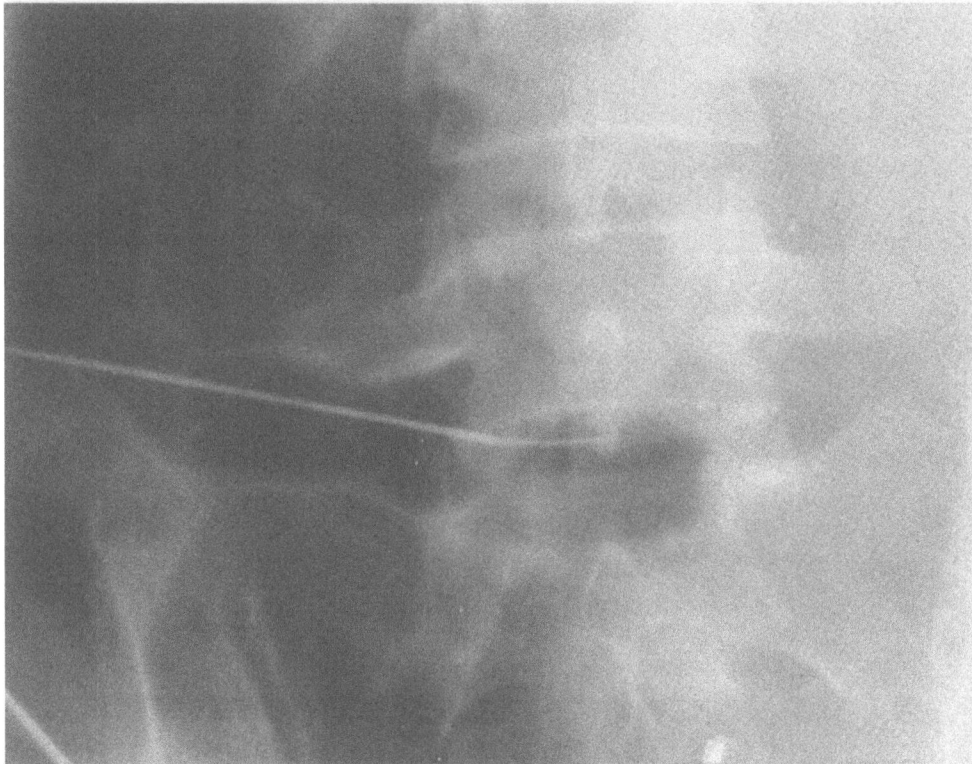

a

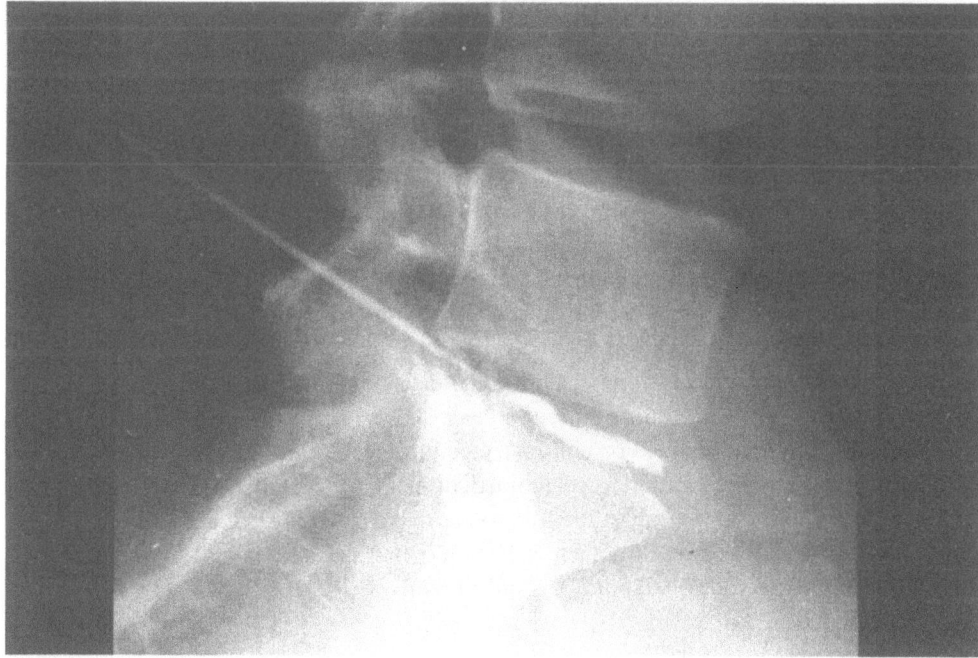

b

Fig. 25. The curved 22-gauge needle is advanced to the center of the L5-S1 disc space. Correct placement of the needle is checked on both AP (**a**) and lateral views prior to perform discography (**b**)

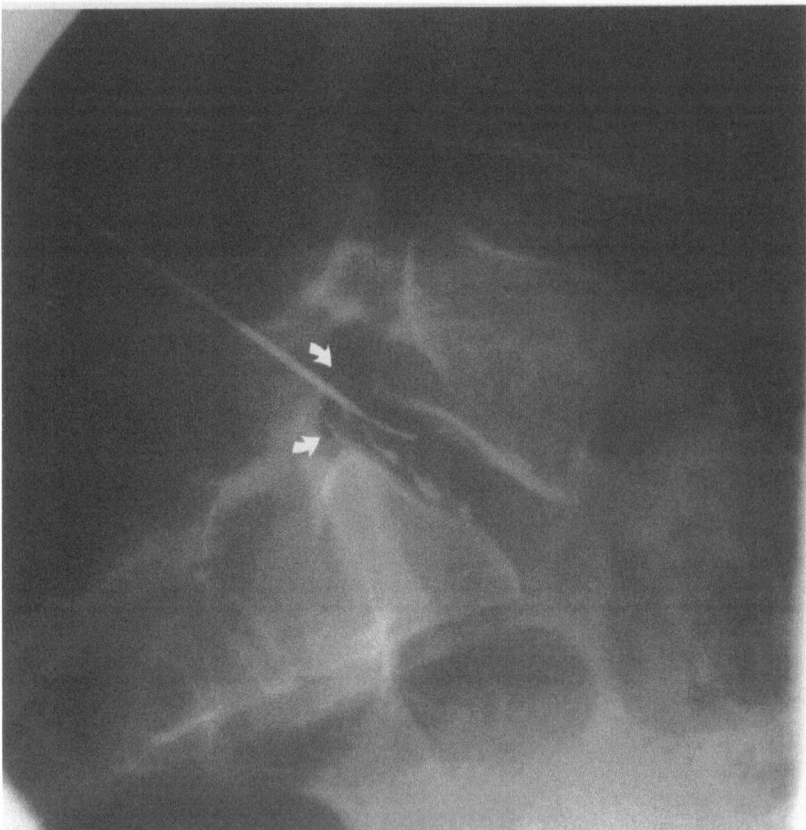

Fig. 26. Mild opacification of the disc herniation (curved arrows) during discography
at L 5-S 1 level

of the contrast media on chymopapain activity. During discography, the
absence of vascular opacification must be verified on the fluoroscopic screen,
especially during the beginning of the injection of contrast material. Vascular
opacification is the only discographic finding which contraindicates enzyme
injection [3, 4, 5, 16]. Dense opacification located in the center of the disc
space on both AP and lateral views ensures the proper placement of the
needle within the nucleus (Fig. 27). In the case of injection of contrast media
within the anulus, contrast material may appear to be located in the center
of the disc on one projection. However the other view demonstrates that
the contrast material is lateral to the center of the disc and to the nucleus
pulposus. Excessive pressure during injection may also indicate incorrect
needle position. Epidural leakage of contrast medium at the time of discography is not a contra-indication to chemonucleolysis [11]. This phenomenon is very common, and the site of the leak aids in placing chymopapain
in contact with the extruded disc fragment [11]. Safety in this case has been
adequately demonstrated in the laboratory and in many series [11].

Enzyme injection

To reduce the possibility of inactivation by a high concentration of radio-
opaque contrast medium, fifteen minutes are allowed to elapse before chymopapain is injected. Two milliliters or 4,000 units of chymopapain are
slowly injected into each disc space to a maximum of 5 milliliters or 10,000

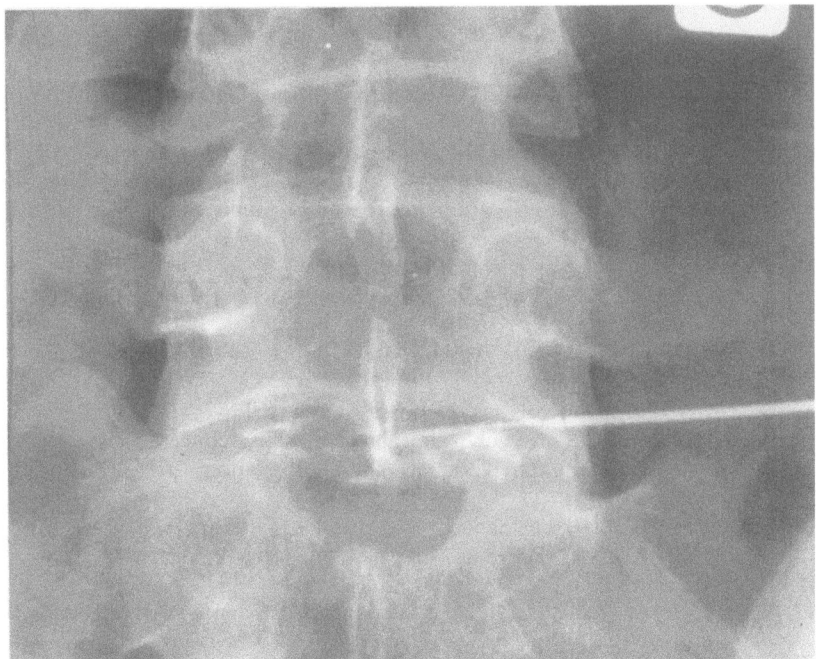

a

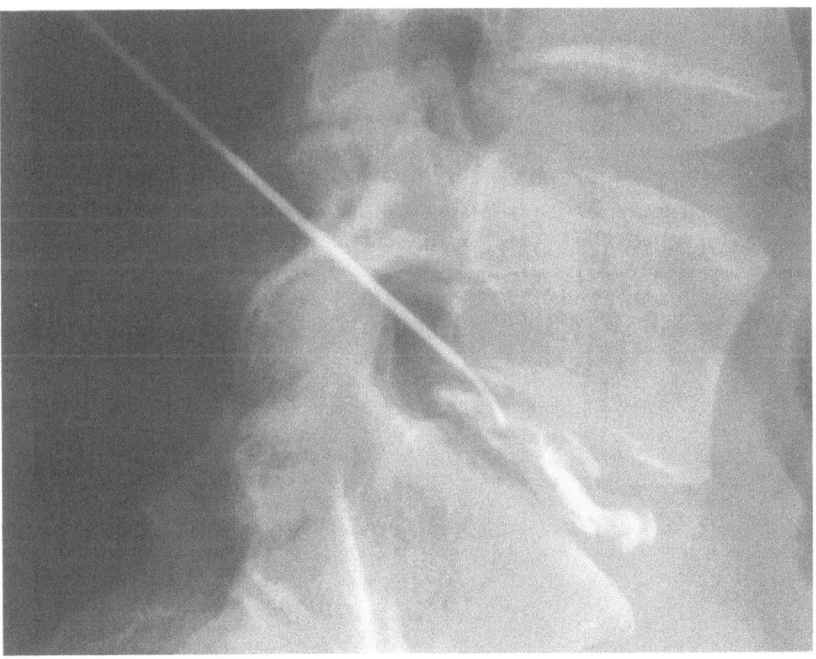

Fig. 27. AP (**a**) and lateral (**b**) views during discography; contrast material must be located in the center of the disc space on both projections

b

units for each patient for multilevel injections [11]. In some cases, the effect of the enzyme can be seen immediately as a milky reflux in the syringe [12]. To reduce leakage of enzyme, a delay of ten minutes is observed after injection and before removal of the needles. After injection of the enzyme the patient is observed for any adverse reactions. Since the procedure has been carried out under local anaesthesia, the patient can signal any abnormal sensations which may mark the beginning of an anaphylactic reaction [11]. After removal of the needle, the patient is placed in a supine position.

References

1. Benoist M, Deburge A, Heripret J, Busson J, Rigot J, Cauchoix J (1982) Treatment of lumbar disc herniation by chymopapain chemonucleolysis: a report on 120 patients. Spine 7: 613–617
2. Brown JE (1969) Clinical studies on chemonucleolysis. Clin Orthop 67: 94–99
3. Crock HV (1983) Practice of spinal surgery. Springer, Wien New York, pp 40
4. Crock HV (1984) Discography and vertebral venography before chemonucleolysis. J Neurosurg 60: 149–150
5. Crock HV (1986) Chymopapain. Lancet 2: 1159
6. Dabezies EJ, Murphy CP (1985) Dural puncture using the lateral approach for chemonucleolysis. Spine 10: 93–96
7. Day PL (1969) Lateral approach for lumbar discogram and chemonucleolysis. Clin Orthop 67: 90–93
8. Erlacher P (1952) Nucleography. J Bone Joint Surg [Br] 34: 204–210
9. Finneson BE (1973) Low back pain. Lippincott, Philadelphia, pp 69–76
10. Keck C (1960) Discography, technique and interpretation. Arch Surg 80: 580–585
11. McCulloch JA (1977) Chemonucleolysis. J Bone Joint Surg [Br] 59: 45–52
12. McCulloch JA (1980) Chemonucleolysis: experience with 2,000 cases. Clin Orthop 146: 35–128
13. McCulloch JA, Waddell G (1978) Lateral lumbar discography. Br J Radiol 51: 498–502
14. Naylor A, Earland C, Robinson J (1983) The effects of diagnostic radiopaque fluids used in discography on chymopapain activity. Spine 8: 875–879
15. Nazarian S (1985) Anatomical basis of intervertebral disc puncture with chemonucleolysis. Anat Clin 7: 23–32
16. Ray CD (1984) Danger of intravenous injection during chemonucleolysis. J Neurosurg 60: 1327
17. Rocolle J, Degott C, Benoist M, Lassale B, Nedjar C, Tardivon A, Deburge A (1985) Etude expérimentale de l'action comparée de deux préparation de chymopapaïne sur le disque interverté et de l'effet de l'injection préalable d'un produit de contraste radio-opaque. Rev Chir Orthop 71 [Suppl 2]: 36
18. Smith L, Brown JE (1967) Treatment of lumbar intervertebral disc lesions by direct injection of chymopapain. J Bone Joint Surg [Br] 49: 502–519
19. Troisier O (1980) Traitement des lombosciatiques par injection intra-discale d'enzymes protéolytiques. 80 observations. Nouv Presse Med 9: 227–230
20. Troisier O (1982) Technique de la discographie extra-durale. J Radiol 63: 571–578

Chymopapain chemonucleolysis in cervical herniated discs

Y. Lazorthes[1], J. Richaud[1], J. C. Verdié[1], and A. Bonafe[2]

[1] University Clinic of Neurosurgery and [2] Department of Neuro-Radiology,
C.H.U Rangueil, Toulouse, France

Since its introduction as a therapeutic procedure for herniated discs, chymopapain chemonucleolysis has been limited to the lumbar spine. Chymopapain chemonucleolysis at the cervical level has been considered a potentially harmful technique. In response to this contraindication, some authors performed intradiscal injection of aprotinin in the treatment of cervical herniated discs [3]. However the effectiveness of this enzyme in the treatment of lumbar herniated discs appears to be doubtful and its use at the cervical level cannot be recommended.

Our report of 15 cases of chymopapain cervical chemonucleolysis published in 1983 (a study in cooperation with the University Hospital of Caen) pointed out the effectiveness and safety of this technique.

The present work is a pilot study, concerning 31 cases, performed in agreement with the Travenol European Division* according to a selective protocol. Indications were highly selective in order to reduce the number of failures that would subsequently require open surgery. Indications were limited to cases of cervical nerve root compression requiring immediate surgical treatment. During the same period of time, open microsurgery, in use in our department since 1968 [1], was performed in 78 cases of cervical herniated discs. In our experience, the immediate and long-term results of cervical chymopapain chemonucleolysis are comparable to those obtained at the lumbar level [11].

Patients

During the past four years, 38 cases of cervico-brachial neuralgia were treated with chemonucleolysis. The present review concerns 31 patients (21 males and 10 females) whose ages ranged from 27 to 55 years (mean 42 years). Seven cases were not included because the follow-up was too short. All patients had cervicobrachial pain resistant to conservative treatment.

The duration of symptoms was superior to 2 months in all patients (2 to 72 months, mean 12 months); it was less than 6 months in 14 cases and exceeded 6 months in 17 cases.

---

* February 17, 1982.

No case of isolated neck pain without brachial symptoms was included in the study. All patients but one had cervical neuralgia (15 moderate and 15 severe) associated with radiculopathy. Signs and symptoms consisted of purely sensory syndromes in 16 cases and sensorimotor radiculopathy in 14 cases (moderate motor impairment in 11 cases and severe motor loss in 3). In one case there was severe motor loss without pain. This finding was not considered a contraindication of this technique.

Associated pyramidal syndrome was observed in 3 cases coexisting with purely sensory syndrome in one case, sensory motor syndrome in another case, and purely motor radiculopathy in a third case. We presently consider associated pyramidal syndrome to be a contraindication.

Pain or sensorimotor radiculopathy (determined in 30 cases) was initiated spontaneously in 13 cases (6 at work) and by acute trauma in 17 cases (6 during sports).

Procedures

The following procedures were used to demonstrate disc herniation:

Plain cervical spine radiographs were performed in all patients, completed by dynamic sagittal views in those patients having a history of trauma, in order to exclude other diseases and cervical spine instability.

Electromyography was used to confirm neuropathy in 19 cases; results were significant in 17 cases and insignificant in 2.

Cervical myelography using iopamidol was performed in 18 patients through direct lateral puncture at C 1–C 2 level under fluoroscopic control and included prone and supine views. Lack of filling of nerve root sleeves and persistent compression of the anterior subarachnoid space were considered indicative of soft disc herniation in 16 cases. In 2 cases, myelography alone was not significant and further investigations were indicated.

Computerized tomography studies were performed in 19 patients, with a CGR CE 10,000 high resolution machine allowing 1 mm thick sections. CT was the only procedure used in 10 patients, and was performed together with myelography in 9 cases. In one case, myelography and CT were used simultaneously. The CT investigation was sufficient to demonstrate disc herniation in 15 cases (Fig. 1), but in 4 cases there was a false negative CT examination. In 2 cases both myelography and CT failed to demonstrate disc herniation that was shown by discography.

In a few cases (during the first part of our study), discography was the only procedure performed to demonstrate soft disc herniation. In 9 cases discography (performed at two levels in two cases) was associated with other neuroradiologic investigation and confirmed uncertain discopathy.

The results of the above neuroradiologic investigations allowed us to demonstrate soft disc herniation in all 31 patients. Double discopathy existed in two cases. The overall number of disc levels treated with chemonucleolysis is 33. The levels treated were: C 4–5 in 2 cases, C 5–6 in 15 cases, and C 6–7 in 16 cases. The two cases with disc herniation at two levels were observed at C 5–6 and C 6–7 levels. No cases with disc herniation at other levels were treated. However, cervical chemonucleolysis can be performed at any level from C 3–4 to C 7–T 1. Associated spinal abnormalities were observed in 5 cases (moderate segmental arthrosis in 3 cases, adjacent cervical congenital block in 2 cases). The appearance of sequestrated or migrated disc fragments

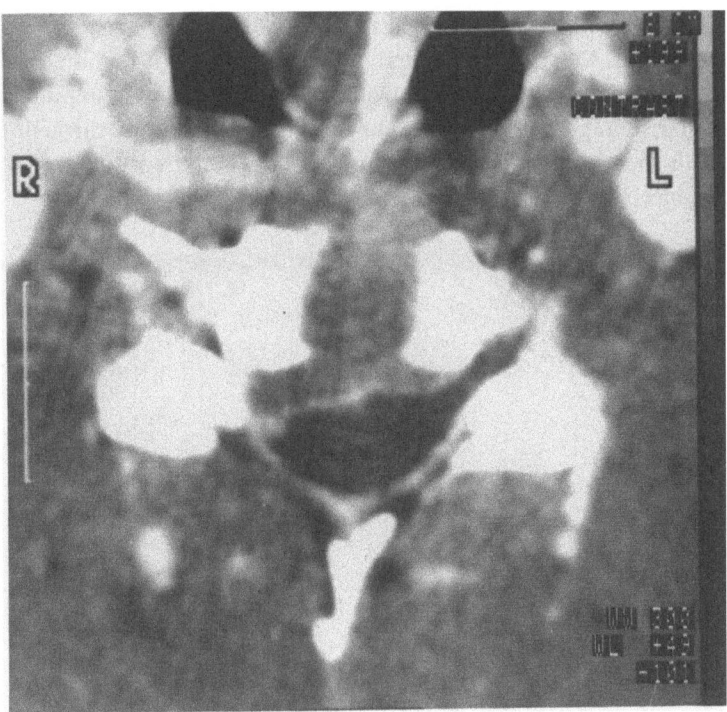

Fig. 1. CT scan at C 5-6 level demonstrating a herniated disc

contraindicated chemonucleolysis and indicated surgical treatment. The follow-up time after chemonucleolysis ranged from 3 to 52 months, with a mean of 8.5 months.

Technique of cervical chemonucleolysis

Cervical chemonucleolysis is usually performed under neuroleptanalgesia as in the latter part of our study (18 cases). General anesthesia was used in the early stage of our study (13 cases).

The procedure of cervical chemonucleolysis consists of three steps: positioning of the needle; discography; chemonucleolysis.

Positioning of the needle

The patient is supine and discography is performed by percutaneous technique using a 22-gauge lumbar puncture needle. The needle is inserted on the right side into the anterior triangle situated between the larynx medially and the right carotid artery laterally (Fig. 2). The central ray is directed with a cephalad tilt until the intervertebral disc is seen in profile. In every case, the intervertebral disc space must be located at right angles to the X-ray beam in order to facilitate the exact positioning of the needle.

Some operators perform the procedure by placing the patient in a supine position with the right shoulder elevated 35° from the horizontal plane and the head turned 45° toward the left side. The needle is advanced toward the center of the disc under C-arm fluoroscopy guidance. Once the outer surface of the disc is reached, the needle is advanced one centimeter further, but no more. Before performing nucleolysis it is mandatory to control needle positioning using antero-posterior and lateral views of the cervical spine; precise positioning in the center of the intervertebral space is crucial. If its tip deviates toward the vertebral end-plates, the needle must be repositioned.

It is essential to avoid penetration of the spinal canal during chemo-nucleolysis. Dural puncture and contamination of the subarachnoid space with enzyme constitute the only life threatening risk of the procedure. The same risk might also exist if a puncture is made laterally through the nerve rooth sheath. However, this complication was not encountered in our experience.

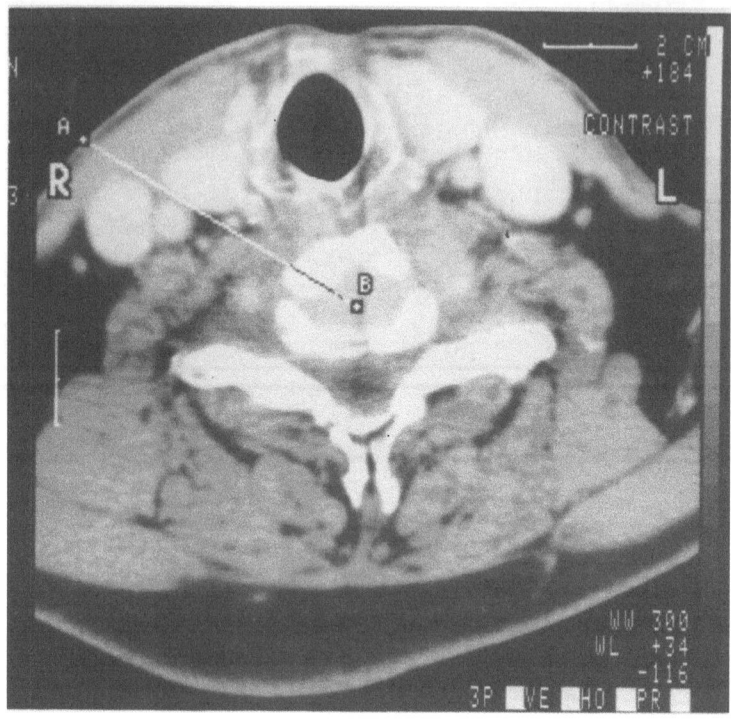

Fig. 2. Anterolateral approach between larynx and carotid artery

Discography

As a part of the operative procedure of chemonucleolysis discography is carried out using one ml of a water-soluble non-ionic contrast media (io-pamidol) at a concentration of 300 mg/ml. In our experience, discography was performed in all cases but one. We now consider this procedure to be obligatory prior to nucleolysis. The contrast media is injected gently under fluoroscopic (lateral view) and clinical controls. In most cases, discography reproduces cervicobrachial pain and confirms the disco-radicular pathologic process. In the present study the discs presented variable resistance to the injection (no resistance in 9 cases and moderate resistance in 24 cases). In one patient with soft disc herniation, the resistance was such that it was impossible to inject chymopapain and surgery was subsequently performed. Two different types of distribution of the contrast media were observed; in 16 cases, subligamentous protrusion was present with no extravasation of contrast media within the epidural space (Fig. 3). Seventeen cases were associated with epidural leak, in which contrast media extended posteriorly to the outer fibers of the annulus. Opacification of the subarachnoid space, contraindicating chemonucleolysis, was never observed.

Chemonucleolysis

Following discography, chymopapain (Discase*) is injected in a dose of 2,000 units (16 cases, first part of our study), 3,000 units (7 cases), or 4,000

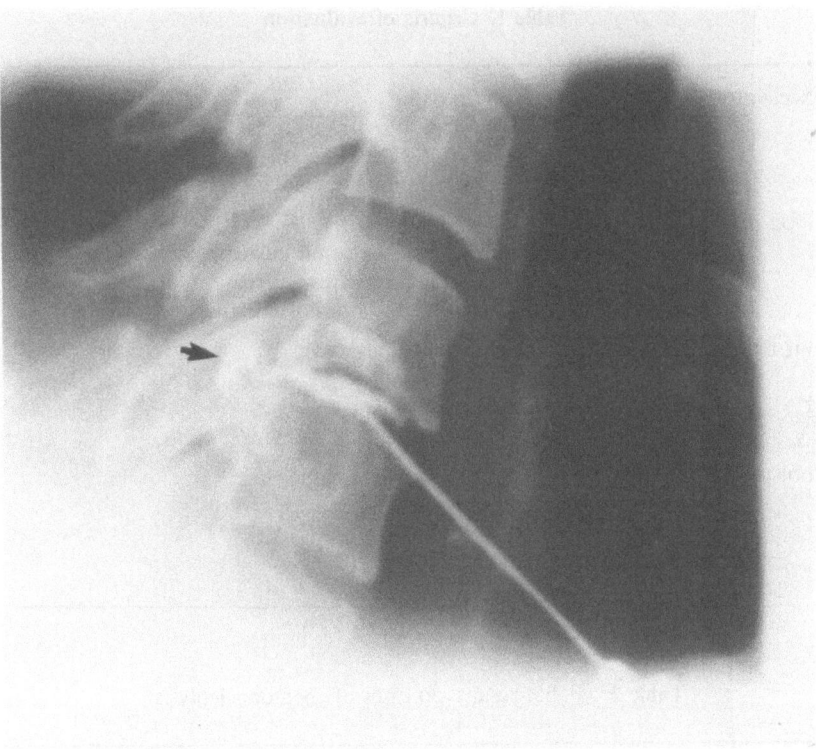

Fig. 3. Discography prior to chemonucleolysis demonstrating posterior extravasation of contrast medium (arrow)

units (9 cases). In two cases with disc herniation at two levels, a double injection (2,000 and 2,000 units in one case, 3,000 and 4,000 units in the other) was performed.

Follow-up

The patients are kept on bed rest for 48 hours without cervical collar, mobilized on the third day, and discharged on the fourth day. After the procedure, patients are followed up clinically, with particular attention to side effects. In some cases transitory exacerbation of neck pain may be observed, requiring administration of analgesics for two or three days. In our experience, a transient and benign allergic skin reaction was noted in one case. No severe anaphylactic reaction was observed and no cases of infectious discitis was encountered.

Results

The criteria of evaluation presented in Table 1 were analyzed by means of a rating scale similar to that developed by MacNab for lumbar disc herniation [12]. The 30 patients (one patient had no chymopapain injection) were evaluated 6 or 8 weeks after the procedure. All 30 patients were also reevaluated at 3 months or more. Results are summarized in Table 2. Excellent or good results were obtained in 25 of the 30 cases (83%). Results were fair or poor in 5 cases (17%). In 3 of these 5 unsatisfactory cases, a surgical procedure was performed with excellent results.

Conclusions

The findings of the present study are in agreement with our previous reports [4, 5, 6, 7, 8, 9, 10] and confirm the effectiveness of cervical chemonucleolysis. No delayed deterioration of the initial excellent or good results was observed

Table 1. Criteria of evaluation

Excellent results	No arm pain
	No neck pain
	No neurological sequelae
	Return to normal activities
Good results	No arm pain
	Moderate neck pain on movement
	No neurological sequelae
	Partial return to normal activities
Fair results	Slight residual arm pain
	Severe neck pain
	Residual neurological sequelae
	No return to normal activities
Poor results	No change in arm pain
	Severe neck pain
	Neurological sequelae
	All activities restricted

Table 2. Global results. 30 cases of chemonucleolysis*

	Excellent	Good	Fair	Poor
Number	18	7	2	3
		25		5
Percent	60%	23%	7%	10%
		83%	17%	

* In 1 case injection of chymopapain was impossible

in cases with extended follow-up. However, the number of patients treated with this technique is still small as compared to lumbar nucleolysis.

The positive and lasting results obtained in this pilot study encourage this new application of chemonucleolysis. However, cervical chemonucleolysis should be restricted to patients who meet several criteria: severe cervicobrachial pain (with or without radicular motor impairment) resistant to conservative treatment for more than two months, radiologic evidence of lateral disc herniation on neuroradiologic investigations, no evidence of disc fragment sequestration or migration or associated osteophytic process, absence of spinal cord compression.

Contraindications to the use of chemonucleolysis are: previous chymopapain treatment or isolated neck pain or moderate brachial pain without clinical or EMG signs, associated pyramidal syndrome, spinal stenosis and spinal instability. The procedure is not contraindicated in allergic patients but must be associated with special premedication.

Finally, this procedure should be considered an alternative to open surgery in selected cases. The safety of the procedure depends on careful discography technique prior to chemonucleolysis. It should be used on a highly selective basis in order to maintain a low rate of failure.

In case of failure, good results can still be obtained with a subsequent surgical procedure. This technique appears to be the last step in conservative treatment of lateral cervical disc herniation.

References

1. Lazorthes Y, Zadeii JO, Lagarrigue J (1970) Les hernies discales cervicales. Rev Méd Toulouse 6: 647–654
2. Lazorthes Y, Theron J, Verdie JC, Houtteville JP, Lagarrigue J, Courtheaux P (1983) Chemonucleolyse discale cervicale. Résultats initiaux dans 15 cas de compression radiculaire. Comm 33ème Congr Annu Soc Neurochirurgie de Langue Française, June 1983
3. Lesoin F, Jomin M, Viaud G, Lozez G, Pruvo JP, Clarisse J (1984) Cervical intradiscal injection of aprotinin. Technical note and preliminary report. Surg Neurol 21: 539–542
4. Lazorthes Y, Richaud J, Verdie JC, Bonafe A (1985) Chémonucléolyse discale cervicale. Résultats préliminaires à propos de 15 cas. 11ème Coll Int Pathologie Locomotrice, Montpellier, March 1–3, 1985
5. Lazorthes Y, Richaud J, Verdie JC, Bonafe A (1985) Chémonucléolyse discale cervicale. Résultats préliminaires à propos de 10 cas. In: Simon L, Leroux JL, Privat JM (eds) Rachis cervical et médecine de rééducation. Collection de pathologie locomotrice, vol 10. Masson, Paris, p 350–354
6. Lazorthes Y, Verdie JC, Richaud J, Theron J, Houtteville JP, Courtheoux P (1985) Chemonucleolysis of cervical discs. A preliminary result in 15 cases of root compression. In: Sutton JC (ed) Current concepts in chemonucleolysis. Royal Society of Medicine, London, p 217–223
7. Lazorthes Y, Richaud J, Verdie JC, Bonafe A (1985) Chemonucleolysis for herniated cervical disc (21 cases). 8th Int Congr Neurological Surgery, Toronto, July 7–13, 1985
8. Bonafe A, Lazorthes Y, Tremoulet M, Verdie JC, Richaud J (1985) Chemonucleolysis for herniated cervical disc. 13th Congr Eur Neuroradiology. Amsterdam, September 11–15, 1985
9. Bonafe A, Lazorthes Y, Tremoulet M, Verdie JC, Richaud J (1985) Chemonucleolysis for herniated cervical disc. Neuroradiology. 22: 319–324
10. Bonafe A, Lazorthes Y, Tremoulet M, Verdie JC, Richaud J, Song MY (1986) Chemonucleolysis for herniated cervical disc. J Intervent Radiol 1: 19–21
11. Lazorthes Y, Richaud J, Rober B, Lagarrigue J, Verdie JC, Bonafe A (1985) La chémonucléolyse dans le traitement des sciatiques chirurgicales. Neurochirurgie 31: 471–493
12. MacNab I (1971) Negative disc exploration. J Bone Joint Surg [Am] 53: 891–895

Percutaneous lumbar discectomy

Percutaneous lumbar discectomy and decompression

P. Kambin

Department of Orthopaedic Surgery, Disc Treatment and Research Center,
University of Pennsylvania School of Medicine, The Graduate Hospital,
Philadelphia, Pennsylvania, U.S.A.

The decrease in the popularity of chemonucleolysis has created a surge of interest in mechanical nuclear decompression and discectomy. Our interest and experience in this unique approach, which permits access to the intervertebral disc without the necessity of violating the spinal canal, dates back to 1973 [3]. Originally, we were utilizing a cannula with the external diameter of 3 mm; however, our subsequent experimental work led us to use a larger diameter sheath to provide an adequate annular opening and permit better access to the intervertebral disc. The intraoperative pressure measurement of a herniated lumbar disc prior to and following the annular fenestration suggests that an adequate fenestration of the annulus, to allow the continuous decompression of the disc, is desirable [2]. A small window in the annulus is easily blocked by blood clots and nuclear debris.

Hijikata [1] has shown that a 4 to 5 mm fenestration of the annulus may remain patent up to 9 months following percutaneous nucleotomy.

The capability of reaching posteriorly with the inserted instruments appears to be an essential step in evacuation and decompression of the protruded discs. The utilization of deflectors and flexible instruments [2] are a step in the right direction in the development of additional tools to achieve this goal.

The use of forced suction allows the reduction of dislodged nuclear fragments in the path of the inserted instruments, thus allowing their evacuation.

The advantages of the percutaneous approach to discectomy include: avoidance of epidural bleeding and perineural fibrosis, elimination of re-herniation through the intraoperatively induced annular fenestration, preservation of spinal stability, establishment of a portal for future herniation away from the neural elements. The percutaneous discectomy is more cost-effective than a laminectomy [6] with decreased O.R. time and the post-operative hospital care. Certainly, future surgical procedures, if deemed necessary, are not compromised.

The complications associated with chemonucleolysis such as sensitivity reaction, anaphylaxis, neurotoxicity, chemical discitis and lateral stenosis associated with rapid disc space narrowing are avoided.

Anatomy

As the lumbar nerve roots leave the foramina, they descend distally and laterally and lay anterior to the transverse processes. Usually, a thin layer of fatty tissue and fibers of the psoas major muscles keep the spinal nerve apart from the annulus at the site of the introduction of instruments which are used in the course of percutaneous lumbar discectomy. Our anatomical study has shown that when the instruments were angulated to reach an intervertebral disc, particularly at the L 5-S 1 level, the chance of injury to the spinal nerve was greater. A far lateral approach may cause serious complications, such as, injury to the content of the peritoneal cavity. When the instruments are inserted at a distance of 9 to 10 centimeters from the spinal processes, they penetrate the skin, subcutaneous tissue, fascia, the fibers of the sacrospinalis, quadratus lumbrum and psoas major muscles, thus entering the annulus dorsolaterally.

The iliac arteries and veins are located anteriorly and are not in the path of the inserted instruments. The sympathetic fibers of the lumbar spine are situated ventral to the vertebral bodies along the medial margin of the psoas major muscle and are not subject to insult in the course of this operative procedure.

Pathology

The nucleus pulposus and annulus fibrosis of the intervertebral disc gradually become more defined and demarcated during the first two decades of life [7]. At this stage, the nucleus consists of loose networks of fibrous tissue in a mucoprotein gel containing various mucopolysaccharides. The degenerative changes of the intervertebral disc may be seen as early as 14 years of age. The changes in the annulus fibrosis always predate nuclear degeneration. The circumferential and vertical tears in the annulus is followed by the ingrowth of vascular tissue through the endplates [8]. This is followed by the decrease of fluid content of the nucleus and its gradual collagenation.

With the advancement of the degenerative process, the peripheral migration of the collagenized nuclear fragment between the torn fibers of the annulus takes place. This is followed by gradual decrease in the height of the intervertebral disc and increase in the bulge or protrusion of the disc [4]. In contrast to herniation of the nucleus pulposus, which shows a well defined and abrupt limited mass, the protruded disc is signified by a gradually contoured soft tissue extradural defect in the CT scan study.

Both of these conditions are capable of producing nerve root compression in the lumbar spine and are treatable by percutaneous lumbar discectomy and decompression.

Patient selection

Proper patient selection is the most important factor which directly influences the final result of percutaneous lumbar discectomy and decompression. We have continued to adhere to the following criteria which was developed in our institution:

(1) Failure of conservative therapy;
(2) Presence of positive Tension Signs;
(3) Positive correlative CT or myelography;

(4) Positive neurological findings or correlative electromyographic evidence of radiculopathy.

All of these patients should have the benefit of a reasonable program of conservative therapy for at least 2 or 3 months. Our conservative treatment includes proper patient's education, which consists of awareness of the pathological changes which have taken place and is felt to be responsible for the patient's pain and the satisfactory results which are often obtained by this conservative therapy. The importance of a post-operative program of exercises and the proper working habits are emphasized.

Bed rest, bracing, anti-inflammatory drugs (which may include short-term steroid therapy), physical therapy modalities and exercises are utilized.

Patients with bony spinal stenosis are not candidates for percutaneous discectomy. Most of these individuals require a laminectomy and posterolateral decompression.

An attempt should be made to diagnose the sequestrated disc. The CT, metrizamide CT evaluation and enhancement CT study are helpful in preoperative screening and the diagnosis of free fragments in the spinal canal. Certainly, all of these individuals require a laminectomy.

The patients with cauda equina syndrome and progressive neurological changes should not be treated by this method. It has been our experience that intervertebral discs which previously have been treated by chemonucleolysis and laminectomy are not good candidates for percutaneous discectomy and decompression.

It appears that the presence of scar tissue and adhesion in the intervertebral disc and annular region prevents adequate evacuation and decompression.

Since the post-operative evaluation of individuals involved in litigation and compensation claims is difficult, surgeons practicing in industrialized nations should exercise extreme caution and remain highly selective in the performance of this operative procedure on these patients.

Operative technique

To maintain a sterile environment, we recommend that this procedure be performed in the operating room.

The instruments which are utilized are as follows (Fig. 1): an 18-gauge needle; a 0.028″ diameter Kirschner wire; cannulated trocar; sheath or cannula; two cutting instruments; deflector and deflector forceps; straight and curved punch forceps.

The procedure can be performed with the patient either in the lateral or prone position; however, a prone position is preferred. This position minimizes movement and rotation of the trunk during surgery and prevents false X-ray imaging in the AP and lateral projections.

To prevent undue pressure on the abdomen and to maintain the hips and knees in flexion, two rolled sheets are placed under the patient, extending from the ilium to the side of the chest wall.

The C-Arm is positioned and covered by a sterile plastic sheet. Avoid blockage of rotation of the C-Arm to assure a reproducible AP and lateral imaging.

Local anesthesia is obtained utilizing 1% Xylocaine solution. The skin, subcutaneous tissue, fascia and superficial muscle layers are infiltrated.

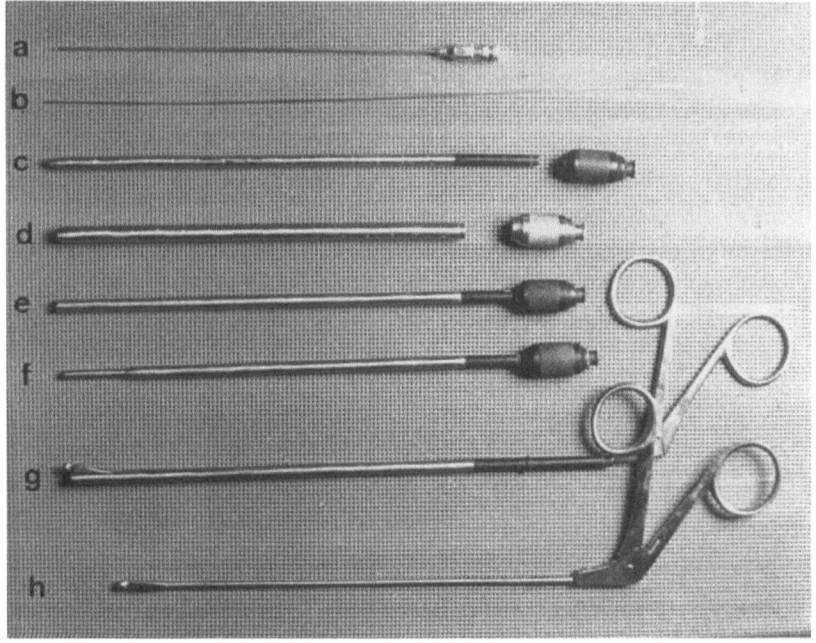

Fig. 1. *a* 18 gauge needle, 15 cm in length; *b* Kirschner wire 22 cm in length; *c* cannulated trocar 4.0 mm outer diameter, 19.0 cm in length; *d* sheath or canula 6 mm outer diameter, 16.0 cm in length; *e* cutting instrument 4.0 mm outer diameter; *f* cutting instrument 2.5 mm outer diameter; *g* deflector tub and flexible end forceps; *h* angled forceps

Avoid periannular infiltration. This will anesthetize the nerve root, thus pre-exposing it to injury.

The point of entry is 9 to 10 centimeters from the mid-line on the patient's symptomatic side. However, a distance of 8 to 9 centimeters may be adequate for slim patients.

With the C-Arm in lateral position, an 18-gauge needle, 6″ in length, is inserted at an angle of 35 to 45 degrees and is advanced to the annular fibrosis at 2 or 10 o'clock position, to the spinal process of the vertebra.

The correct positioning of the 18-gauge needle is the single most important step affecting the final outcome of percutaneous discectomy.

One should avoid far lateral or too close to the mid-line entry. When the instruments are introduced too close to the mid-line, they bypass the nucleus, thus, preventing adequate evacuation of the disc material.

A far lateral approach greatly enhances the risk of bowel rupture and entrance to the peritoneal cavity.

As we insert the needle, we have to simultaneously rely on our sense of touch as well as on the radiographic appearance.

Resistance should be encountered when the needle reaches the annulus.

If no resistance is encountered and the lateral X-ray projection demonstrates that the needle has bypassed the annulus, it indicates vertical insertion. This requires withdrawal of the needle and reinsertion at a reduced angle with respect to the horizontal plane.

In contrast, if the needle has been inserted horizontally, one usually encounters resistance. The lateral X-ray projection, in this case, demonstrates that the tip of the needle has not reached the annulus. This requires withdrawal and reinsertion in a more vertical direction.

The needle should be inserted parallel to the vertebral plates. This allows for better evacuation of the disc material and minimizes the chance of neural injury. This needle may be utilized for the introduction of local anesthetics into the intervertebral disc.

Following the proper positioning of the needle, the stylet of the needle is replaced by a 0.028″ Kirschner wire. The needle is then withdrawn (Fig. 2).

At this time, the cannulated blunt trocar is passed over the Kirschner wire.

Watch for possible migration of the guide wire in the course of introduction of the cannulated trocar.

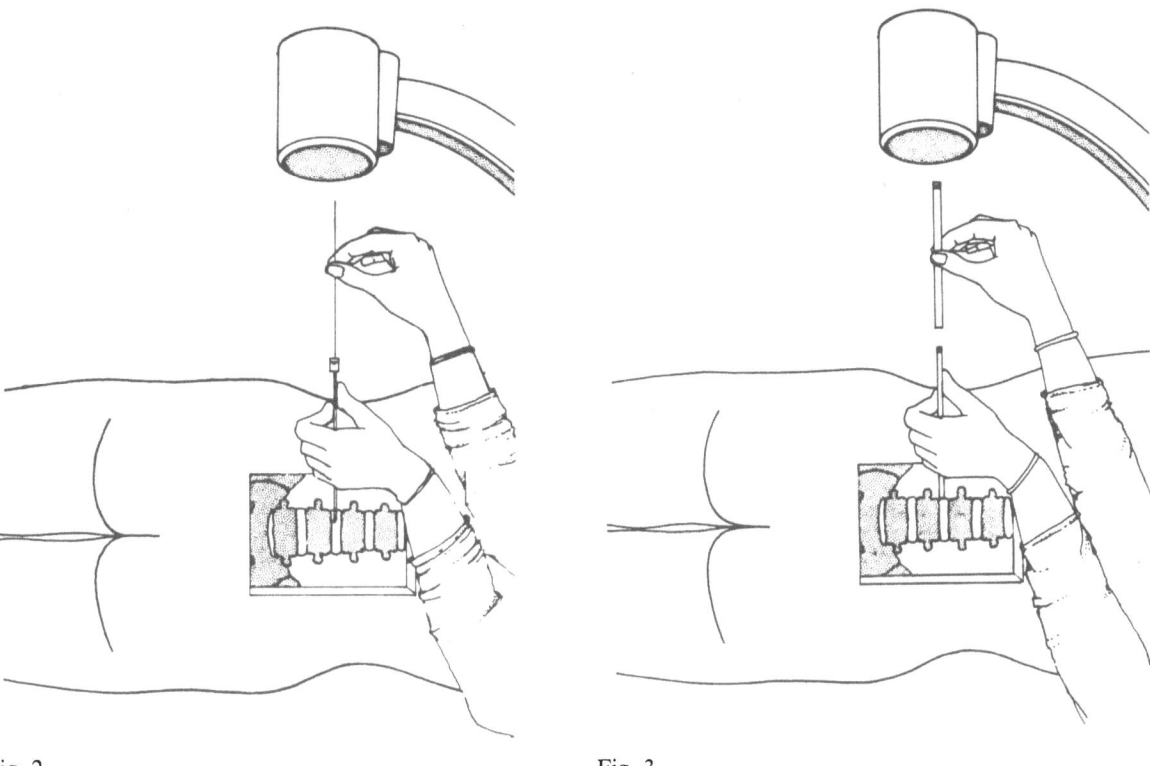

Fig. 2

Fig. 3

Fig. 2. The stylet of the needle is replaced by a 0.028″ wire to accommodate and direct the cannulated trocar

Fig. 3. The 4.9 mm sheath is passed over the cannulated trocar

As soon as the direction of the cannulated trocar is established, the guide wire should be removed. This is a precautionary measure. By doing so, the blunt end of the trocar will bypass the nerve roots, preventing undue damage to the nerve fibers.

A sheath with an internal diameter of 4.9 mm is passed over the trocar until it reaches the annulus fibrosis (Fig. 3).

At this time, one should check for possible entrapment of the spinal nerve. The sheath is held firmly against the annulus to prevent central or periannular migration. The central migration of the sheath will cause serious complications. It permits deep penetration of the inserted instruments causing abdominal or vascular injuries. The correct position of the sheath is radiographically monitored.

A simple needle test provides great assistance in the detection of nerve root entrapment under the sheath. The needling of the annulus by moving around the internal diameter of the sheath should not produce radicular pain. However, if the nerve is trapped, the introduction of the needle will cause severe radicular symptoms.

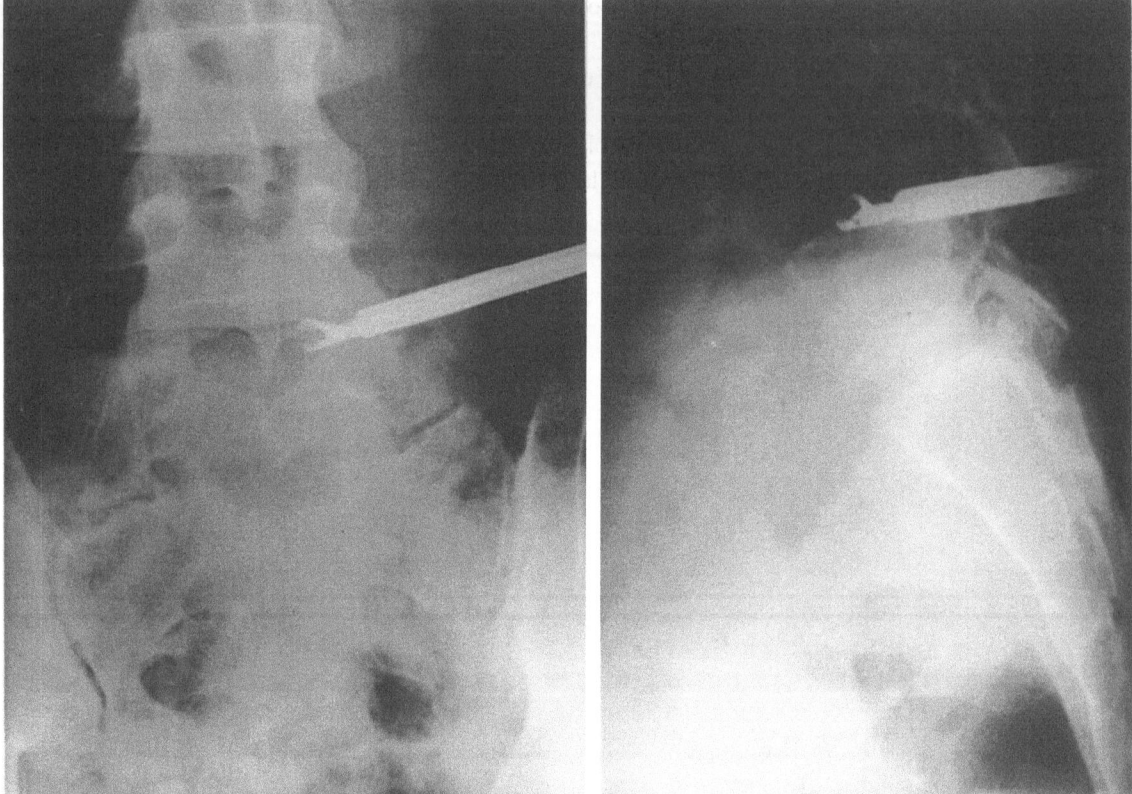

Fig. 4. AP and lateral intraoperative roentgenographic evaluation demonstrating the introduction of the forceps into the L 4–L 5 intervertebral disc for the evacuation of the nuclear fragments

The fenestration of the annulus is accomplished first with a small and then a larger cutting instrument.

The instrument design does not permit more than 2 centimeters of penetration.

The windowing of the annulus is done by a firm rotary movement maintaining the downward pressure.

This step of the operation is painful usually. The patient may experience radicular pain. At this time, additional analgesics are administered by the anesthesiologist.

Following the fenestration of the annulus, the disc material is evacuated with forceps and suction (Fig. 4).

The cutting instrument or the deflector tube is inserted into the sheath and attached to a 50 cc syringe. Forced suction is introduced to the center of the disc, which is followed by further evacuation utilizing straight or curved forceps.

Since a straight probe limits the site of evacuation to the tip of the inserted instruments, we feel that the use of the deflector tube, angled and flexible forceps, are essential.

These instruments allow us to reach posteriorly, close to the site of the herniation, and facilitates better decompression and disc extraction.

One can utilize automated instruments inside the sheath. These instruments permit simultaneous suction, irrigation and evacuation.

We have used instruments with rotary cutting capabilities as well as ones with oscillation or forward and backward movements.

In general, these instruments deprive the surgeon of the capability to distinguish between attached, semi-attached, or loose fragments which are being extracted. Also, it limits the site of evacuation to the center of the disc space.

For these reasons, we feel that they should be used only in association with the manual instruments.

Certainly, powered instruments are necessary and have a place in performing percutaneous interbody fusion.

Results

In our experience, the percutaneous discectomy and decompression has been effective. Our satisfactory results have remained above 80% (modified MacNab's standard) [5]. We have not encountered any neurovascular complications or intestinal injuries. We have reported 2 patients who developed inguinal and anterior thigh pain following percutaneous disc evacuation and decompression. This was attributed to intraoperative bleeding from the bone and veins at the site of annular fenestration. The symptoms on both patients subsided with bed rest and the use of analgesics.

The post-operative utilization of hemovac, when bleeding is apparent, has eliminated this complication.

References

1. Hijikata SA (1987) Percutaneous nucleotomy. Int Symp Graduate Hospital, University of Pennsylvania, Philadelphia, March, 1987
2. Kambin P, Brager MD (1987) Percutaneous posterolateral discectomy. Anatomy and mechanism. Clin Orthop 223: 145
3. Kambin P, Gellman H (1983) Percutaneous lateral discectomy of the lumbar spine: a preliminary report. Clin Orthop 174: 127
4. Kambin P, Nixon JE, Chait A (1987) Bulging annulus pathophysiology and roentgenographic findings. Scientific Program, Am Acad Orthopaedic Surgeons, San Francisco, January, 1987
5. Kambin P, Sampson S (1986) Posterolateral percutaneous suction − excision of herniated lumbar intervertebral discs: report of interim results. Clin Orthop 207: 37
6. Kambin P, Sampson S (1984) Laminectomy vs. percutaneous lateral discectomy: a comparative study. Orthop Trans 8: 408
7. MacNab I (1986) Disc degeneration and low back pain. Proc R Coll Phys Surg Can, 1952. Clin Orthop 208: 3
8. Ritchie TH, Fahrni WT (1970) Age changes in the lumbar intervertebral disc. Canada J Surg 13: 63

Percutaneous automated discectomy

G. Onik

Allegheny-Singer Research Institute, Pittsburgh, Pennsylvania, U.S.A.

Herniated lumbar discs are a major health problem throughout the world. Over 200,000 back operations are performed each year in the United States alone. The traditional procedure for herniated lumbar discs, surgical removal through laminectomy, while benefiting most patients, carries the risk of soft tissue injury of both joints and neural structures and can have a prolonged recovery period following surgery. Because of the potential problem associated with surgery the trend in the treatment of this problem is moving toward the use of more conservative treatment modalities prior to resorting toward surgery. Prior to chemonucleolysis those patients who had failed conservative means such as bed rest, traction, physical therapy, and epidural steroids had laminectomy as their only alternative. Chymopapain raised the hopes of both patients and physicians that a relatively noninvasive treatment could be used instead of surgery in this patient population. The fact that 70,000 chymopapain injections were made within 6 months of introduction of its use in the United States attests to this desire on the patient's part not to have to undergo back surgery. The use of chymopapain, however, has been greatly curtailed due to its association with major complications which include anaphylaxis, subarachnoid hemorrhage, disc space infection, and transverse myelitis with associated paraplegia.

Percutaneous discectomy by mechanically decompressing the center of the disc space has the advantages of chymopapain without the associated risks. In contrast to laminectomy there is no need for general anesthesia and since the procedure does not violate the spinal canal there is no risk for postoperative epidural fibrosis. A number of procedures have been developed in order to percutaneously remove lumbar disc material as described by Hijikata, Kambin, Suezawa, and Friedman [1–6]. What all of these techniques have in common is that they effect disc decompression percutaneously by removing the disc material with grasping forceps by hand. Consequently, large instrumentation is needed to gain access to the disc, which increases soft tissue damage as well as the potential for nerve injury. It is for this reason that acceptance of these procedures has been slow.

In 1985, Onik [7–9] described an automated percutaneous discectomy procedure in which he used a reciprocating suction cutter that cut separate pieces of disc material up to 180 times per minute. This enabled the procedure to be completed in a reasonable period of time while minimizing the size of the introduction cannula to 2.8 mm thereby reducing soft tissue damage and the possibility of nerve injury. In addition, the small flexible probe could be placed through a curved cannula which has allowed access to the L 5-S 1 disc space which had not been previously possible with the other techniques. The instrumentation with the procedure has now been released for sale and at the time of this writing over 200 surgeons in the United States are practicing the procedure and have performed the operation over 2,000 times. The purpose of this chapter is to describe this procedure, the selection of patients for the procedure, and the current results and complications associated with percutaneous automated discectomy.

Patient selection

The key to performing percutaneous automated discectomy and obtaining a high percentage of good results is correct patient selection. The most important factor that needs to be considered is that this is not a procedure for patients with acute or chronic low back pain but is designed to treat patients whose predominant complaint is leg pain (sciatica) secondary to a herniated lumbar disc who have failed other means of conservative therapy. The patient's leg pain should be greater than their back pain and confined to one leg, with the pain preferably radiating below the knee. A history of paresthetic discomfort in a specific dermatomal distribution that corresponds with the patient's radiographic pathology should also be sought. In general, the patient who is appropriate for the procedure will have physical findings that are consistent with a herniated disc, which include a positive straight leg raising and/or bow string sign, and the presence of neurologic findings to include wasting, weakness, sensory alteration, and reflex changes. The radiographic evaluation of patients for percutaneous lumbar discectomy is critical in choosing patients for this procedure. Every attempt must be made to exclude patients who have extruded or free fragments of disc that have migrated from the disc space or who have other associated lumbar pathology such as central spinal stenosis, lateral recess stenosis, or severe degenerative facet disease.

In general we do not obtain plain lumbar films in evaluating patients for percutaneous automated discectomy. In females over the age of 40 we will, however, obtain a flexion and extension views of the lumbar spine to exclude any evidence for segmental instability, since this problem is common in this patient population and it is a problem that could be made worse by removing disc material.

It should be stated at this time that we do not feel that a myelogram alone is adequate evaluation for these patients. We feel that either MRI or CT needs to be obtained to fully evaluate the patient for lateral recess stenosis and central canal stenosis as well as for visualizing free fragments of disc. The CT scan should be obtained so that all areas of the lumbar spine from the midbody L 3 to the sacrum are covered. A scan that has slices only at the disc spaces which leave gaps between the discs is inadequate since migrated free fragments or significant bony pathology can be missed.

At this point we have found MRI a useful adjunct to CT in a number of cases, however, its role in choosing patient for percutaneous automated discectomy is not yet clearly defined.

There are a number of cases in which CT and MRI still leaves doubts as to which level it is causing the patient's symptoms, i.e., multiple herniations are shown or whether the herniation is a free fragment. In these instances we have resorted to doing discography looking for the patient's pain response to the disc injection. If the patient has reproduction of his symptoms upon disc injection, that level is operated upon and if his pain is not reproduced the case is cancelled.

Technique

The procedure itself is conducted in the operating room or the radiology department under C-arm fluoroscopic control. Strict sterile technique is followed, and anesthesia staff is usually on standby. The procedure is carried

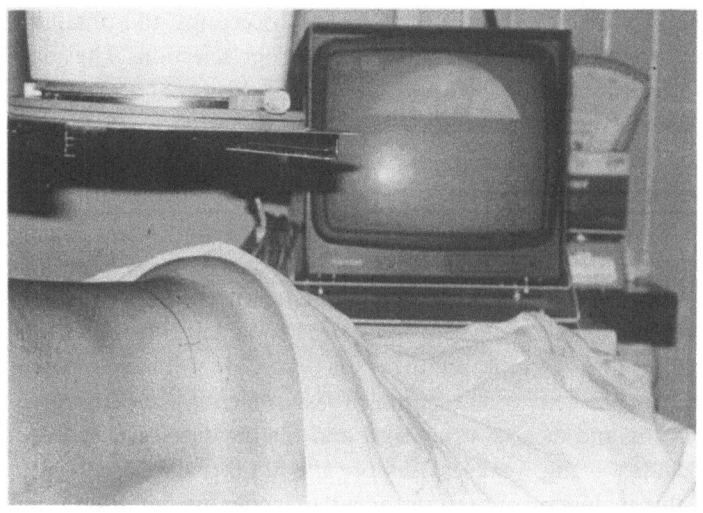

Fig. 1. Prior to draping, the patient is shown in the lateral decubitus position. The dotted line represents the iliac crest. The vertical line represents the plane of the L 4-5 disc. The horizontal line represents the measured distance from the midline as calculated from the CT scan. Where the two lines cross is the needle entry point

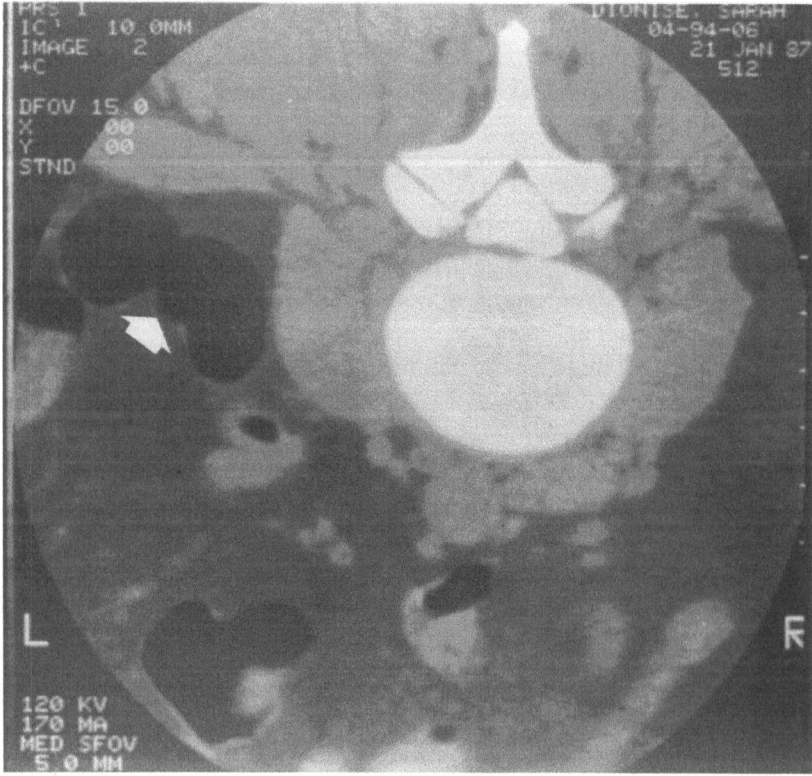

Fig. 2. CT scan of the lumbar spine showing posteriorly placed bowel. The white arrow indicates bowel just lateral to the psoas muscle

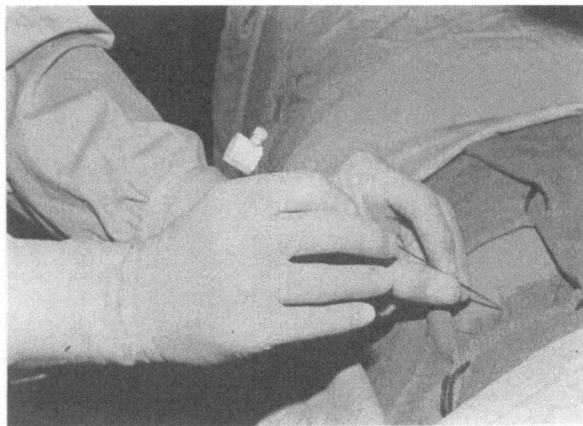

Fig. 3

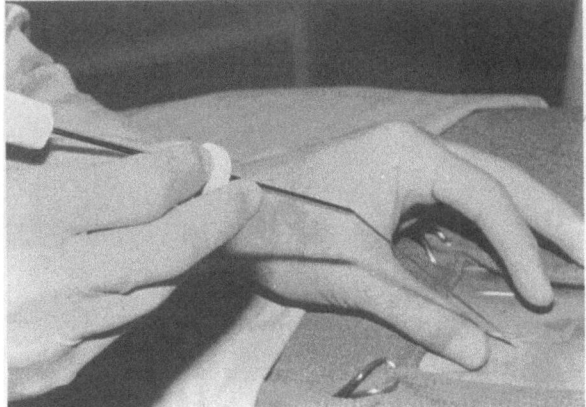

Fig. 4

Fig. 3. 18-gauge trocar being placed. The white hub is removable

Fig. 4. The trocar has been placed and confirmed in position. The hub has been removed and the cannula and dilator are placed over the trocar, down to the annulus

Fig. 5. The cannula is down to the annulus, the biopsy stop has been brought down to the skin, and the dilator has been removed. The trephine is now placed over the trocar and through the cannula to incise the disc

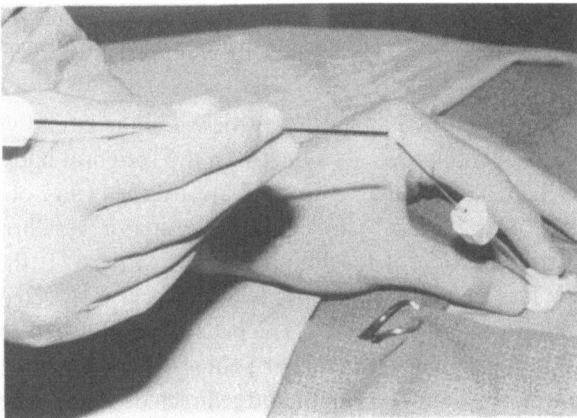

Fig. 5

out under local anesthesia; it is contraindicated to perform the procedure under general anesthesia due to the risk of neural injury. The procedure is performed in either the lateral decubitus position or in the prone position. When performed in the prone position, the patient is placed over a bolster on a Collis table in order to open up the disc spaces posteriorly. When performed in the lateral decubitus position, attention again must be paid to opening up the disc spaces posteriorly as well as making sure the patient does not rotate out of the straight lateral position (Fig. 1). The entry point for the procedure is chosen from a preoperative localization CT scan obtained through the disc slice of interest on the side of the patient's symptoms. This nontargeted full view of the abdomen CT scan, obtained in the prone position aside from choosing the exact entry point for the posterolateral approach to the disc space, rules out the presence of any retroperitoneal structures, such as bowel that may be in the path of the trocar (Fig. 2). Recent radiologic literature indicates that marked posterior displacement to the colon, which could interfere even with this posterolateral approach, can occur in a relatively high percentage of patients while in the prone position. By means of this posterolateral approach, first a 22 gauge Green biopsy needle (Cook, Inc.) is used to localize the exact projectory to the disc. An 18-gauge needle with a removable hub is then placed beside this needle in a tandem fashion down to the disc (Fig. 3). The trocar placement

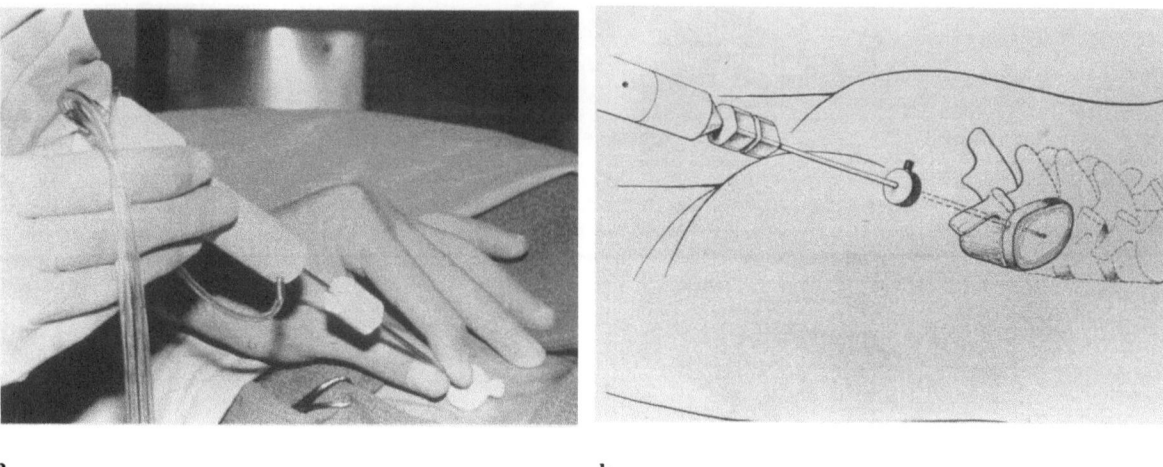

a   b

Fig. 6. a The Nucleotome is placed through the cannula after removing the trocar and the trephine. **b** Diagramatic representation of the nucleotome within the disc

is monitored in a lateral X-ray view. On the lateral view, the trocar should be parallel to and midway between the vertebral body endplates and the tip of the trocar should be directed toward the center of the disc. The point of the trocar should be touching the posterior vertebral body line when the gritty sensation of touching the annulus is felt. If the trocar is anterior to this line when the annulus is felt the trocar is directed too anteriorly and is withdrawn and redirected. The patient is monitored continually for any sign of radicular pain. If radicular pain is experienced by the patient, the trocar is withdrawn its full length and redirected. When the annulus is felt and the tip of the trocar (on the lateral view) is at the posterior vertebral body line,

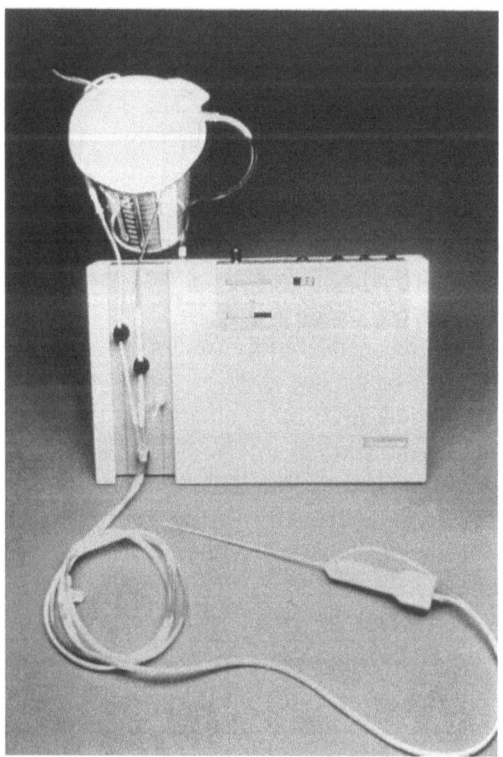

Fig. 7. The Nucleotome system as it stands today. The console controls the fluid, aspiration and compressed air that drives the probe

the AP view is then checked to confirm that the tip of the trocar is not medial to a line connecting the medial aspect of the pedicles. If the trocar tip is lateral to this line, the trocar is then advanced into the disc until the tip is in the center of the disc on the AP view. The lateral view is then checked to confirm the central placement of the trocar. The cannula, with a tapered dilator in place is inserted over the trocar (Fig. 4). The cannula and dilator are advanced down to the wall of the annulus, and their position is confirmed radiographically in 2 views. Once confirmed to be in place, the tapered dilator is removed from the cannula leaving the trocar and cannula in place. The circular trephine is placed over the trocar and through the cannula (Fig. 5). The fluoroscopic unit at this point is perpendicular to the cannula and the trephine, confirming that they are actually against the annulus before the trephine is used. After the disc is incised, the trephine and trocar are removed leaving the cannula in place and the Nucleotome (Surgical Dynamics Inc., San Leandro, California) aspiration probe is placed into the disc (Figs. 6 and 7). After the position of the Nucleotome is confirmed, the instrument is turned on and the disc material is aspirated. The port of the instrument is first placed in the direction of the herniation, and after no more disc material can be removed, the direction of the port is changed to remove disc material from other areas of the disc. After no more disc material can be removed from these areas, the rigid cannula can be angled to change the position of the Nucleotome within the disc. As much disc material as possible is removed through a single puncture site. The disc material, however, is aspirated for at least 40 minutes in each case. While the instrument is actively removing disc material, firm pressure, directed toward the annulus is applied to the cannula as a precaution to keep it against the annulus. When no further disc material can be removed, the Nucleotome is turned off and pulled back into the cannula and both are removed. The puncture site is then covered with a bandage. In most cases the patient is held for 3 hours and then discharged the same day of surgery. When the surgery is performed later in the afternoon or the patient does not have a relative or friend to accompany them home, the patient is held over night.

Results

A multi-institutional study is now currently being conducted to evaluate percutaneous automated discectomy. The following criteria were used in entering a patient into this prospective study: a) the patient's major complaint had to be sciatica (leg pain must be unilateral, leg pain greater than back pain); b) a history of paresthetic discomfort in a specific dermatomal distribution; c) positive findings on a straight leg raising test, cross over pain, or a positive bow string sign; and d) the presence of 2 of 4 positive neurological findings (wasting, weakness, sensory alteration, and reflex alteration). To be included in this prospective study, the patient had to satisfy at least 2 of the 4 above criteria. In addition, all patients underwent a computed tomography scan, or MRI scan which showed a herniated nucleus pulposus in an area consistent with the clinical findings. The patients must have had at least 6 weeks of conservative therapy without success and must otherwise have been a candidate for a laminectomy. Patients were excluded from this study if they had any of the following: a) history of previous lumbar surgery; b) previous chymopapain injection; c) Workmen's compensation claims; d) any other cause of back pain as revealed on the CT, such as severe degenerative facet disease, lateral recess stenosis, or evidence of a free fragment; and e) any contraindication to local anesthesia.

The success of the procedure was judged on criteria that considered the need for further intervention and the patient's pain and functional status. Therapy was judged successful if the radicular pain was moderately to totally improved; if the patient was no longer taking narcotic analgesics; if the patient's functional status was improved (when the patient's functional status was impaired before the procedure); and if the patient and physician were both satisfied with the results. If any of these criteria were not met, the treatment was considered unsuccessful.

At the time of this writing, 120 patients have met the prospective criteria for the study and have had the procedure and have at least 6 weeks follow-up. Ninety-three of these patients have a follow-up of 6 months or greater. Using the criteria for success as outlined for the study, 74% of patients with at least 6 weeks follow-up have had successful results. Those patients with at least 6 months follow-up, 71% have met all of the above criteria for success. Of the failures, in which surgical results are available, 8 had demonstrated free fragments at operation. Seven had herniations under the posterior longitudinal ligament or a bulging disc; one patient had lateral recess stenosis. Fifteen patients have had no surgical results at this time.

Complications

In 120 patients, no serious complications had been reported. There has been no nerve damage or great vessel injury associated with the procedure. There was one instance of vasovagal reaction associated with the procedure and one presumably psoas hematoma which resulted in 4 days of paresthesia in distribution of the lateral femoral cutaneous nerve on the side of the procedure with no subsequent sequelae. This hematoma was of no hemodynamic significance and the patient was still discharged the same day of the procedure. In November of 1986 the instrumentation for the procedure was released for sale and since that time over 2,000 procedures have been completed outside of this study protocol, again, with no serious complications reported [10].

This is a remarkable record of safety for a new operation in which all of the operators are still on the steep portion of the learning curve. I think the safety of the procedure, as it now stands, can be attributed to a number of factors. Firstly is that of the CT planning of the procedure. Without this safety factor there would undoubtedly eventually be bowel perforations as well as increased incidence of infection. The second and possibly the most important safety factor, which markedly decrease the chance of a nerve injury is the fact that the procedure is performed under local anesthesia with the patient able to respond to any manipulation of his nerve root. In addition careful radiologic checks have been incorporated into the procedure to insure that the Nucleotome is within the disc space and the cannula is up against the annulus before the Nucleotome is activated. Once within the disc space the Nucleotome which is blunt ended cannot be inadvertently pushed through the annulus and since the Nucleotome can cut only material that can be deformed into its side port, only nucleus and not annulus can be cut by its action. This assures that the Nucleotome cannot cut its way out of the inside of the disc. In our experience once the Nucleotome is inside the disc it is unusual for the Nucleotome to have to be removed therefore decreasing the chances for infection, as compared to a grasping forcep that must be continually removed and replaced within the disc.

Conclusion

In conclusion, automated percutaneous lumbar discectomy has now been performed in over 2,000 patients without serious complications. In a multi-institutional study the success rate is approximately 70%. Its lack of complications such as anaphylaxis and transverse myelitis make it a safer procedure than chymopapain. The fact that it can be performed on an outpatient basis with local anesthesia, without violating the spinal canal and producing

 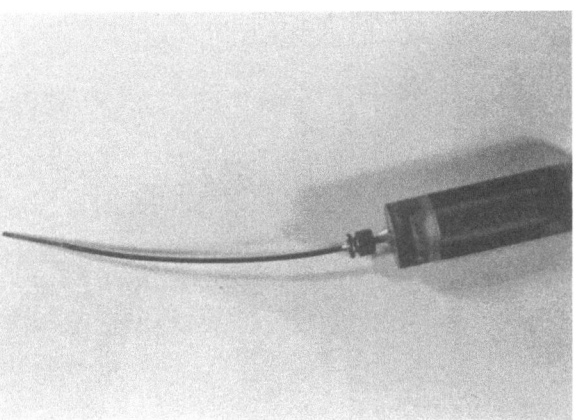

a   b

Fig. 8. a Nucleotome is shown with the curved cannula for the approach to the L 5-S 1 disc space. **b** The flexible probe takes the curve of the cannula so it can be redirected back into the plane of the disc

epidural fibrosis, make it an attractive alternative to a traditional laminectomy. The small size of the instrumentation needed to do the procedure, thereby decreasing the chance of soft tissue and nerve injury, as well as its ability to approach the L 5-S 1 disc space make it a more attractive alternative than the previous percutaneous discectomy techniques (Fig. 8).

Certainly more cases need to be performed and long-term follow-up is needed to fully evaluate the results of this procedure, however, the preliminary data indicates that percutaneous automated discectomy should replace chymopapain in the appropriate patient population and has the potential for replacing many surgical laminectomy procedures, thereby decreasing hospital cost, patient morbidity, and post surgical rehabilitation time.

References

1. Hijikata S, Yamagishi M, Nakayama T, Oomori K (1975) Percutaneous diskectomy: a new treatment method for lumbar disc herniation. J Toden Hosp 5: 5–13
2. Hijikata S, Nakayama T, Yamagishi M, Ichihara M (1978) Percutaneous nucleotomy for low back pain. Presented at the 14th World Congr Soc Int Chirurgie Orthopedique et de Traumatologie, Kyoto, Japan, October 15–20, 1978

3. Hijikata S (1981) Percutaneous nuclectomy for low back pain: the second report. Presented at the 15th World Congr Soc Int Chirurgie Orthopedique et de Traumatologie, Rio de Janeiro, August 30–September 4, 1981

4. Kambin P, Gellman H (1983) Percutaneous lateral diskectomy of the lumbar spine. Clin Orthop 174: 127–132

5. Schreiber A, Suezawa Y (1986) Trans discoscopic percutaneous nucleotomy in disc herniation. Orthoped rev 15: 75–78

6. Friedman WA (1983) Percutaneous diskectomy: an alternative to chemonucleolysis. Neurosurgery 13: 542–547

7. Onik G, Helms C, Ginsburg L et al (1985) Percutaneous lumbar diskectomy using a new aspiration probe: porcine and cadaver model. Radiology 155: 251–252

8. Onik G, Helms C, Ginsburg L, Hoaglund F, Morris J (1985) Percutaneous lumbar diskectomy using a new aspiration probe. AJNR 6: 290–293

9. Onik G, Maroon J, Helms C et al (1987) Automated percutaneous discectomy: Initial patient experience. Radiology 162: 129–132

10. Written Communication, Ron Allan Surgical Dynamics, May 1987

The lateral percutaneous approach to discectomy

W. A. Friedman[1] and S. L. Kanter[2]

[1] Department of Neurological Surgery, University of Florida, Gainesville, Florida
[2] Scott-White Clinic, Temple, Texas, U.S.A.

Percutaneous approaches to lumbar disc disease have evolved as a response to the morbidity generated by the variety of open procedures available. Two basic approaches have been pursued: a posterolateral approach similar to that used for vertebral biopsy or chymopapain injection [3]; and a straight lateral approach. The following text concerns the latter approach and the results of its utilization at the University of Florida [1, 2].

Technique

The lateral percutaneous approach to the lumbar disc was developed by Robert Jacobson, who has now performed over two hundred such procedures [personal communication, 1985]. It is easily performed under local, regional, or general anesthesia. The patient is placed in the lateral decubitus position, with the painful leg down. A roll is placed under the dependent flank to rotate the superior iliac crest out of the projected surgical path. A C-arm fluoroscope with image intensification is positioned for lateral lumbar spine radiography and the field is prepared and draped. A one-inch incision is made in the skin immediately above the iliac crest, and a specially designed (nasal-type) speculum is advanced, under X-ray control, through the retroperitoneal soft tissues and psoas muscle, to the midpoint of the lateral surface of the desired interspace (Fig. 1). The speculum is gently opened and a 40 French chest tube, with trocar in place, is inserted. The speculum and trocar are removed, leaving the chest tube in position at the lateral edge of the anulus fibrosis. An 18-gauge Kirschner wire is passed into the chest tube and popped through the anulus, where it remains for the duration of the procedure, to prevent migration of the chest tube. Utilizing specially lengthened instruments and intermittent fluoroscopic monitoring the anulus is incised with a #15 blade and the nucleus pulposus is removed with pituitary-type rongeurs in piecemeal fashion (Figs. 2 and 3). With the pathological side down, an angled rongeur easily reaches into the posterolateral aspect of the disc space, thus completing the discectomy (Fig. 4). Total operative time is between 15 and 30 minutes.

Results

Fourteen patients underwent this procedure at the University of Florida in 1982 and 1983. Twelve were male and two were female, with ages ranging from 19–63 years. No patient had a prior lumbar spine operation. All patients presented with sciatica (with or without back pain), positive mechanical findings, and appropriate radiographic findings on myelography and computed tomographic spine imaging. Thirteen patients had herniated discs at the L 4–5 level while one had an L 3–4 disc herniation. No patients with L 5-S 1 ruptures were included in this series because of the potential risks identified in the anatomical study described below. Follow-up evaluation of these patients was conducted by telephone interview in January, 1986. Response rate was 100%, with follow-up times ranging from 2.25–3.33 years. Patients were questioned regarding current activity level, pain, analgesic requirement, work status, and overall satisfaction with the procedure. At the time of follow-up, 12/14 (86%) patients were satisfied while 2 (14%) were not. Of the 12 satisfied patients, 11 had good results from percutaneous discectomy and one required a second, open procedure to remove a free fragment. Of the 2 dissatisfied patients, one was initially satisfied but developed recurrent left sciatica two years after the percutaneous procedure; the other presented with back pain, bilateral sciatica, and a large central disc herniation at L 3–4. The symptoms were not helped by the operation. All satisfied patients experienced minimal or no pain, returned to work, and reported few or no restrictions to activity. No patient required narcotic medication at follow-up.

Vogel, after learning the procedure from Jacobson, performed approximately 100 lateral percutaneous discectomies. He utilized the aforementioned technique to enter the L 4–5 interspace, but removed a dowel of bone from the iliac crest to effect a less oblique approach to the L 5-S 1 interspace [personal communication, 1986]. He noted six cases of temporary dysesthesiae in the femoral nerve distribution on the side of the incision. This complication has also been noted by Jacobson, who now recommends the use of a nerve stimulator prior to incising the anulus fibrosis [personal communication, 1987].

Discussion

Kanter and Friedman conducted an anatomical study using radiographic data to assess the risk of damage to bowel or vascular structures impeding the path of the surgical instruments employed in this technique [4]. Abdominal computed tomographic scans were analyzed at the L 4–5 and L 5-S 1

Fig. 1. Artist's rendition of the lateral percutaneous discectomy technique. The speculum is inserted to the lateral aspect of the selected disc space. A forty French chest tube is then positioned

Fig. 2. After the trocar and speculum are removed, the disc is incised. The previously inserted K-wire is not shown

Fig. 3. The nucleus pulposus is removed in piecemeal fashion with pituitary rongeurs

Fig. 4. A modified "up-biting" rongeur is used to reach into the opposite postero-lateral disc space

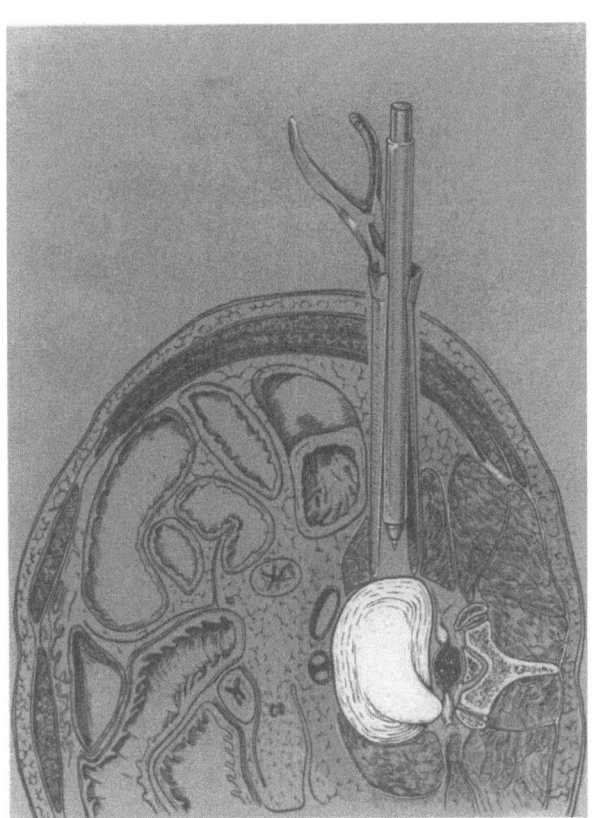

Fig. 1

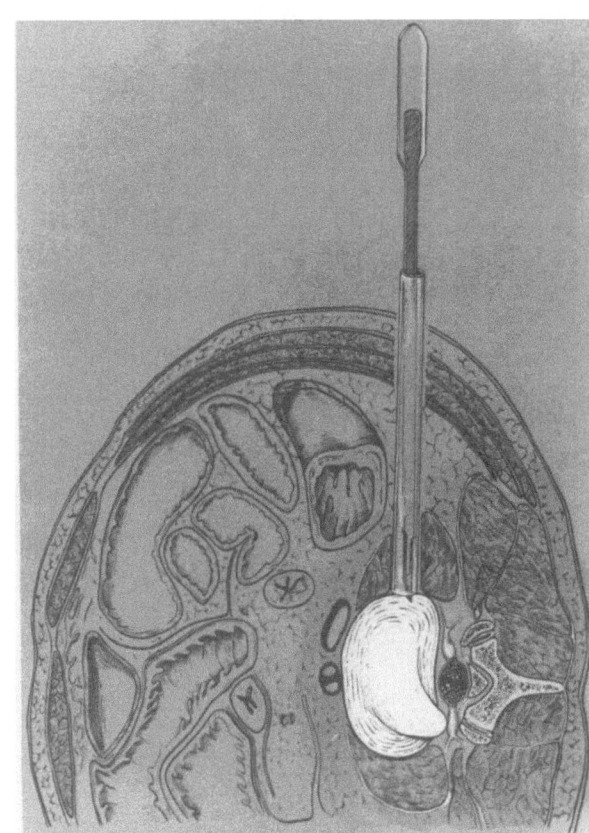

Fig. 2

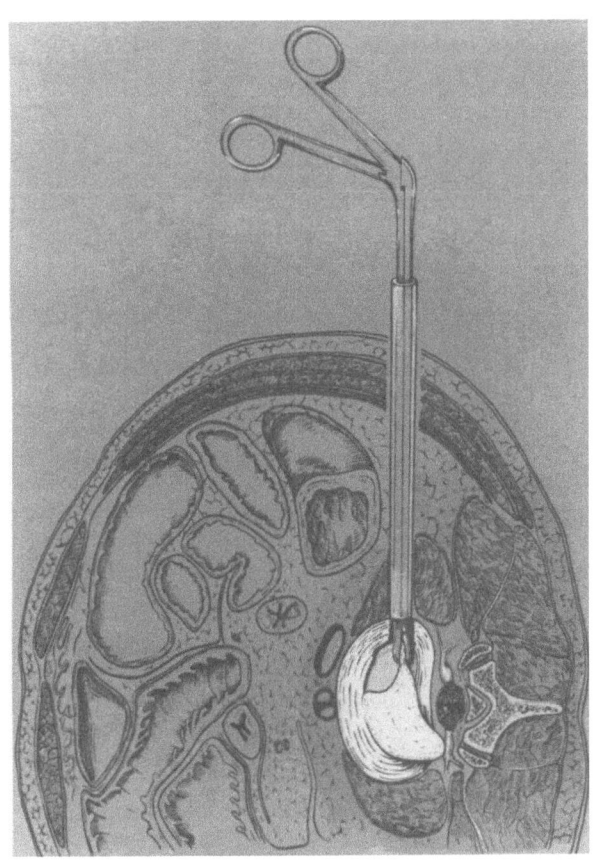

Fig. 3

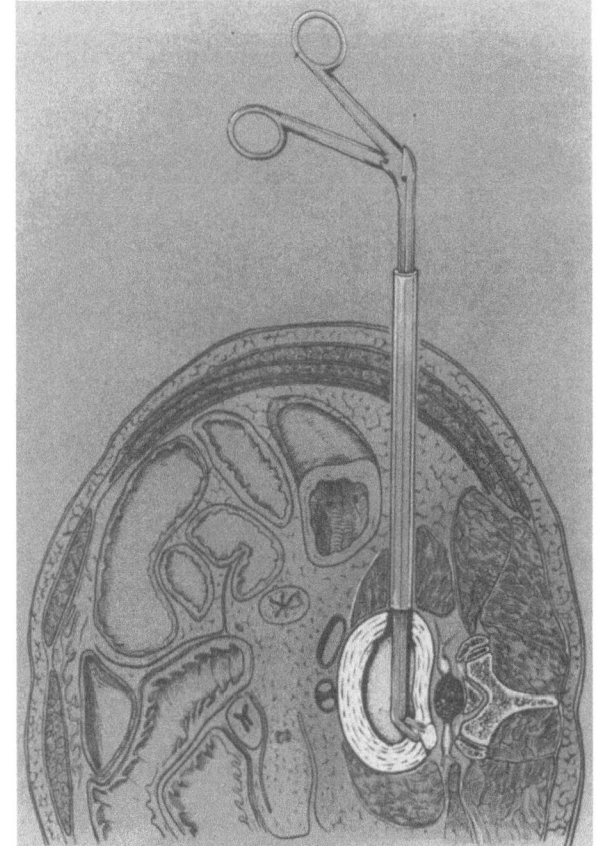

Fig. 4

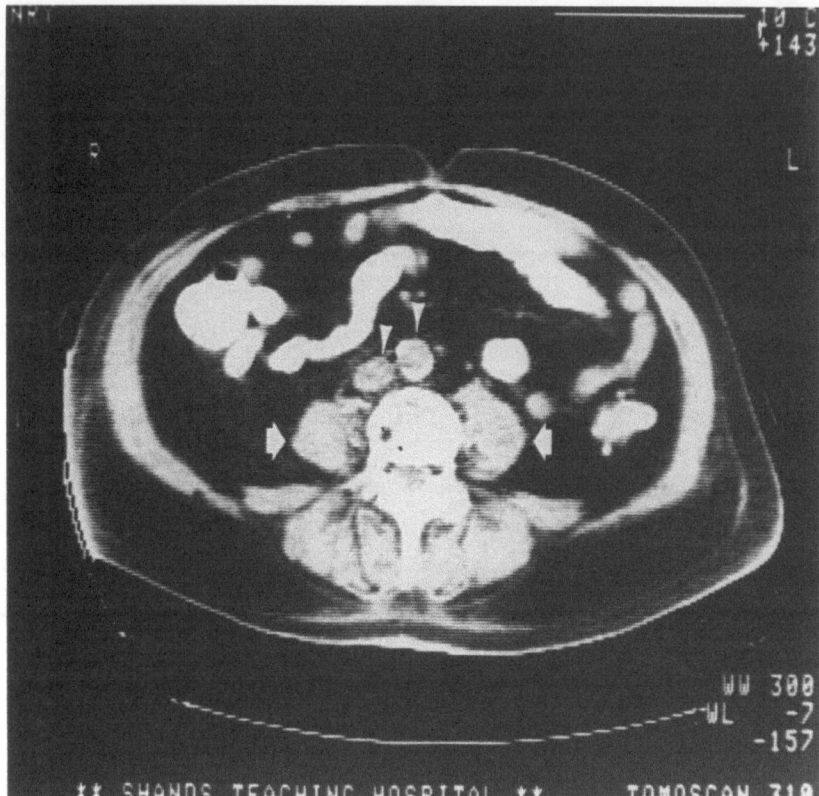

Fig. 5. Abdominal computed tomography, immediately above the level of the iliac crest passing near the L 4-5 interspace, obtained with the patient in the supine position. The aorta and vena cava (arrowheads) are ventral to the vertebral body. They present no obstruction to a lateral approach through the psoas muscle to the disc space

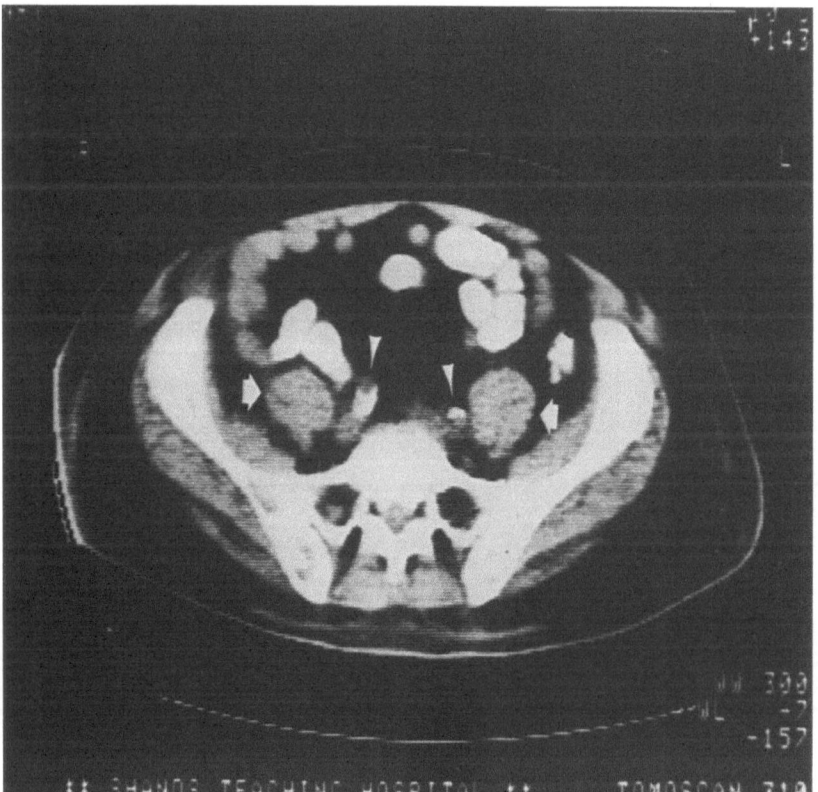

Fig. 6. Abdominal computed tomographic scan, approximately at the level of the L 5-S 1 interspace, obtained with the patient in the supine position. The iliac vessels (arrowheads) run along the medial aspect of the psoas muscles (wide arrows) and obstruct the lateral percutaneous approach to the disc space

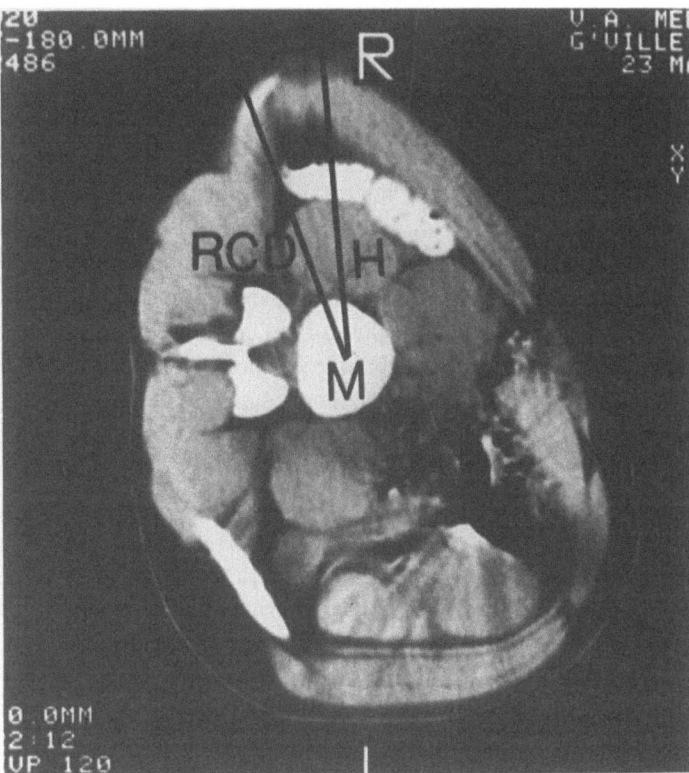

Fig. 7. Gastrograffin abdominal computed tomographic scan in the surgical position clearly demonstrates a loop of bowel in the surgical path

levels. Since the surgical instruments pass throught the psoas muscle, the overlap of vascular structures and the psoas muscle was of interest. Little or no overlap between these structures was noted at the L4–5 level, while significant overlap was appreciated at the L5-S1, as is clearly depicted in Figs. 5 and 6. Gastrograffin abdominal computed tomographic scans were obtained in the surgical position on several operative candidates. Approximately one-third of the patients who were otherwise suitable candidates for percutaneous discectomy, were found to have segments of bowel obstructing the path through which the chest tube would pass (Fig. 7).

The lateral percutaneous discectomy technique is severely restricted in terms of patient selection. Based on the anatomic findings of Kanter and Friedman, it is probably not safe to approach the L5-S1 interspace from the strictly lateral direction due to potential damage to vascular structures which lie immediately medial to the psoas muscle at that level. In addition, of patients who are otherwise suitable candidates for discectomy, approximately one-third were found to have bowel impeding the projected path of the surgical instruments. Although these authors did not experience a complication with major hemorrhage or bowel injury in their series, anecdotal reports from others who have performed the operation are alarming. Consequently, this operation may be safely recommended only for those patients that have herniated discs at the L4–5 or L3–4 levels, without suggestion of free fragment, and without bowel impeding the surgical path.

Conclusions

Percutaneous approaches to lumbar discectomy offer several distinct theoretical advantages over more standard techniques, including no lumbar incision, muscle stripping, or bone removal. The epidural space is never violated and the nerve root is never directly manipulated. Percutaneous discectomy can be performed under local anesthesia and may conceivably be accomplished as an outpatient procedure. Moreover, there is no risk of anaphylaxis or transverse myelitis as has been reported with chemonucleolysis.

Disadvantages of the lateral percutaneous approach include blind removal of relatively normal disc material hoping that the herniated fragment will thus be decompressed. In addition, other structures, including bowel, major blood vessels, or lumbar plexus branches may be at risk. These disadvantages have led these authors to currently recommend microdiscectomy as the lumbar disc technique of choice. The further development of "automated" posterolateral methods, however, may lead to a favorable reappraisal [5].

Acknowledgements

Figures 1–4 are reprinted from Friedman WA (1983) Percutaneous discectomy: an alternative to chemonucleolysis? Neurosurgery 13: 542–547. Figures 5–7 are reprinted from Kanter SL, Friedman WA (1985) Percutaneous discectomy: an anatomical study. Neurosurgery 16: 141–147.

References

1. Day AL, Friedman WA, Indelicato PA (1987) Observations on the treatment of lumbar disc disease in college football players. Am J Sports Med 15: 72–75
2. Friedman WA (1983) Percutaneous discectomy: an alternative to chemonucleolysis? Neurosurgery 13: 542–547
3. Kambin P, Gellman H (1983) Percutaneous lateral discectomy of the lumbar spine: a preliminary report. Clin Orthop 174: 127–132
4. Kanter SL, Friedman WA (1985) Percutaneous discectomy: an anatomical study. Neurosurgery 16: 141–147
5. Maroon JC, Onik GM (1987) Percutaneous automated discectomy: a new method for lumbar disc removal. J Neurosurg 66: 143–146

Facet joints percutaneous diagnostic and therapeutic procedures

Facet joint arthrography and steroid injection

M. Wybier[1] and J.-D. Laredo[2]

[1] Department of Bone and Joint Radiology, Hôpital Cochin, and
[2] Department of Bone and Joint Radiology, Hôpital Lariboisière, Paris, France

Technique, indications, and results

Degenerative changes in lumbar facet joints are a very frequent finding, especially with computed tomography (CT) [2]. These changes are suspected in many cases to be the cause of low back pain, especially when radiological procedures fail to demonstrate impingement of the nerve root by a herniated disc. However, it is always difficult to determine that pain actually originates in the facet joints. Pain may originate in facet joints with normal radiological appearance. Conversely, significant radiological degenerative changes in facet joints may remain asymptomatic. Facet arthrography with therapeutic injection is the only radiological procedure which can help to ascertain to what extent low back pain may originate in the lumbar facet joints.

Synovial cysts originating from the lumbar facet joints are another valuable indication of facet arthrography. These cysts may compress the nerve roots within the central canal or lateral recess. They are easily recognized on CT. Facet arthrography allows opacification and therapeutic injection of synovial cysts. The procedure also helps to determine which cysts should be excised.

Facet arthrography is used mostly at the lumbar level. However, it may also be performed at the thoracic or cervical levels.

Anatomy of lumbar facet joints

The facet joint space resembles a vertical gutter with a variable degree of concavity [18]. This concavity is oriented dorsally and medially. However, this orientation varies from 20° to 60° with respect to the sagittal plane [16], depending on patient build, lumbar level and side (asymmetry of the facets is frequent) [9, 13]. The ventro-medial and cephalic aspects of the facet joint are in close relation to the nerve roots within both the spinal canal and the foramen. The facet joint is a synovial joint with two major and minor capsular recesses. The superior recess is located in the intervertebral foramen. The inferior recess is at the caudal end of the joint. Superior and inferior recesses bulge during extension and flexion of the lumbar spine, respectively [18]. The two minor capsular recesses are the ventro-medial recess beneath the ligamentum flavum, and the dorso-lateral recess (Fig. 1). The capsule is lined with a thin synovium containing villi and fat pads, especially in the two

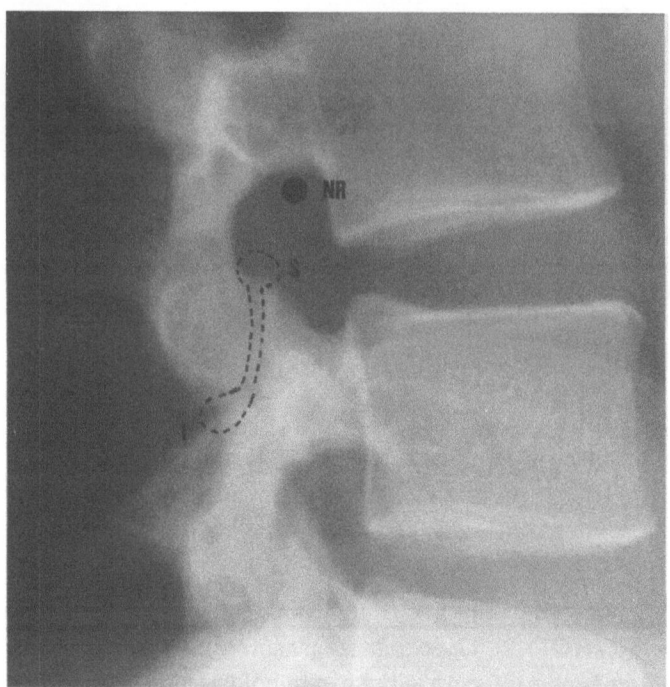
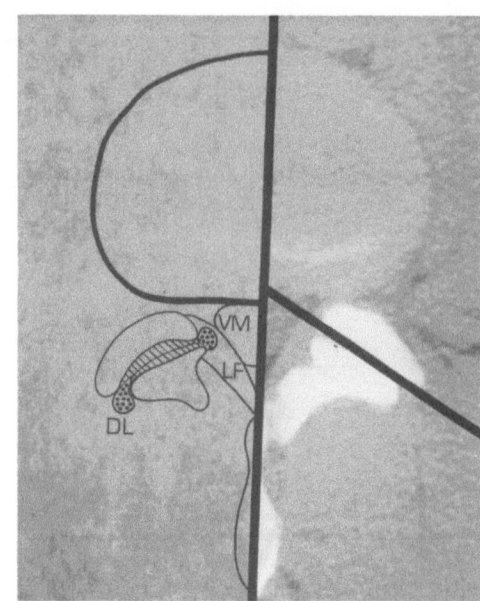

a   b

Fig. 1. Anatomy of the facet joint. **a** The superior recess (*S*) protrudes into the spinal canal and the foramen and is close to the nerve root (*NR*). The inferior recess (*I*) projects out of the neural space. **b** The ventromedial recess (*VM*) projects toward the dural sac and lies beneath the ligamentum flavum (*LF*). The dorsolateral recess (*DL*) projects out of the neural space. Facet orientation is about 45° with respect to the sagittal plane

major recesses. Meniscus-like villi project from the superior recess into the articular space (Fig. 2) [8]. Adipose tissue of the joint is continuous with that of the epidural space. The facet joint capsule is richly innervated by the dorsal ramus of the lumbar spinal nerve [8]. A single dorsal ramus supplies at least two facet joints, so that each facet joint has plurisegmental innervation [17].

Fig. 2. Meniscus-like synovial villus (★)

Facet arthropathy
Degenerative disease

Facet arthritis is infrequent before the age of 30 years [11] and almost constant in patients over 45 years [13]. It involves at least 3 contiguous lumbar levels in 50 percent of cases.

Two groups of predisposing abnormalities must be considered:

(1) Static disorders [12]: they include lumbar lordosis, which accounts for most facet degenerative disorders [16], and lumbar scoliosis which jeopardizes the facet joints located in its concavity.

(2) Disc diseases: in patients over 45 years, facet arthritis is almost as frequent as degenerative disc disease. According to the 3-column concept of the spine [12], a mechanical disorder in one spinal column may lead to mechanical imbalance in the two other joints. Thus, narrowing of the disc space – whatever its origin (spinal osteochondrosis, chemonucleolysis, surgical excision or spinal infection) – increases mechanical stress on neighboring facets, which may result in capsuloligamentous laxity and facet joint instability [3].

Chronic mechanical stress on facet joints may cause synovial effusion and expansion, sometimes resulting in synovial diverticuli (Fig. 3) or cysts, which may threaten the nerve roots either in the lumbar canal or in the foramen. Further degenerative changes are joint-space narrowing, subchondral bone sclerosis and cysts, and osteophytes. These osseous changes are easily diagnosed on plain radiographs. At that final stage, active synovitis is usually greatly reduced.

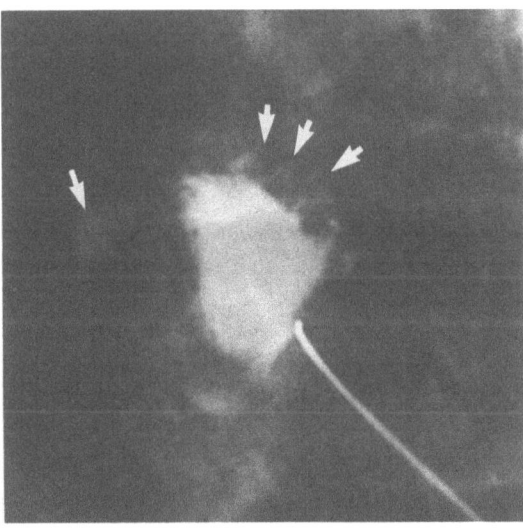

Fig. 3. Degenerative synovial outline with several diverticuli (arrows)

It has also been postulated that the synovial villi may become inflamed or trapped between the articular processes, especially in the case of facet joint instability, resulting in acute lumbar pain and blockage [5, 6].

Other diseases

A wide range of diseases may involve the facet joints. Trauma with rotation of the spine may cause facet joint strain [16, 18]. Facet joints are commonly involved in seronegative spondylarthropathy. Septic facet arthritis is exceptional. These disorders can produce acute symptoms or lead to facet joint degeneration.

Technique of facet joint arthrography

Facet arthrography is a relatively non-invasive procedure which can be performed on an outpatient basis.

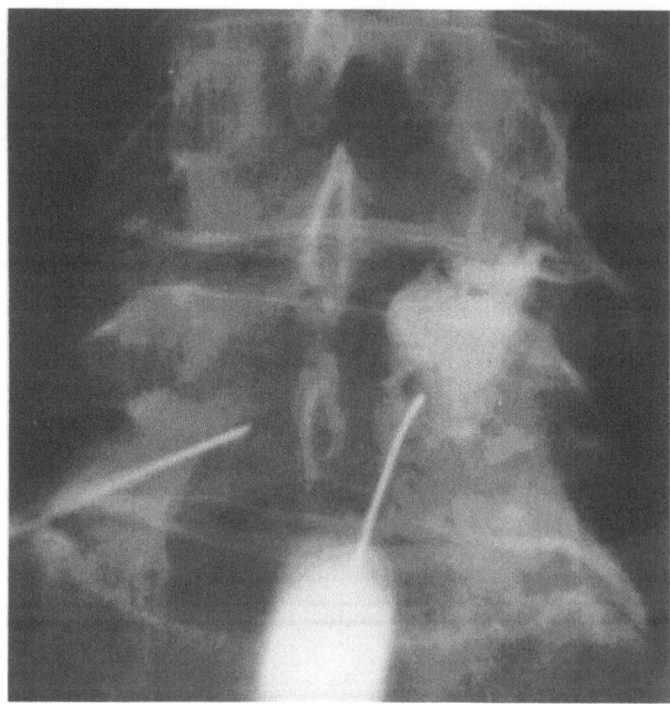

a

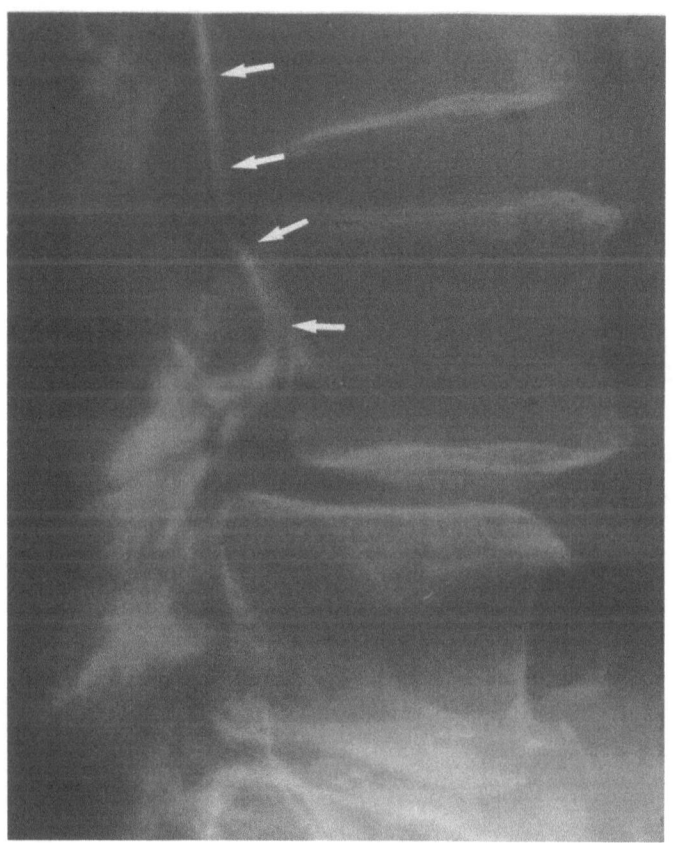

b

Fig. 4. a The joint capsule may rupture during filling. This leads to opacification of the epidural space (arrows) (**b**)

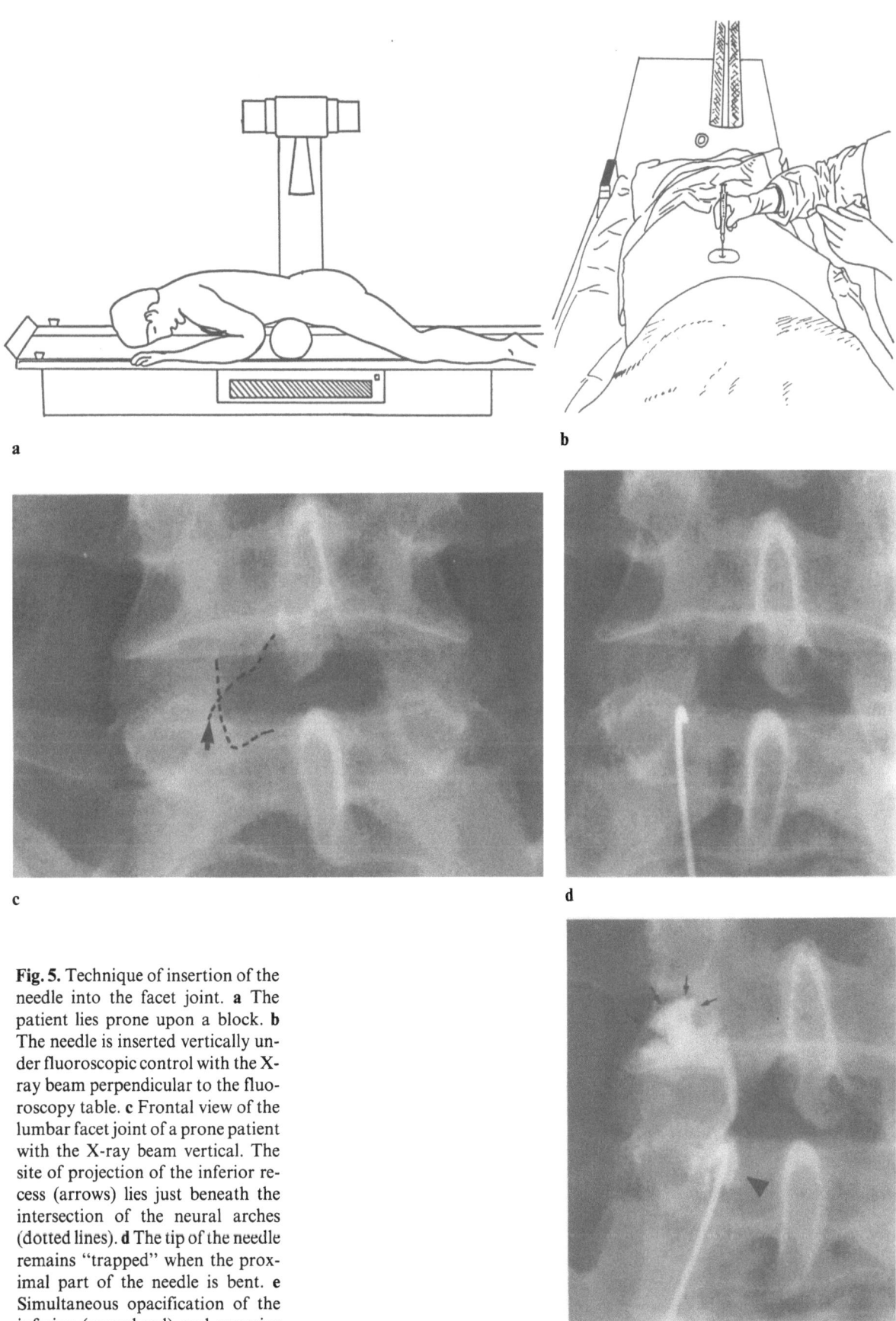

Fig. 5. Technique of insertion of the needle into the facet joint. **a** The patient lies prone upon a block. **b** The needle is inserted vertically under fluoroscopic control with the X-ray beam perpendicular to the fluoroscopy table. **c** Frontal view of the lumbar facet joint of a prone patient with the X-ray beam vertical. The site of projection of the inferior recess (arrows) lies just beneath the intersection of the neural arches (dotted lines). **d** The tip of the needle remains "trapped" when the proximal part of the needle is bent. **e** Simultaneous opacification of the inferior (arrowhead) and superior (arrows) recesses

Material

— A standard X-ray table with single plane fluoroscopy;
— 20-gauge, 8 cm-long disposable spinal needles;
— 5 ml of water-soluble contrast material. Non-neurotoxic material (Io-pamidol, Iohexol) is more suitable because leakage of contrast material into the epidural space is not infrequent (Fig. 4);
— sterile gloves;
— sterile compresses and skin disinfectants;
— in case of therapeutic injection, 5 ml of 0.5% Xylocaine and 1 vial of injectable long-acting corticosteroid.

Injection technique at the lumbar level

The needle is inserted into the inferior recess of the facet joint, which is large, superficial and far from the nerve roots. This recess is approached through a direct posterior route with the patient prone, according to the technique described by Chevrot [18] (Fig. 5). A plain postero-anterior view of the facet joint is performed with the patient placed prone with a block under the abdomen so that the lumbar spine is flexed and the inferior recess of the facet joint bulges. The central ray is vertically oriented over the lumbar level to be injected. The disc space is usually not seen in profile on this vertical projection. On this view, the inferior recess is located beneath the lower margin of the inferior articular process. The needle tip must reach the recess immediately beneath the intersection of the neural arches as shown in Fig. 5. Performing this technique is still possible in patients who have undergone laminectomy (Fig. 6). Lateral and oblique plain views of the facet joint are also obtained to facilitate further interpretation of the arthrographic views.

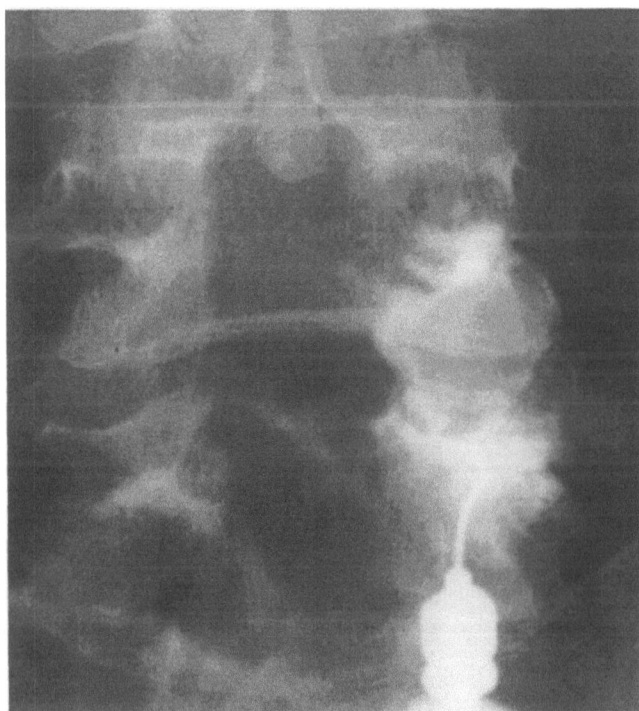

Fig. 6. Facet arthrography in a patient previously operated on for L 4-5 disc herniation. Facet mechanical disorders are frequent after disc curettage and may induce facet syndrome

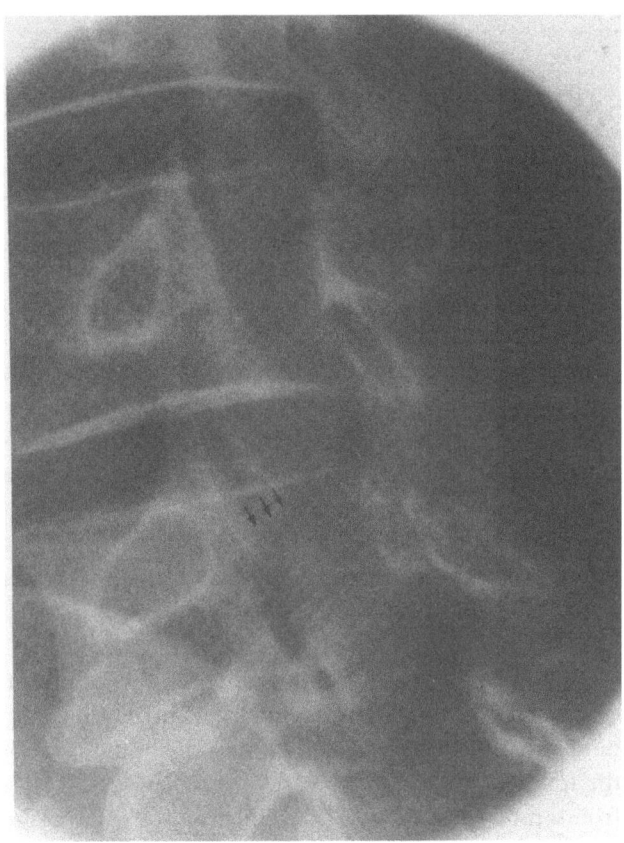

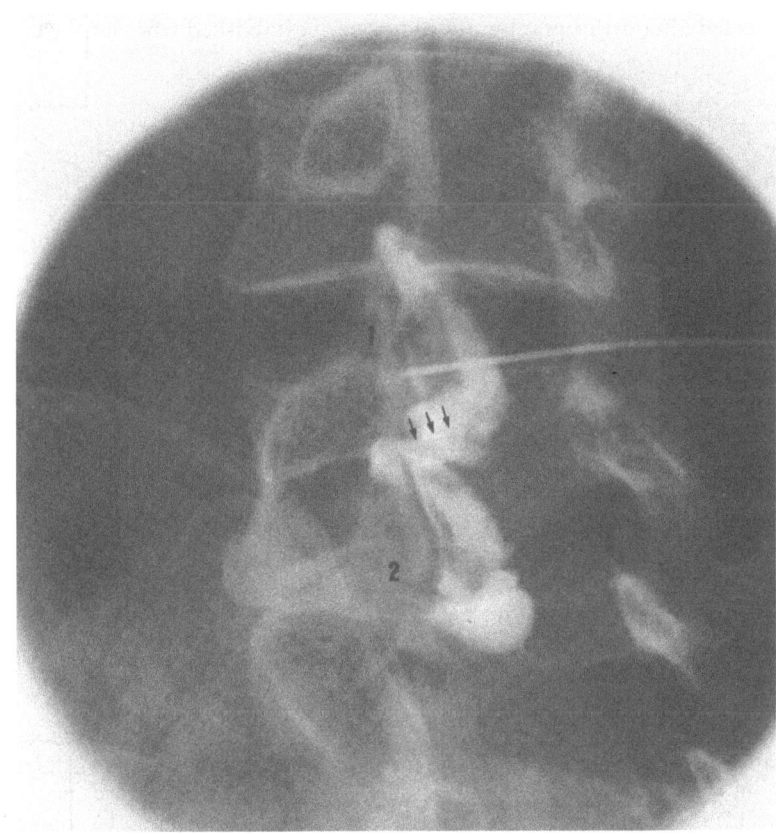

Fig. 7. In the case of spondylosis (arrows) (**a**), injection of a facet joint (1) may lead to injection of the underlying joint (2) through the bone defect (arrows) (**b**)

The procedure is gentle enough to make prior local anesthesia unnecessary, unless major osteoarthritis prevents easy fluoroscopic guidance. The overlying skin is prepped. The needle is inserted vertically under fluoroscopic control parallel to the X-ray beam toward the inferior recess until the bone is reached. A correctly oriented needle will appear as a dot superimposed on the point described above. If placed correctly, the needle tip usually seems to be "trapped" in the joint. The injection of a very small amount of contrast material immediately induces concurrent opacification of the superior recess, confirming correct intra-articular needle position (Fig. 5 e). Depending on the size of the capsule, 1 to 3 ml of contrast material can be slowly injected. Excessive pressure on injection must be avoided. In case of spondylolysis located above or below the injected facet joint, concurrent opacification of the over- or underlying facet joint is usual (Fig. 7). In case of Baastrup's disease the facet joint may communicate with an interspinous neocavity and through it with the contralateral facet joint. In such cases 5 to 10 ml of contrast media may be necessary to obtain full opacification.

Infrequently, the site of the inferior recess is difficult to identify or reach, especially when facets are sagittally oriented, which is usual at the upper lumbar levels. In such cases, one may attempt a direct oblique approach to the joint space. The joint space is profiled by slowly rotating the patient from a prone position into a slightly prone oblique position with the relevant side up. However, as the joint space is curved, it is sometimes impossible to penetrate the joint even when adequately profiled (Fig. 8).

When a therapeutic injection is indicated, a vial of injectable long-acting corticosteroid, with or without additional instillation of 0.5% Xylocaine, is injected after arthrography. It has not been established whether the capsule

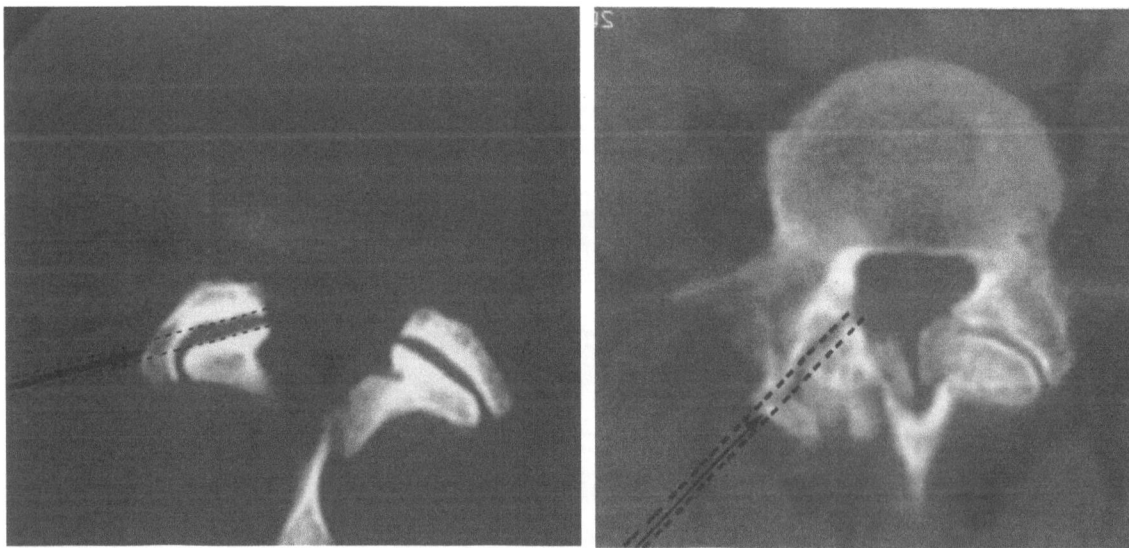

a   b

Fig. 8. Pitfalls to the oblique approach of the facet joint space. **a** Because the joint space is curved, the needle (arrow) may not penetrate the joint, even though the joint space is seen in profile (dotted lines). **b** Apophyseal osteophytes may not be visible on an oblique view profiling the joint space (dotted lines) but prevent the needle (arrow) from being inserted

should be ruptured during injection so that the drugs act both in the joint and in the periarticular tissues. During injection of contrast material and drugs, one must note whether the injection reproduces the patient's pain.

Injection technique at the cervical and thoracic levels

At these levels, the puncture is easier with the spine in lateral projection in which the joint space of the cervical facet joints is seen in profile. The point of skin puncture is determined under fluoroscopic control. The patient remains seated. Optimal distance from the midline of the spinous processes is determined on the AP view by projecting a lead marker placed on the skin on the articular processes. Then the patient is turned to a lateral position to determine the level of the puncture and the optimal cephalad tilt (Fig. 9).

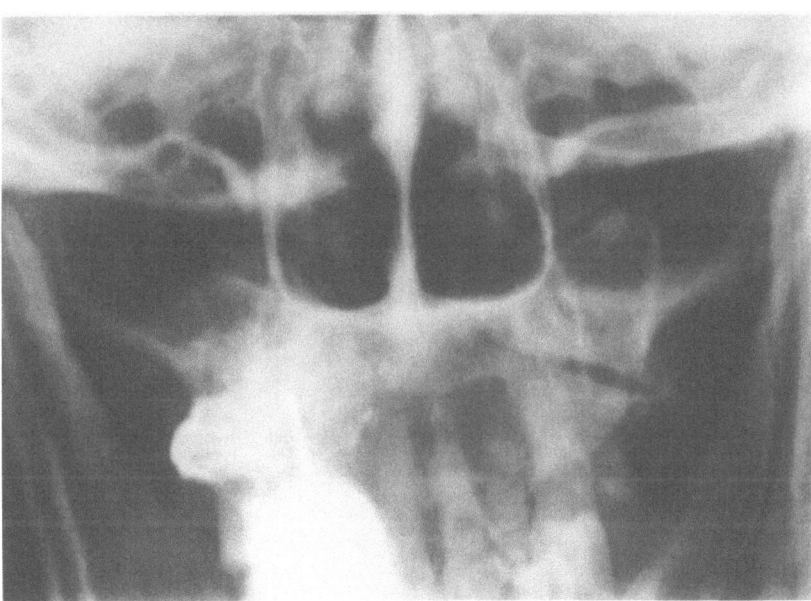

Fig. 9. Arthrography of C 1-C 2 right facet joint

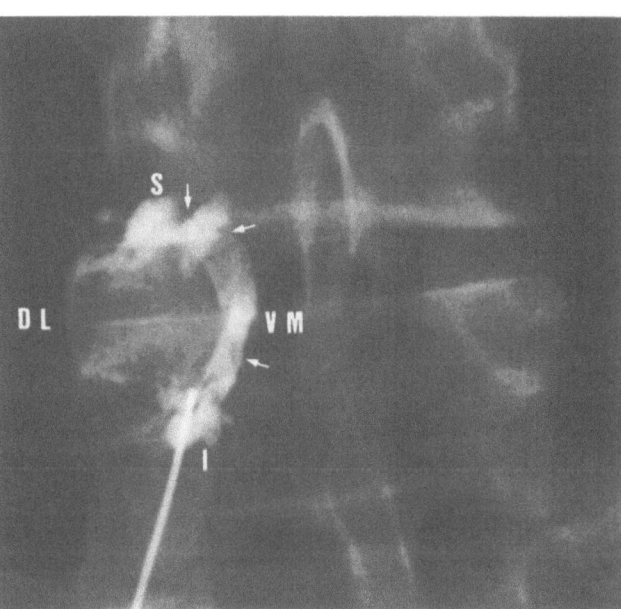

Fig. 10. Normal frontal arthrogram with filling of the superior (S), inferior (I), ventromedial (VM), and dorsolateral (DL) recesses. Slightly lobulated outline and small filling defects (arrows) are normal features

Arthrographic views

Four projections are necessary. The frontal and lateral views demonstrate the relationship between the synovial recesses and the lumbar canal and foramen (Figs. 10 and 11). The anterior oblique position with the affected side up allows good visualization of joint space and cartilage (Fig. 12 a).

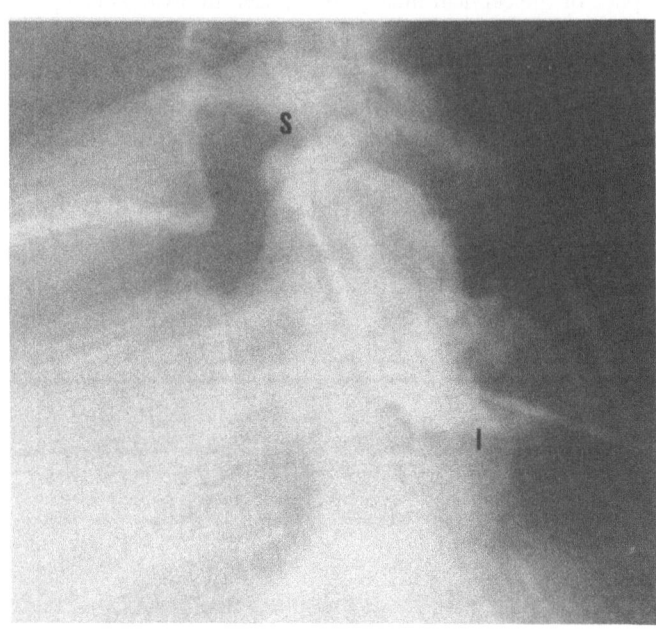

Fig. 11. Normal lateral arthrogram with the superior (*S*) and inferior (*I*) recesses

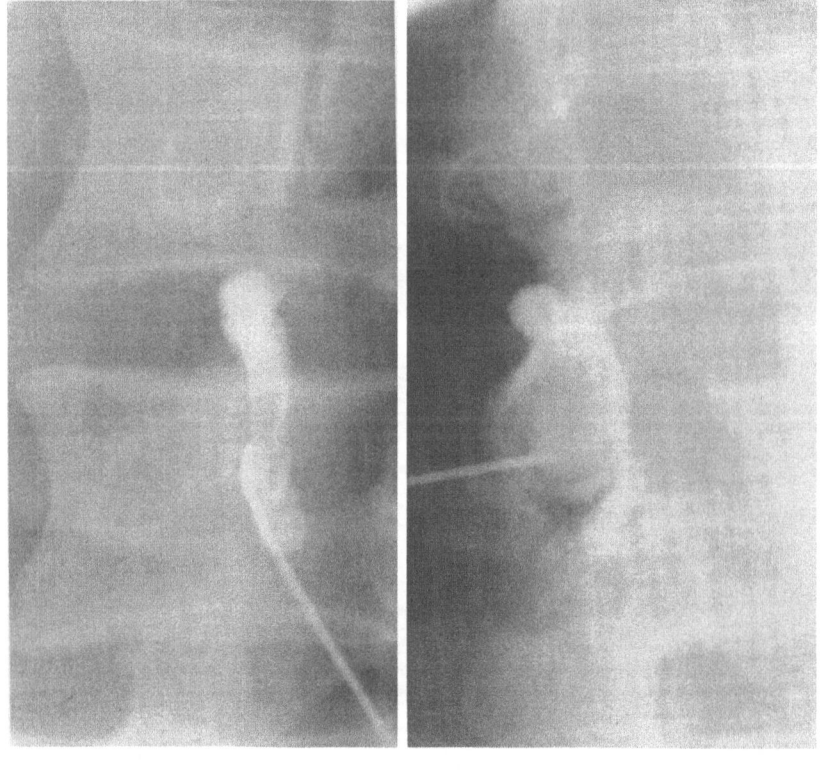

a   b

Fig. 12. Normal anterior oblique arthrogram. (**a**) With the affected side up the joint capsule appears S-shaped. (**b**) With the contralateral side up

The contralateral oblique projection provides a true frontal view of the facet joint. The size and contours of the joint are well analyzed on this projection (Fig. 12 b).

Normal arthrographic appearance [4, 5, 18]

The contours of the facet joint capsule are smooth, or slightly lobulated when large synovial villi or fat pads are present. Villi and fat pads also induce normal filling defects. The capsule varies in shape, depending on facet orientation and radiologic projection. It is usually oval on AP views (Fig. 10) and S-shaped on oblique views (Fig. 12). The size of the capsule must be evaluated with respect to that of the facets. Normally, the synovial recess does not project into the site of the nerve roots. Both the superior and the inferior recesses can easily rupture during the injection of contrast material, unless the joint is filled very gently. Rupture of the superior recess induces opacification of the epidural space. Opacification of either lymphatic or venous drainage is unusual, occurring as a result of inflammation or rupture of the joint capsule.

Indications

Facet arthrography can be performed for diagnostic and therapeutic purposes.

Degenerative diseases

The first step is to determine those patients whose symptoms actually originate from facet degeneration. There is no clinical or radiologic gold standard to implicate the facet joints as a cause of low back pain, but rather, a set of clinical and radiologic criteria. Table 1 summarizes the signs and symptoms of the two clinical syndromes related to facet degeneration [1, 7, 15, 16, 21]. Tables 2 and 3 indicate the range of investigations that help to determine which facet joints may cause pain and should be injected.

Table 1. Signs and symptoms of the two principal syndromes due to facet degeneration at the low lumbar level

---

Facet syndrome

Frequent low back blockage of short duration, generally provoked by anteflexion of the trunk.

Chronic lumbosacral pain with either uni- or bilateral pain referred to the buttock or the dorso-lateral aspect of the thigh. Generally no radiation beneath the knee.

Pain provoked by hyperlordosis and changes in position, especially by rising from a low sitting position.

Pain reproduced by pressure over the facet joint.

Absence of signs and symptoms of disc pressure on nerve roots such as severely diminished straight leg raising test, severe reflex impairment, sensory loss or motor weakness in a specific dermatome.

Nerve root pressure syndrome

Intermittent painful claudication suggesting degenerative lumbar stenosis.

True sciatica, with signs and symptoms of nerve root involvement.

Table 2. Facet joint arthrography in facet syndrome determination of the level of injection

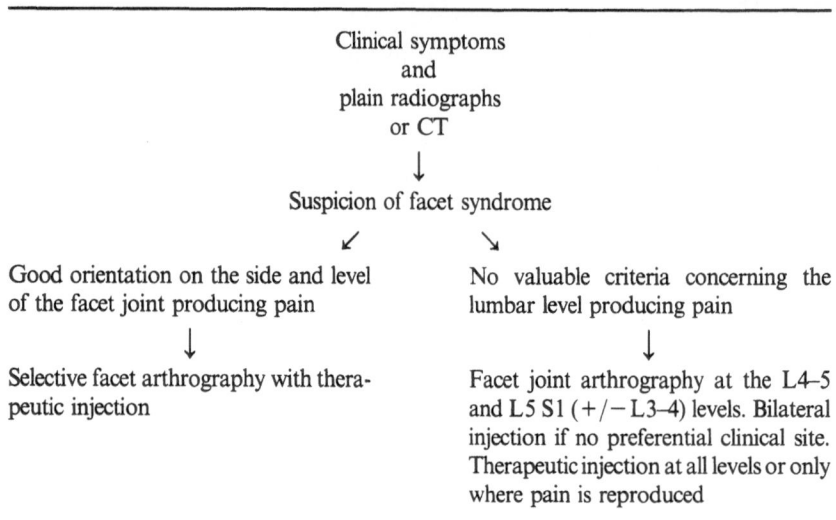

Table 3. Radiological work-up in case of nerve root pressure syndrome at the low lumbar level

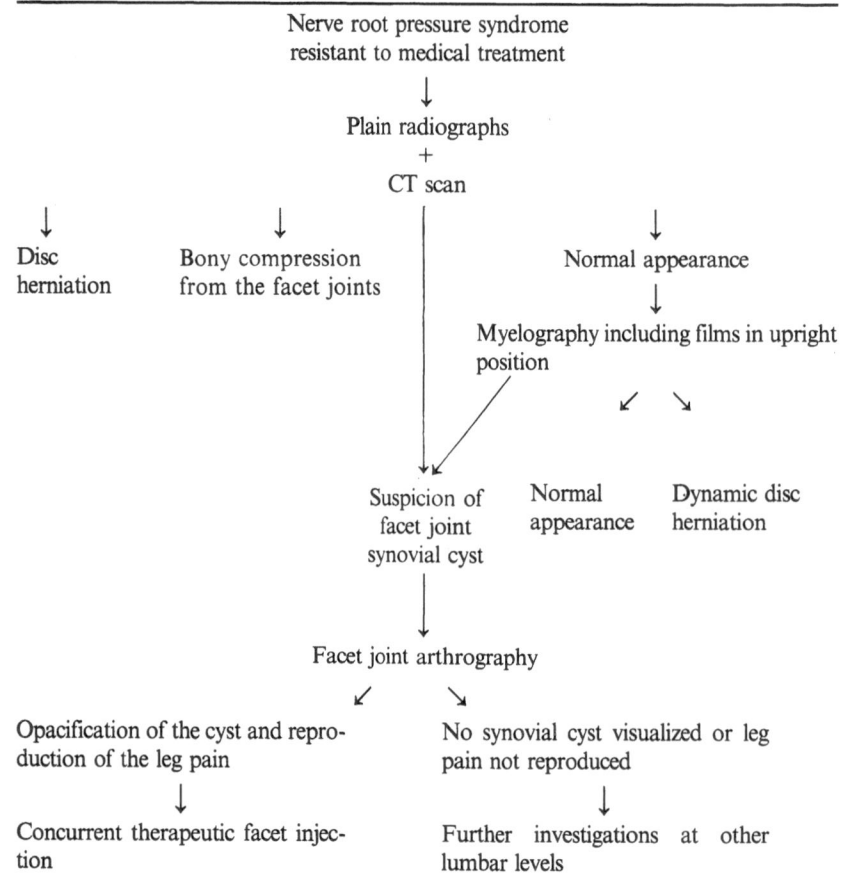

In the case of facet syndrome, the level of facet joint arthrography is determined on clinical and radiological grounds. However, it is always difficult to establish a relationship between symptoms and a specific apophyseal joint for several reasons. As mentioned above, pain originating in facet joints may be associated with a normal roentgenographic appearance while, conversely, significant radiological degenerative changes in facet joints may remain asymptomatic. Moreover, the arthrographic appearance of facet joints has relatively little value in diagnosing facet syndromes. Even the exacerbation of pain during capsular filling is not completely reliable. Reproduction of the patient's pain by facet joint injections at several levels performed during the same procedure is not infrequent. Experimental studies also demonstrated that intra- or even extra-articular injections at several

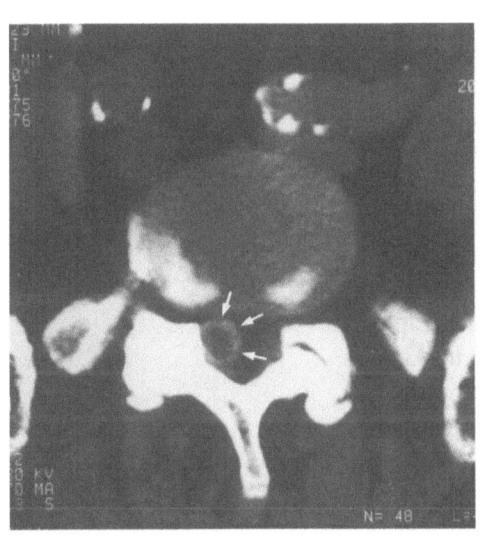

a

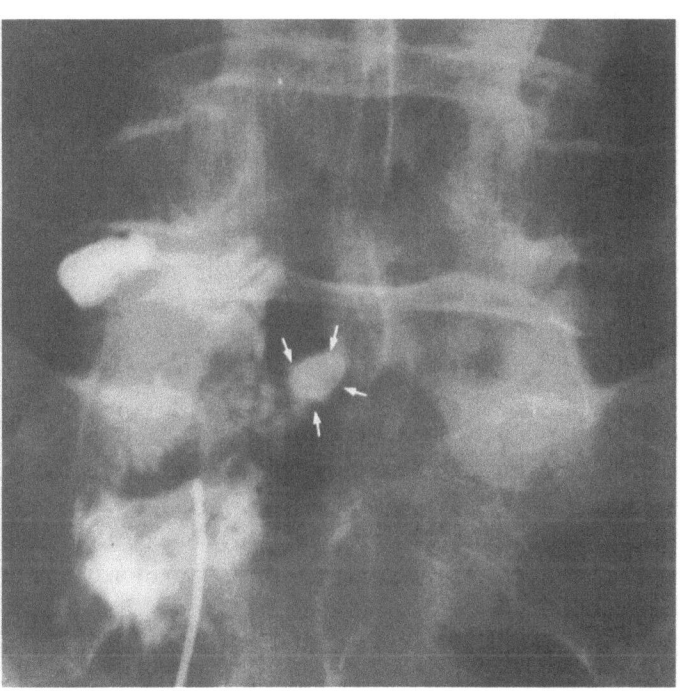

b

Fig. 13. Synovial cysts: CT and arthrography. **a** They usually appear on CT as focal nodules within or through the ligamentum flavum, with a thick, hyperdense wall (arrows), which may enhance after intravenous injection of contrast media. **b** Facet arthrography confirms that the cyst (arrows) projects from the ventromedial recess. Filling the cyst reproduced the patient's leg pain. **c** Cysts may sometimes expand from the dorsolateral recess into the dorsal soft tissues (arrows). Such cysts might induce referred pain

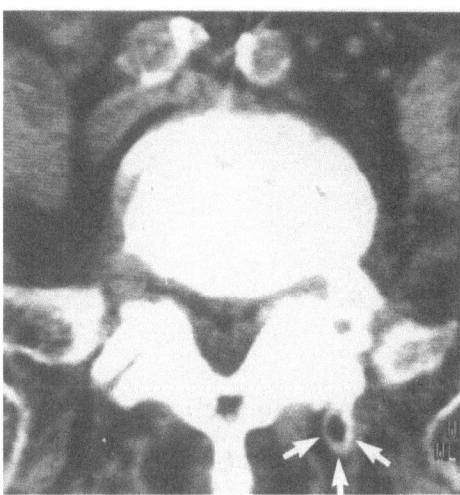

c

different lumbar levels may result in the same painful response [14]. The fact that a single dorsal ramus of the lumbar spinal nerve supplies two or three different facet joints may explain these findings. Finally, relief of pain obtained by steroid injection into the facet joint appears to be the best diagnostic test available in facet syndrome. For these reasons, facet joint arthrography with therapeutic injection is widely indicated in suspected facet syndrome even when facet joint appearance is normal. In this case, it is wise to perform injections at two or three levels during the same procedure. In the absence of referred pain radiating to one side, the injections should be bilateral. Following the injection of corticosteroids within the facet joints, pain relief may be complete. However, complete pain relief is usually of short duration and after two to four weeks, pain usually recurs, but residual pain is usually milder than before the intraarticular injection. Furthermore, this relatively innocuous procedure may be repeated. A positive response is a prerequisite for those patients who are to undergo percutaneous rhizolysis of the dorsal ramus or surgical neurotomy [11, 19, 20, 22].

In the case of nerve root pressure syndromes, the clinical and radiologic findings are usually unequivocal. Synovial cysts or diverticuli are not infrequent. Those expanding from either the superior or the antero-medial recesses can project into the neural canal and compress the nerve roots (Figs. 13 and 14). Compression of a sacral root in the sacral foramen by a synovial diverticulum from the inferior recess of an L 5-S 1 facet joint is infrequent (Fig. 15). The purpose of therapeutic injection is to relieve pain. Such pain relief is predictive of a good response to cyst excision. Attempts to burst the cyst during facet injection are usually unsuccessful because the cyst has a thick, fibrous, sometimes calcified wall. The contents of the cyst may be gelatinous and therefore impervious to contrast material. Facet arthrography is sometimes the only way to detect a synovial expansion projecting into the foramen (undetected on myelography) or which appears only in the upright position (undetected on CT). Surgical excision of the cyst is generally successful.

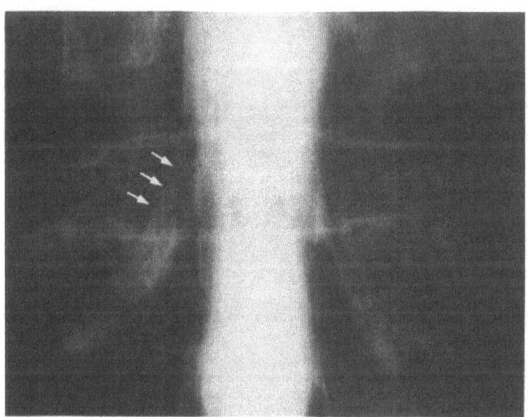 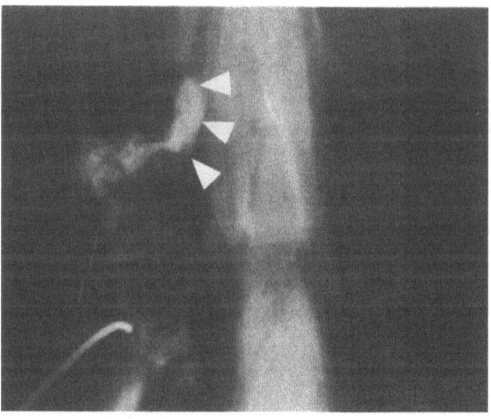

a   b

Fig. 14. Synovial cysts: myelography combined with facet joint arthrography. **a** The right L 5 nerve root is widened on myelography (arrows). **b** The concurrent arthrogram of the right L 4-5 facet joint reveals a synovial cyst (arrowheads) originating from the superior recess and compressing the nerve root. Filling the cyst reproduced the sciatica

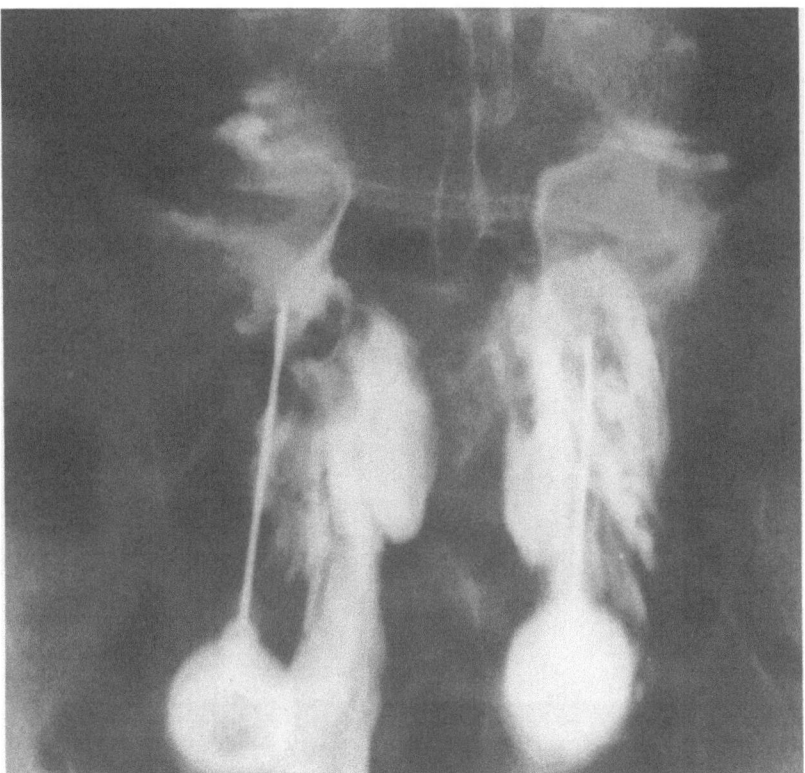

a

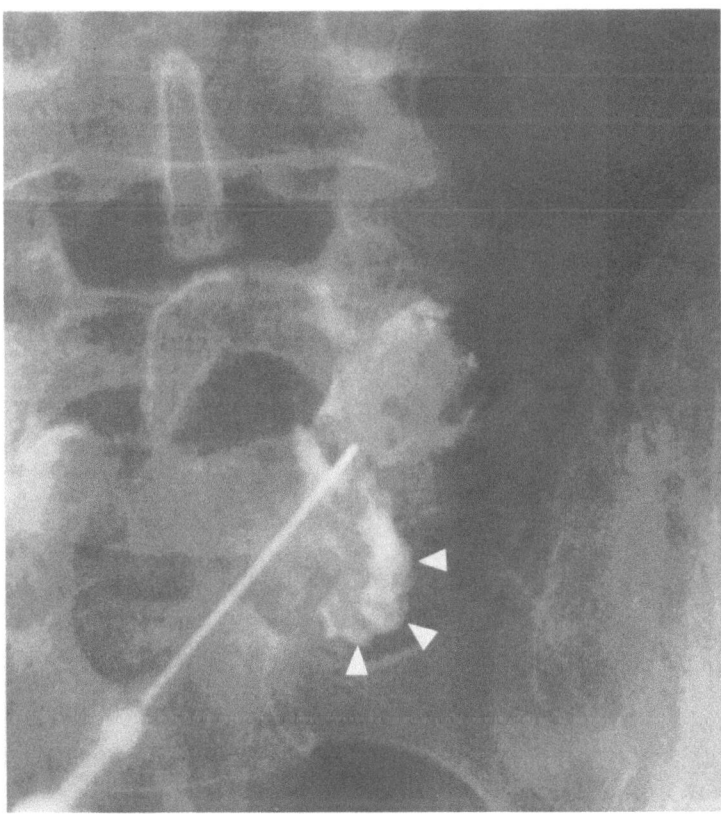

Fig. 15. Dorsal extracranular L 5-S 1 synovial cysts. Large synovial expansions from the inferior recess are not infrequent at the L 5-S 1 level (**a**). They may induce local pain or facet syndrome. Infrequently, such expansions (arrowheads) can project into a sacral foramen (**b** and **c**) and compress the sacral root (arrow) as demonstrated by CT (**d**)

b

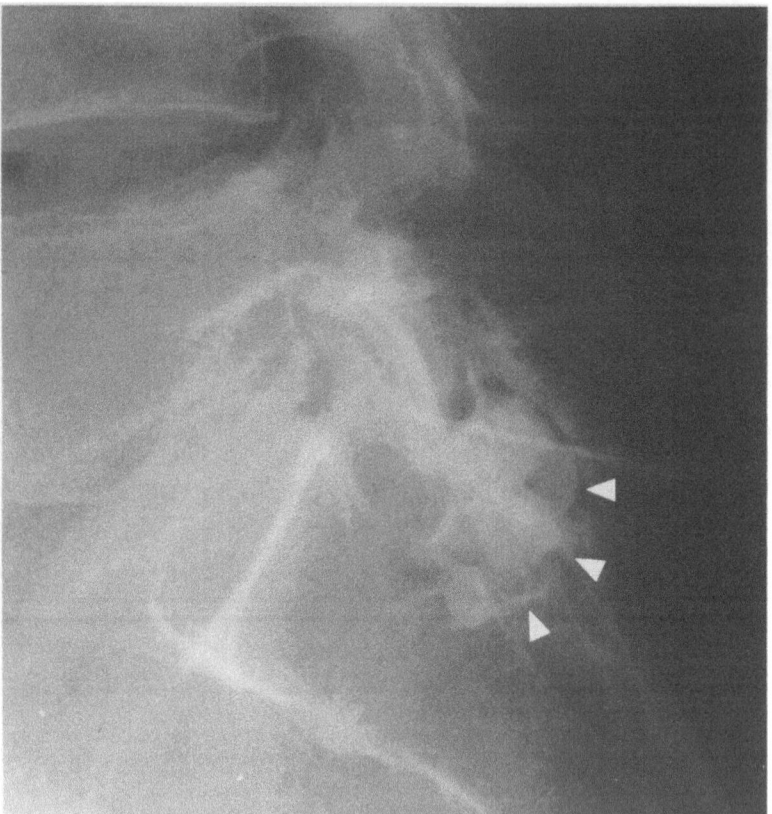

Fig. 15 c

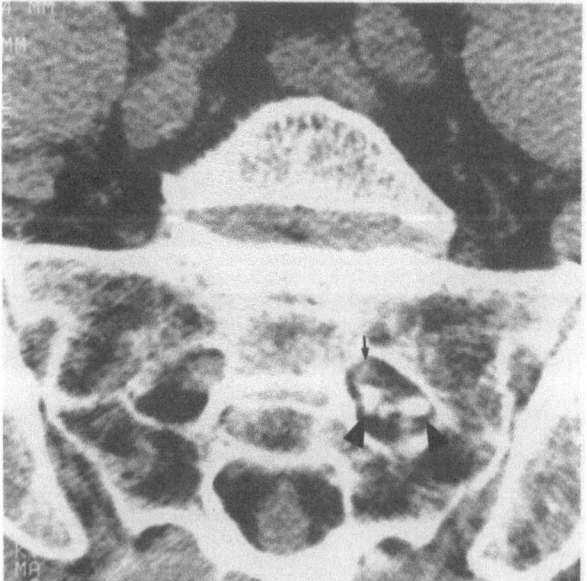

Fig. 15 d

Other diseases

Inflammatory spondyloarthropathy: multiple facet joint therapeutic injections may help relieve pain in the affected spinal segment.

Septic arthritis: facet arthrography allows sampling of joint effusions and reveals abcesses, if any.

Table 4. Possible indications of facet arthrography

Facet syndrome

Nerve root compression by a facet joint synovial cyst

Narrowing of the lumbar canal caused primarily by degeneration of the facet joint

Pre-rhizolysis survey

Spondylolysis with nodule protruding from the bone defect, which could compress a nerve root

Inflammatory spondylarthropathy

Septic facet joint arthritis

References

1. Amor B, Dougados M (1986) Orientation générale du diagnostic des lombalgies. Rev Prat (Paris) 36: 705–710
2. Carrera GF, Williams AL, Haughton VM (1980) Computed tomography in sciatica. Radiology 137: 433–437
3. Crock HV (1981) Normal and pathological anatomy of the lumbar spinal nerve root canals. J Bone Joint Surg [Br] 63: 487–490
4. Dory MA (1981) Arthrography of the lumbar facet joints. Radiology 140: 23–27
5. Destouet JM, Gilula LA, Murphy WA, Monsees B (1982) Lumbar facet joint injection: indication, technique, clinical correlation and preliminary results. Radiology 145: 321–325
6. Emminger E (1972) Les articulations interapophysaires et leur structure méniscoide vue sous l'angle de la pathologie. Ann Med Phys 15: 219–237
7. Ghormley RK (1933) Low back pain with special reference to the articular facets with presentation of an operative procedure. JAMA 101: 1733–1777
8. Hadley LA (1961) Anatomico-roentgenographic studies of the posterior spinal articulations. Ann J Radiol 86: 270–276
9. Kenesi C, Lesur E (1985) L'orientation des apophyses articulaires L4, L5 et S1. Leur rôle possible dans la pathologie discale. Anat Clin 7: 43–47
10. Lewin T (1964) Osteoarthritis in lumbar synovial joints. A morphologic studies. Acta Orthop Scand 73: 107–112
11. Lora J, Long D (1976) So-called facet denervation in the management of intractable back pain. Spine 1: 121–126
12. Louis R (1985) La stabilité vertébrale selon la théorie des trois colonnes. Anat Clin 7: 33–42
13. Maslow GS, Rothman R (1975) The facet joints: another look. Bull NY Acad Med 51: 1294–1311
14. Mc Call IW, Park WM, O'Brien JP (1979) Induced pain referral from posterior lumbar elements in normal subjects. Spine 4: 441–446
15. Mooney V, Robertson J (1976) The facet syndrome. Clin Orthop 115: 149–156
16. Naveau B, Laredo JD (1986) Pathologie mécanique des articulations vertébrales postérieures lombaires. Rev Prat (Paris) 36: 725–733
17. Pedersen HE, Blunck CFJ, Gardner E (1956) The anatomy of lumbosacral posterior rami and meningeal branch of spinal nerves (sinu-vertebral nerves). J Bone Joint Surg [Am] 38: 377–391
18. Sellier N, Vallee C, Chevrot A, Frantz N, Revel M, Wybier M, Gires F, Pallardy G (1986) Arthrographie articulaire vertébrale postérieure lombaire. J Radiol 67: 487–506
19. Shealy CN (1976) Facet denervation in the management of back and sciatic pain. Clin Orthop 115: 157–164
20. Shealy CN (1975) Percutaneous radiofrequency denervation of spinal facets. Treatment for chronic back pain and sciatica. J Neurosurg 43: 448–451

21. Shealy CN (1974) The role of the spinal facets in back and sciatic pain. Headache 14: 101–104
22. Theron J, Blais M, Casasco A, Courtheoux P, Adam Y, Derlon JM, Houtteville JP (1983) La radiologie thérapeutique du rachis lombaire – Chémonucléolyse discale, infiltrations et coagulation des articulations postérieures. J Neuroradiology 10: 209–230

Percutaneous radio frequency lumbar facet denervation: rhizolysis

B. Lavignolle[1], J. Senegas[1], J. L. Honton[2], J. Guerin[3], J. M. Caillé[4]

[1] Departments of Anatomy and of Spine Surgery and Rehabilitation, [2] Department of Orthopedic Surgery, [3] Department of Neurosurgery, [4] Department of Radiology, University Hospital Bordeaux, Bordeaux, France

Surgical denervation of lumbar facets was first suggested by Rees [49] in 1971, with a percutaneous scalpel technique. Shealy [53] introduced the procedure in North America, but changed to percutaneous radio frequency denervation of spinal facets, as an alternative to treatment of chronic back pain. The simple thesis was that back pain and referred pain did not originate from the intervertebral discs, but rather from painful degenerated facet joints. Thus, severing the sensory nerve supply from these joints would cure the pain instantly. In spite of the obvious overenthusiasm of the innovators of the procedure, the theory of facet denervation was attractive and interest in this technique was aroused in the patients who had not responded to the usual conservative therapy, before considering lumbar fusion.

Why facet denervation?

Five main advances in the last ten years have led to the development of rhizolysis.

(1) Better knowledge of the so-called facet syndrome was developed by Goldthwait [18], Putti [46], Ghormley [17], Bagdley [2], Steindler [54], de Sèze [50], and Maigne [34] who defined clinical signs; explanations of referred pain were offered by Leriche [27], Galetti and Procacci [see 4], Kellgren [22], Mooney and Robertson [39], Perl [45], Feinsten, Langton, and Jameson [14], and Hirch [21].

The exact mechanism is speculative, but it may be related to the contiguous depolarization of adjacent nerve cells of the anterior primary ramus as they lie juxtaposed to those of the posterior primary ramus in the spinal ganglia, resulting in antidromic conduction into the symptomatic leg.

(2) The pathogenesis of segmental lumbar instability was approached by MacNab [32], Farfan [12], Cauthen]8], Vernon Roberts [57], Yong Hin and Kirkaldy Willis [60]. Biomechanical studies by Adams and Hutton [1], Miller and Schultz [38] demonstrated that the higher contact pressures with disc narrowing during extension could possibly damage the facet joints and produce pain from the subchondral bone and soft tissues nipped between the facets by way of branch nerves of the posterior ramus. Pain in the richly

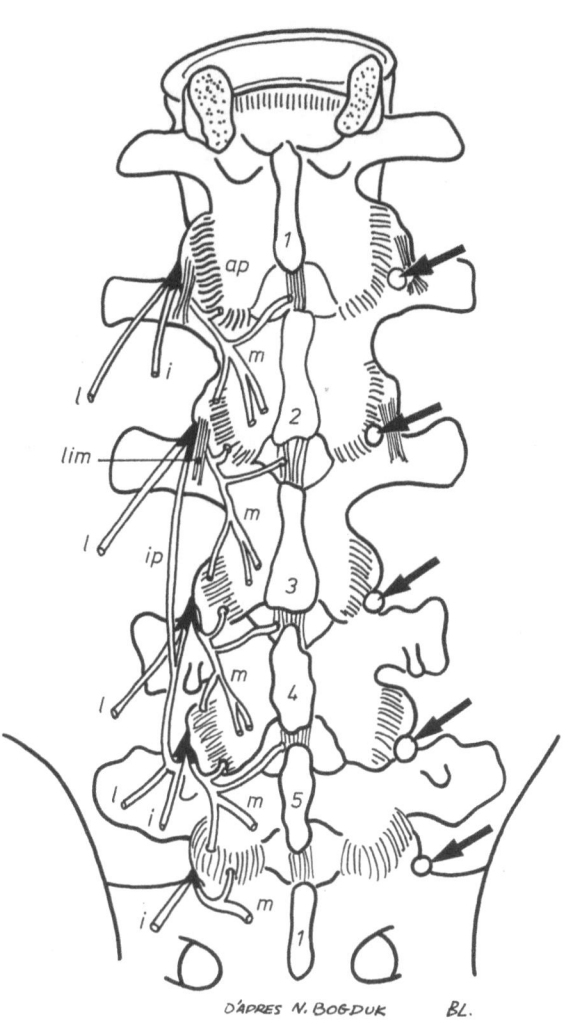

ap

Fig. 1

DʹAPRES N. BOGDUK   BL.

Fig. 2

Fig. 1. Branches of the dorsal rami (Bogduk [3]). *ap* Zygapophyseal posterior joints; *m* medial branches with articular and interspinous branches; *i* intermediate branches; *ip* intermediate plexus; *l* lateral branches; *lim* ligament mamillo accessory; ○← target point for rhizolysis

Fig. 2. Origins of the various rami innervating the lumbar zygapophyseal joints (Auteroche [31]). *1* Dorsal ramus of spinal nerve; *2* medial branch of dorsal ramus; *3* lateral branch of dorsal ramus; *4* ventral ramus of spinal nerve; *5* spinal nerve trunk prior to its bifurcation into ventral and dorsal rami. *T* Spinal nerve trunk; *A* anterior branch of spinal nerve; *M* medial branch of dorsal ramus; *L* lateral branch of dorsal ramus; *S* superior facet joint; *I* inferior facet joint

C and A delta innervated membrane then triggers a reflex paraspinous muscle spasm.

(3) Anatomy of the posterior ramus was explored by Danforth and Wilson [9], Lockhart [29], Lazorthes [24, 25], Pedersen, Blunck and Gardner [44], Fox and Rizzoli [15] neatly summarized by McCulloch [35] and in recent studies by Bogduk and Don Long [3], Paris [43], and Louis and Auteroche [31]. The functional unit of facet joint innervation spans at least three vertebral levels (Fig. 1). Radicular topography seems to contrast markedly with the metameric concept when applied to innervation of the lumbar facet joints (Fig. 2).

(4) Rhizolysis technique was precisely described and modified by McCulloch and Organ [36], Burton [6], Bogduk [3], Lora and Don Long [30], Ray [48], Rashbaum [47], and Hickey [20].

(5) The management of patients with mechanical low back pain syndrom was developed by Mooney and Cairns [40], Burton [7], Frymoyer [16], White [58, 59], Lidstrom and Zachrisson [28], and Nachemson [41], who evaluated both patient disability through a multidisciplinary approach and the parameters to consider in what may be termed "reasonnable treatment", avoiding multiple operations which do not increase the success rate in controlling pain.

Nevertheless, a review of the literature revealed that no one was sure that a percutaneous technique guaranteed that all the nerve supply to a joint would be sectioned. The facet joint receives branches from more than one root and thus a single lesion does not totally denervate the joint.

Many structures in the lumbar region may be the source of pain, thus single facet denervation may sometimes be insufficient. In addition, patient selection for rhizolysis is difficult, because backache is a symptom and patients with this complaint may be manifesting effects of physical and emotional stress or reacting to situational pressures. Further complicating the problem of patient selection, there is a tendency among physicians to dismiss these complaints as being functional in nature ...

Patient selection

Clinical features

The procedure is limited to patients with a history of mechanical low back pain syndrome (MLBS) with referred pain, who evoke segmental mechanical instability, worsened by activity and relieved by rest. Such pain is seen in all age groups after adolescence, but is common in the 30s through 40s.

The lumbar pain usually has two components; there is a constant low-grade backache aggravated by bending with lifting and prolonged sitting position or car travel. This is usually associated with a sensation of stiffness or loss of flexibility most pronounced in the morning on arising. The second painful component is in the form of intermittent acute episodes of incapacitating back pain with "sciatic" scoliosis, lasting from a few days to a few weeks. Pain is not aggravated by coughing and sneezing.

Physical examination demonstrates loss of flexibility in the lumbar region as manifested by loss of full flexion, maintenance of lumbar lordosis on flexion, pain on leg raising. Pain is relieved by lying down in "knee-chest position" or by the "cow-boy crouch" position. Examination shows only posterior ramus nerve root irritation with posterior skin tenderness and reactive spasm. Referred pain rarely radiates below the knee, is often bilateral, is more vague than radicular pain of sciatic nerve root involvement and is not associated with paresthesia.

Passive Straight Leg Raising (SLR) is possible up to 60°, but bilateral active SLR causes a painful response in the back. Neurological examination is normal.

Rhizolysis is limited to patients with chronic MLBS for at least one year and failure of standard conservative treatment measures, including bed rest, medication, bracing, acupuncture, physiotherapy, manipulations, exercises to strengthen back muscles, and infiltrations of short acting steroids.

Thus, patients with the following will be excluded:

Clinical evidence of nerve root involvement and myelography or CT scan showing a herniated disc or a lumbar stenosis;

inflammatory disease in the lumbar spine and metabolic disease (diabetes, osteoporosis ...);

lumbar scoliosis, spondylolisthesis, and congenital lumbo-sacral anomalies;

previous lumbar fusion, except for facet syndrome above the level of fusion;

more than two operations in the lumbar spine;

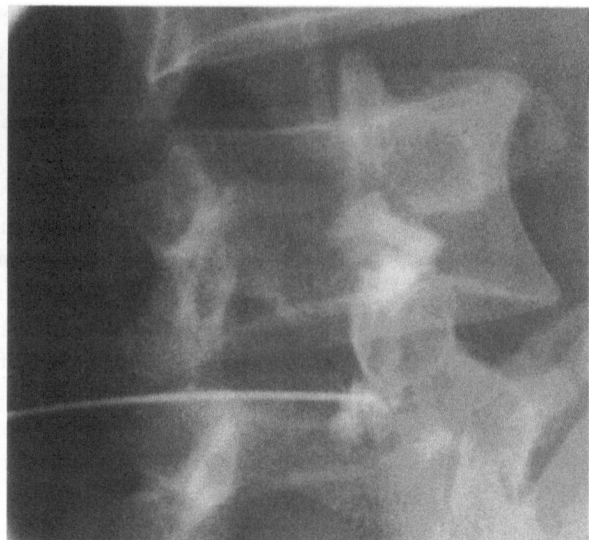

Fig. 4

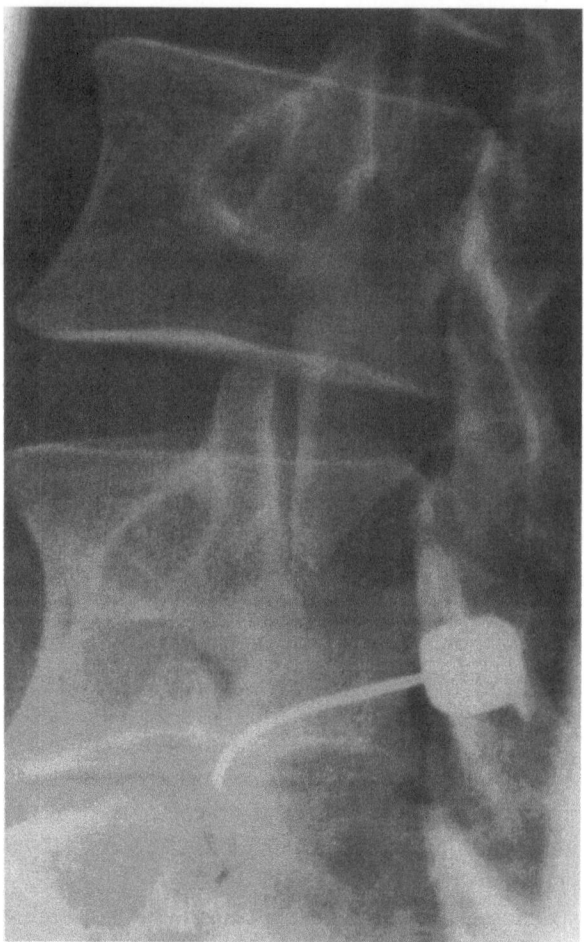

Fig. 3

Fig. 3. Needle approach during facet joint arthrography

Fig. 4. Facet joint arthrography. Oblique view

patients with inappropriate responses on physical examination, suggesting a non-organic component to their disability, as for example in McCulloch and Waddell [37]: psychogenic spinal pain or situational pain in cases of litigation passing through the workmen's compensation board. A pain questionnaire as developed by Mooney [40] is necessary to establish a percentage of disability with regard to the patient's way of life and activities. "Pain drawing" test is especially of interest in detecting patients with hysteria or hypochrondria, who subsequently should consult a psychiatrist.

Radiographic examination

Except to exclude more serious pathology, X-rays have few indications in patients with MLBS. However, in our patients, 80% had signs of segmental instability, which has been described by MacNab [33], Knutsson [23], Drum [10], and Hadley [11].

48% of these patients had major signs of disc narrowing and facet abnormalities with subluxation of joints in 60% of younger and about 80% of older patients.

CT scan shows neither disc herniation, nor stenosis, but only facet arthrosis, asymmetrical posterior joints and often muscle degeneration with increase of fat. In cases of doubt with a degenerative bulding disc, *discography associated with CT scan* will exclude disc herniation. Likewise, in such cases, we use *intradiscal pressure recording* after injection of a constant volume of 1,5 ml of isotonic saline solution. Discomanometry with intradiscal isotonic saline solution usually shows, in these cases, very low pressure related to degenerative discs and is entirely painless even when disc is injected with more than 3 ml of isotonic saline.

Facet arthrography usually shows enlargement of the articular space, or more rarely cystic expansion, with pain reproduction relieved by local anaesthetics and steroids. Similarly, with the injection test of the posterior ramus as described by Mooney and Robertson [39], pain must be relieved for at least one hour by nerve block with marcaine® 0,5% (Bupivacaine), performed under fluoroscopy with an oblique approach on the external side of the superior facet. Immediately after the injection, patients perform motions that normally would precipitate pain and their pain is recorded on a graph.

In our experience, about one third of patients were excluded from rhizolysis, because of an unsuccessful nerve block or facet injection after arthrography (Figs. 3–4).

EMG of paraspinous muscles

Applied to verify posterior ramus nerve root syndrome as reported by Oudenhouven [42], the EMG examination shows signs of muscular denervation, but has no significance as to level involved, because lumbar muscles receive innervation at several levels. This examination is not as selective for diagnosis as facet injection or nerve block.

Selection score

In our experience, we tried to define a score for rhizolysis indications with predictive results (Table 1).

Table 1. Selection score

Positive		Negative	
Mechanical pain on extension with loss of full flexion	+ 10	Diabetes, obese and deficient muscle condition	− 10
Clinical signs of facet syndrome	+ 20	Tension syndrome (myofascial pain)	− 10
Typical referred pain	+ 10	Post menopausal fibrosis	− 10
Positive facet injection or nerve block	+ 20	Osteoporosis	− 10
Posture flexion improving or relieving pain	+ 10	Multilevel spondylosis	− 10
Sedentary work	+ 10	More than one operation	− 10
No psychological or social factors	+ 10	Psychogenic back pain	− 20
Exercise program performed once daily	+ 10	Situational back pain (off work for more than one year)	− 20
	+ 100		− 100

Total
up to 80 = good
up to 60 = fair
lower than 50 = bad

Operative procedure
McCulloch [36]

Anaesthetic protocol

The patient requires a premedication one hour before the procedure for sedation, such as oral Fortal® (Pentazocine) or oral Tranxene® (Dipotassic chlorazepate). However, the patient must remain mentally alert and capable of identifying and describing the response to various stimuli given during the testing period prior to making the lesion. The following protocol is one which has been found to work well.

Preparation

Instrumentation used for percutaneous lumbar facet joing denervation is shown in Table 2. The procedure is done in the operating room under sterile conditions. The patient is placed in the prone position with the lumbo-sacral

Table 2. Surgical instrumentation for percutaneous lumbar facet joint denervation

---

Syringe and disposable needles for local anesthetic infiltration

Standard supplies for cleansing of skin and draping the lumbosacral region

A facet rhizolysis kit in custom stainless steel autoclaving case
The OWL* temperature sensing probe and the radiofrequency electrode, insulated except for 5 mm bare tip,
or the radionics** type SRK-TM temperature monitoring electrode for stimulating and creating lesions,
a set of four 3 1/2 inch 14 gauge stainless steel needles with matching solid stylets, lead wires

The instrument for providing electrical stimulation and for making radio frequency lesions
the OWL* R.F. generator-stimulator, model RFS 1
or the Radionics** model RFG-3 AV lesion-generator system

---

* OWL Instrument LTD, 61 Alness Street, Downsview, Ontario, Canada M3J 2 H2. Dr. Richard Weiss, Reichpietschufer 20, D-1000 Berlin 30. Sophysa, BP n 3 Guipel, F-35440 Montreuil sur Ille, France.
** Radionics, Inc., 76 Cambridge Street, Burlington, MA 01803, U.S.A.

region on a radiolucent platform, using radiographic sagittal and lateral plane fluoroscopic control. The patient should not have any bolstering and specifically there should be no bolstering under the chest to increase the lumbar lordosis. The patient is prepared and draped, showing the lumbar area with the C-arm fluoroscopy apparatus positioned for anterior-posterior viewing (Fig. 5).

Surgical procedure

1% Xylocaine® (Lidocaine) is used to anesthetize the skin and soft tissues over the facet joint. However, the anaesthetic agent is not placed around the facet joint for fear of interfering with nerve root conduction and patient response. A 3 1/2 inch 14 gauge needle is inserted down to the lateral edge of the facet joint aiming for the region just superior to the transverse process. This area is within an angle seen on anterior posterior radiography formed

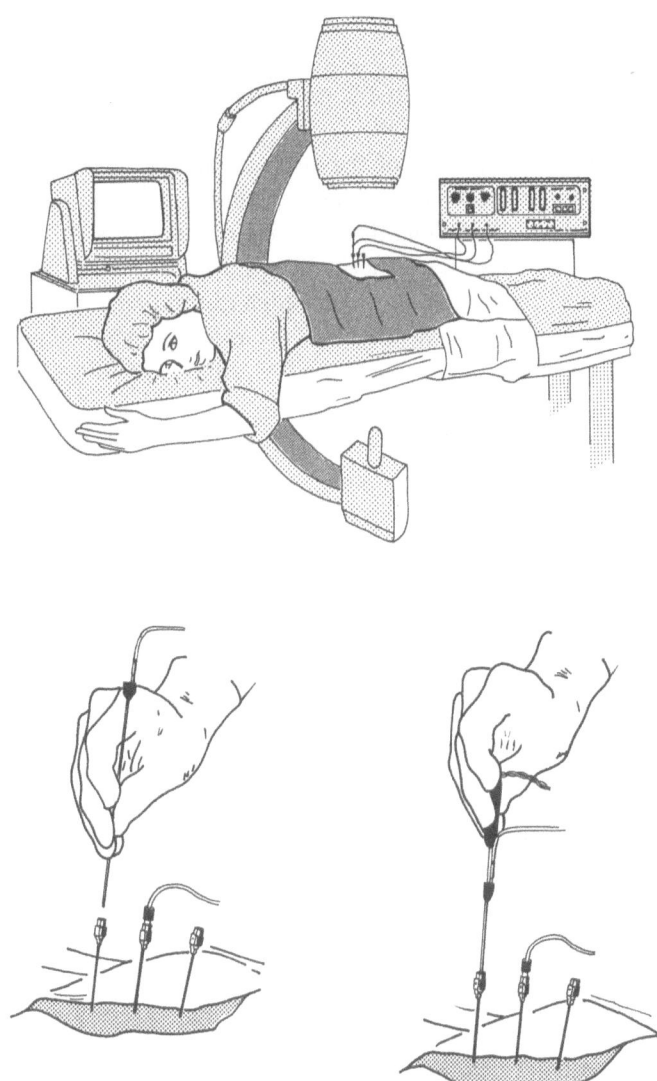

Fig. 5. Patient position and needle placement for rhizolysis

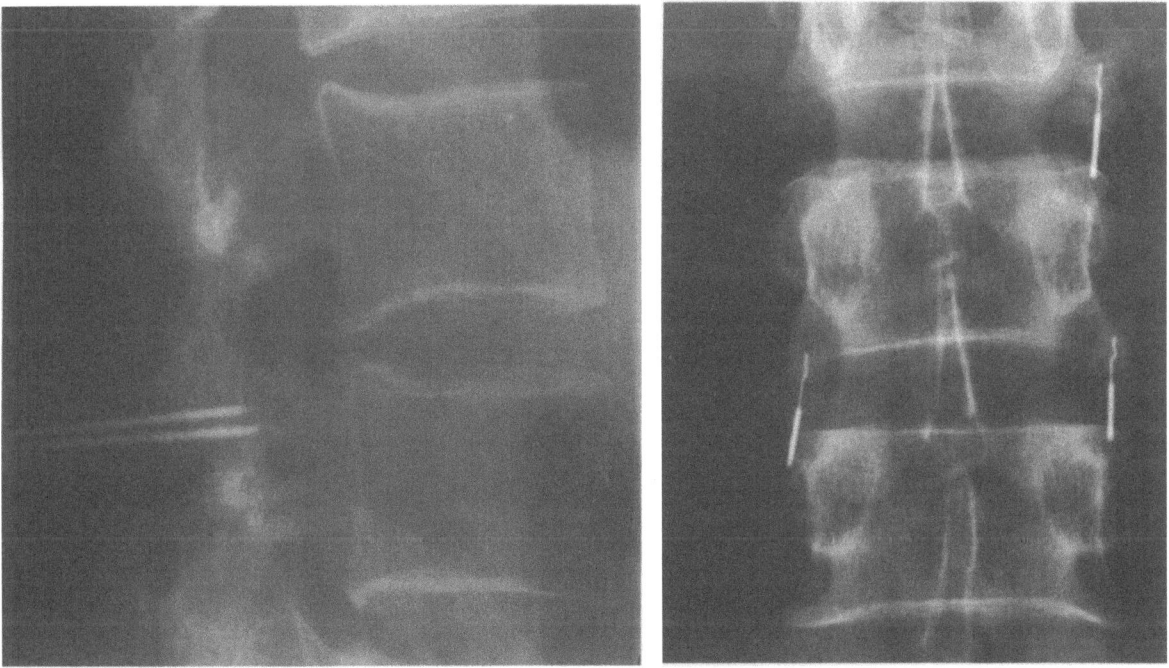

a   b

Fig. 6. Needle placement on lateral (**a**) and frontal (**b**) views

by the lateral edge of the superior facet and the superior border of the transverse process (Fig. 6).

Following removal of the guide needle stylet, the RF lesion electrode with a 5 mm exposed tip is inserted through the 14 gauge needle and the needle is retracted along the electrode beyond the skin to avoid skin burn. The C-arm image intensifier is then moved to the lateral position and the electrode is advanced ot a depth edge of transverse process. The electrode tip reaches its proper depth just as the intertransverse ligament is penetrated. With the blunt tipped electrode, this is palpable to the surgeon. The temperature sensing probe is inserted within the lesion electrode. Needles are placed at the levels where the rhizolysis is planned. Usually, this is lumbar 3, 4, 5 or lumbar 4, 5 and lumbosacral level. Sometimes this is thoracic 12 and lumbar 1, 2. Needle placement is done on one side only or bilaterally, according to clinical signs and facet injection tests. The reference electrode with metal plus is inserted in unoccupied needle.

Stimulation

Stimulation is used to confirm that the electrode tip is properly positioned adjacent to the posterior primary ramus and away from the anterior primary ramus. Again, ensure that the bare tip of the electrode is beyond the tip of the needle. Otherwise, the electrode will short circuit the needle.

Two stimulating frequencies are used:

2 hertz stimulation for observing motor effects. Usually, a slowly increase of 0,5 to 5 volts is required until the patient feels local back contraction and discomfort. Twitching in the lower extremity means the electrode is too close to the anterior ramus and *must be removed and placed more posteriorly*;

100 hertz stimulation for producing sensory effects. Sensory response is elicited with a slow increase of the stimulus voltage, almost always between 0,5 to 5 volts.

The response to this stimulation will be variable. If the stimulus is turned

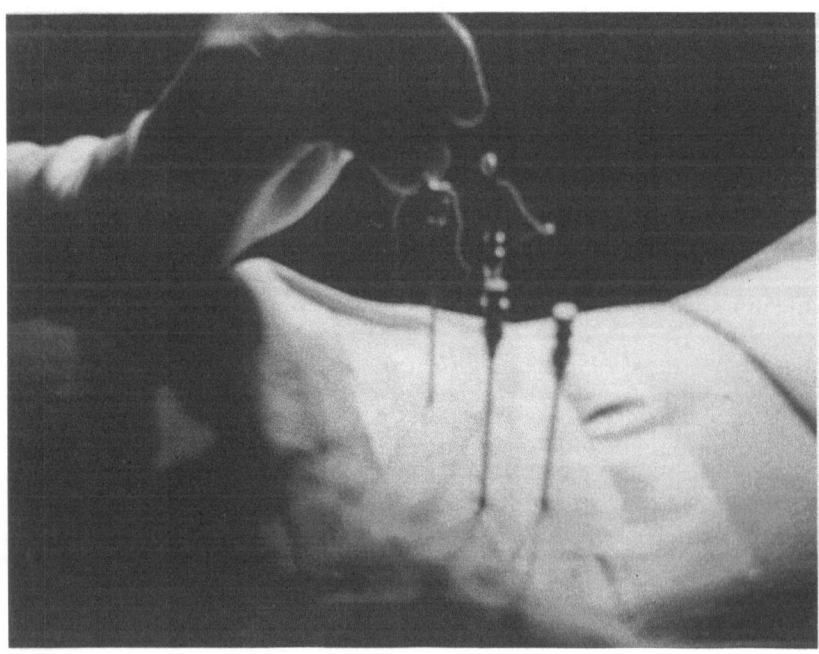

Fig. 7. Lesion making

high enough, the patient may describe the discomfort in terms of his clinical symptoms.

Lesion making (Fig. 7)

The patient is informed that during lesion making, pain, sometimes intense, will be felt in the low back and may be referred down the leg. A temperature of 85 °C is desired for about 50 to 60 seconds. Usually, between 100 to 150 milliamperes of RF current is required, with 25 to 30 volts for a probe tip temperature of 85 °C. When 85 °C has been reached, within 60 to 90 seconds the temperature will increase above 90 °C, followed shortly by a rise in lesion voltage and a decrease in miliamperes. Decreased current and increased voltage indicate a rise in electrical resistance of the tissue surrounding the probe tip, produced by tissue coagulation. If the surgeon has his hand on the electrode, he can feel the lesion being completed by the delivery of a crackling sensation.

On removing the electrode, there should be some degree of charring at the end to signify that a lesion has been produced. Lesions are completed on one side before moving to the contralateral side.

The selection of levels for lesion making is decided on the basis of the clinical presentation, the response to facet joint blocks and the response to motor and sensory stimulation.

Postoperative care

The patient is discharged from the hospital three hours following the procedure. Antalgic medication usually suffices for control of discomfort. This is followed in a few days by occurrence of a different type of pain in the area where the lesions have been produced. This is the rising inflammatory reaction for healing the coagulum.

This non-mechanical discomfort does not usually interfere with work capabilities and disappears within ten days to three weeks. Most patients are in their final recovery state six weeks from the date of the procedure. Those patients who notice initial relief tend to be the ones with a good response at six weeks and those patients who notice no initial relief are the ones who have treatment failure at six weeks.

Results

Overall experience

In all known series, after selection and positive facet injection or posterior nerve block, 60 to 80% of patients obtained good or excellent results at a mean follow-up time of one year (Table 3).

In all published series, there is a tendency toward deterioration of pain relief with time; in the French experience, 40% of patients achieved satisfactory relief with rhyzolysis at a mean follow-up of 3 years.

Rhizolysis raises serious questions about the efficiency of the procedure compared to placebo. Stovall King and Lagger [55] performed a randomized study with a placebo procedure. At six months, the results showed a high level of statistical difference (PS to chi square analysis).

Table 3. Published results

		Patients (number)	Successes (percentage)
N. Shealy [53]	1975	235	80%
J. A. McCulloch [36]	1976	82	67%
C. Burton [6]	1976	126	67%
J. Lora, D. Long [30]	1976	82	79%
R. C. Oudenhouven [42]	1979	66	74%
B. Fassio [13]	1980	30	60%
J. C. Verdie [26]	1982	97	59%
R. F. Rash-Baum [47]	1983	100	68%
French experience [61]	1985	927	60%
U.S.A. experience [62]	1985	1,020	65%

Personal experience

We performed rhizolysis in 537 patients over a period of 9 years according to the above criteria and technique. Mean follow-up time was 4 years (2 to 9 years). The group consisted of 46% females and 54% males. The age range was from 35 to 71 years and the average age was about 50 years.

These patients were manual workers (43%), sedentary workers (42%), and housewives or retired people (15%). The duration of low back pain was more than one year in 49% of cases, and more than two years in 51%.

Work was related to back injury in 30% of cases and patients were unable to work for 6 months to one year in 52% of cases, and for more than one year in 33% of cases.

Previous surgery had been performed (one operation) in 32% of cases. 3% of patients had chemonucleolysis and 7% had had 2 operations.

Results were classified as:

excellent: pain free, return to work;

good: pain reduction without need of drug (pain relief recording up to 50%), functional increase, normal social or professional way of life;

bad: pain unchanged, medication, no return to work.

An overall 58% of patients achieved satisfactory relief at one year and 44% at the mean follow-up of 4 years.

First group
(ideal candidates)

Virgin back, sedentary work, selection score superior to 60; 70% of the 325 patients filling those criteria had successes at one year and 58% at 4 years.

Second group

Patients with previous surgery, heavy work, selection score lower than 60 (N = 212); only 47% of those patients had good results at one year and 30% at the mean follow-up time of 4 years.

A comparison of all successes and failures at one and 4 years follow-up demonstrates *a limited time effect of rhizolysis*. Regeneration of nerves following facet denervation probably occurs and no one is certain that this percutaneous denervation guarantees that all the nerve supply to facet joints can be sectioned. 22% of patients required repeated rhizolysis after one year, performed only if they had been improved during this period.

A chi square analysis with comparison of successes at the mean follow-up time of 4 years between the first group with a good score and the second group with a bad score is highly significant (PS) and shows the *importance of initial evaluation*.

Factors without significant influence on failures

Sex, age, presence or absence of referred leg pain radiographic changes of facet joints and disc narrowing.

Factors with significant influence on failures

Duration of work lay-off (more than one year), heavy work, postoperative group (more than one operation with failed back surgery syndrome and presence of arachnoiditis or epidural fibrosis), multilevel spondylosis, psychological factors and neuroleptic treatment, litigation in progress and non-organic components of patient's disability, spondylolysis with olisthesis, chronic low back and radicular pain due to lateral stenosis which is sometimes difficult to differentiate from facet syndrome, but may be recognized by CT scan, EMG, and nerve root infiltration prior to surgery.

Complications

There have been no serious complications in the 2000 patients of all known series [3, 6, 11, 13, 26, 30, 36, 42, 47, 53, 55, 56, 61]. The only reported complications have involved 2 generalized reactions to local anaesthesia, 2 superficial infections, and 3 skin burns from dispersive electrodes.

No mortalities are known at this time. No patient's state was worsened by rhizolysis. Sensory or motor deficits were not found, but five patients described leg paresthesia and it is possible that the spinal root ganglia was injured and the resulting complaint of pain was secondary to denervation dysesthesia.

No neuropathic joint and no radiographic changes relative to disc space or facet integrity was observed at the final control following rhizolysis. In our series, no intervertebral instability occurred at the treated levels in the postoperative period. However, in 3 patients, the initial degenerative process progressed to a segmental instability which later required surgical fusion.

Conclusion

Percutaneous RF lumbar facet denervation seems to be a valid method of therapy in selected patients who do not respond to standard conservative treatment for typical facet or mechanical low back pain syndromes (MLBS).

Rhizolysis is not the final solution to MLBS. But facet denervation seems to be useful in a limited group of patients and appears to be of at least short-term benefit to patients with organic pain and without psychogenic components or litigation in progress. This procedure was successful in our experience at the mean follow-up time of 4 years in 58% of the selected group and in patients who had not previously undergone more than one operation. Approximately 70% of patients have an immediate reduction in MLBS; three years later, these results degrade somewhat so that about 50%

of the cases will have a permanent good result. This represents a limited role for rhizolysis in the management of low back pain. However, facet denervation used in conjunction with a back reeducation program (William's postural exercices) may reduce the need for repeated back surgery in MLBS, if instructions given in the accompanying program are followed. This symptomatic treatment provides relief of pain in selected patients with major disability when treatment has failed.

The procedure should be considered prior to lumbar fusion and has low risk on out-patients.

Nevertheless it should never be considered a substitute for the time — honored treatment of rest, immobilization, analgesia, weight reduction, graduated exercizes and spinal education, and facet injections of steroids or sclerosing material (1 ml solution with 2% Phenol, 25% Glycerol, 25% Dextrose added for 1 ml of 1% Xylocaine).

Rhizolysis has no role in the group of patients with herniated disc or lumbar stenosis manifested by sciatica, and with MLBS of discogenic origin.

References

1. Adams MA, Hutton W (1983) The mechanical function of the lumbar apophyseal joints. Spine 8: 327–330
2. Badgley CE (1941) The articular facet in relation to low back pain and sciatic radiation. J Bone Joint Surg [Am] 23: 481–496
3. Bogduk N (1983) The innervation of the lumbar spine. Spine 8: 286–292
4. Boureau F, Willer JC (1979) La douleur. Masson, Paris
5. Bouvier JP, Privat JM, Fuentes JM (1980) Facet syndrome. In: Simon L (ed) A sciatique. Masson, Paris, pp 182–189
6. Burton C (1976) Percutaneous radio frequency facet denervation. Appl Neurophysiol 39: 80–86
7. Burton C (1981) Conservative management of low back pain. Post Grad Med. Low back Pain, Vol 70, no 5
8. Cauthen C (1983) Lumbar spondylosis and stenosis. In: Cauthen JC (ed) Lumbar spine surgery. Wilkins, Baltimore, pp 5–19
9. Danporth MS, Wilson PD (1925) The anatomy of the lumbosacral region in relation to sciatic pain. J Bone Joint Surg [Am] 7: 160–167
10. Drum DC (1970) An introduction to the study of posture and spinal mechanics. Published by the author, Toronto [quoted by Haldeman S (1975) in: Goldstein M (ed) NINCDS Monograph Series 15, Bethesda, MD, pp 217–226]
11. Drevet JG (1983) Rhizolyse percutanée et syndrome articulaire postérieur lombaire. In: Simon L (ed) Lombalgies et médecine rééducation. Masson, Paris, pp 325–328
12. Farfan HF (1973) The mechanical disorders of the low back. L Fibiger, Philadelphia
13. Fassio B (1981) Denervation articulaire postérieure percutanée. Rev Chir Orthop [Suppl 11] 67: 131–136
14. Feinstein B, Langton JNK, Jameson R (1954) Experiments on pain referred from deep somatic tissues. J Bone Joint Surg [Am] 36: 981–997
15. Fox JL, Rizzoli HV (1973) Identification of radiologic coordinates for the posterior articular nerve of Luschka in the lumbar spine. Surg Neurol 1: 343–346
16. Frymoyer JW (1980) Epidemiologic studies of low back pain. Spine 5: 419–423
17. Ghormley RK (1933) Low back pain with special reference to the articular facets with presentation of an operative procedure. JAMA 101: 1773–1777
18. Goldthwait JE (1911) The lumbo-sacral articulation. An explanation of many cases of "lumbago" "sciatica" and paraplegia. Boston Med Surg 64: 365–372
19. Hadley LA (1936) Apophyseal subluxation. Disturbances in and about intervertebral foramen causing back pain. J Bone Joint Surg [Am] 18: 428–433
20. Hickey RF (1977) Denervation of the spinal facet joints for the treatment of chronic low back pain. N Engl J Med 85: 96–99

21. Hirsch C (1963) The anatomical basis for low pain. Acta Orthop Scand 33: 1–17
22. Kellgren JH (1938) Observations on referred pain arising from muscle. Clin Sci 3: 175
23. Knutsson F (1944) The instability associated with degeneration in the lumbar spine. Acta Radiol 25: 593–609
24. Lazorthes G (1956) Innervation des articulations interapophysaires vertebrales. 43rd CR Assoc Anat (Lisbonne), pp 488–494
25. Lazorthes G (1972) Les branches postérieures des nerfs rachidiens et le plan articulaire vertébral postérieur. Ann Med Phys 15: 192–202
26. Lazorthes Y, Verdie JC (1976) Thermocoagulation percutanée des nerfs rachidiens à visée analgesique. Neurochirurgie 22: 445–453 [Nouv Press Med 11: 2131–2134 (1982)]
27. Leriche R (1940) La chirurgie de la douleur. Masson, Paris
28. Lidstrom A, Zachrisson M (1970) Physical therapy in low back pain and sciatica. Scand J Rehab Med 2: 37–42
29. Lockhart RD (1938) Human anatomy. Churchill Livingstone, London (see [49])
30. Lora J, Long D (1976) So-called facet denervation in the management of intractable back pain. Spine 1: 121–126
31. Louis R, Auteroche P (1983) Innervation of the zygapophyseal joints of the lumbar spine. Anat Clin 5: 17–28
32. MacNab I (1977) Backache. Williams and Wilkins, Baltimore
33. MacNab I (1969) Pathogenesis of symptoms in discogenic low back pain. In: American Academy of Orthopedical Surg, Symposium on the Spine, vol 6, pp 97–110
34. Maigne R (1974) Douleurs d'origine vertébrale et traitement par les manipulations. Expansion Scientifique, Paris [A new approach to vertebral manipulations. C Thomas, Springfield, OH]
35. McCulloch JA (1970) Anatomy of lumbar posterior rami. Can Med Ass J 116: 30–32
36. McCulloch JA, Organ LW (1976) Percutaneous radiofrequency lumbar rhizolysis. Appl Neurophysiol 39: 87–96 [Can Med Assoc J 116: 30–32 (1977)]
37. McCulloch JA, Waddell G (1980) Non organic physical signs in low back pain. Spine 5: 117–125
38. Miller JA, Schultz AB, Haderspeck AK (1983) Posterior element loads in lumbar motion segments. Spine 8: 331–337
39. Mooney V, Robertson J (1976) The facet syndrome. Clin Orthop 115: 149–156
40. Mooney VM, Cairns D (1978) Management in the patient with chronic low back pain. Orthop Clin North Am 9: 543–557
41. Nachemson A (1970) Intravital dynamic pressure measurements in lumbar discs. A study of common movements, Maneuvers and exercices. Scand J Rehab Med [Suppl] 1: 37–42
42. Oudenhouven RC (1977) Paraspinal EMG following rhizotomy. Spine 2: 229–304
43. Paris SV, Nyberg R, Mooney VT (1980) Three level innervations of the lumbar facet joints. Int Soc Study lumbar spine, New Orleans
44. Pedersen HE, Blunck CFJ, Gardner E (1956) The anatomy of lumbosacral posterior rami and meningeal branches of spinal nerves. J Bone Joint Surg [Am] 38: 377–391
45. Perl ER (1975) Pain spinal and peripheral nerve factors. In: Goldstein M (ed) The research status of spinal manipulative therapy. US Department of Health, Bethesda, MD (NINCDS Monograph Series 15, pp 175–181)
46. Putti V (1927) New conceptions in the pathogenesis of sciatic pain. Lancet 2: 53–60
47. Rashbaum RF (1983) Radiofrequency facet denervation. Orthop Clin North Am 14: 569–575
48. Ray BS (1943) The management of intractable pain by posterior rhizotomy. Res Publ Assoc Res Nerv Ment Dis 23: 391–407
49. Rees WES (1971) Multiple bilateral subcutaneous rhizolysis of segmental nerves in the treatment of the intervertebral disc. Syndrome Ann Gen Prac 26: 126–127

50. Seze S de (1960) Arthrose lombaire posterieure et hyperlordose lombaire. Rev Rhum 27: 73–81
51. Senegas J, Lavignolle B, Guerin J (1983) Dégénérescence discale, syndrome des facettes et rhizolyse lombaire percutanée. In: Simon L (ed) Lombalgies et médecine rééducation. Masson, Paris, pp 313–324 [Ann Med Phys 25: 255–262 (1982)]
52. Senegas J, Lavignolle B (1983) Approche globale du patient lombalgique. In: Simon L (ed) Lombalgies et médecine rééducation. Masson, Paris, pp 342–353
53. Shealy CN (1975) Percutaneous radiofrequency denervation of spinal facets. An alternative approach to treatment of chronic back pain and sciatica. J Neurosurg 43: 448–451
54. Steindler A (1948) The interpretation of sciatic radiation and the syndrome of low back pain. J Bone Joint Surg [Am] 22: 28–34
55. Stovall King J, Lagger R (1976) Sciatica viewed as a referred pain syndrome. Surg Neurol 5: 46–50
56. Theron J, Louis AX (1980) Syndrome douloureux des facettes articulaires lombaires et electrocoagulation transcutanée. Encycl Med Chir App Locom 54: 25–28
57. Vernon Roberts B (1979) The pathology and interrelations of intervertebral disc lesions, osteoarthrosis of the apophyseal joints, lumbar spondylosis and low back pain. In: Jayson MIV, Dixon A (eds) Lumbar spine and back pain, vol 4. St J Pitman, London, pp 83–114
58. White AW (1966) Low back pain in men receiving workmen's compensation. Can Med Assoc J 95: 50–56
59. White A (1982) Workshop on idiopathic low back pain. Spine 7: 141–149
60. Yong Hing K, Kirkaldy Willis WH (1983) The pathophysiology of degenerative of the lumbar spine. Orthop Clin North Am 14: 491–504
61. Lavignolle B, Senegas J (1985) French experience of percutaneous RF lumbar rhizolysis. Int Congr Altern Spinal Surgery, Paris, June 18, 1985. Omnis Travenol, Deerfield, Ill., U.S.A.
62. Ray CD (1985) USA experience of facet rhizolysis. Int Cong Altern Spinal Surgery, Paris, June 18, 1985. Omnis Travenol, Deerfield, Ill., U.S.A.

Interventional vascular radiology
in musculo-skeletal lesions

Technique of angiography and embolization

F. Gelbert, D. Reizine, and J.-J. Merland

Department of Neuroradiology and Therapeutic Angiography, Hôpital Lariboisière,
Paris, France

Several improvements have contributed to the increasingly widespread use of vascular investigations. Low osmolality quickly eliminated contrast media (iopamidol, iohexol) have recently been developed. They can be used in fragile or elderly patients and those presenting cardiac or renal insufficiency. Rapid elimination and better tolerance has resulted in modification in anesthetic procedures. General anesthesia has been abandoned in favor of neuroleptanalgesia. The patient remains conscious during the examination. Timely medication is administered by the anesthesiologist, who is present during the examination. The patient awakens immediately after the procedure. The entire exploration, including pre-anesthetic examination, can be performed during a 24-hour hospitalization. If, however, the patient presents a past history of allergy, premedication is given before the examination.

Major improvements have also been made in x-ray examination. Digital angiography provides rapid and precise subtraction imaging. New catheters has also contributed to the improvement of these examinations. New thin, flexible polyethylene catheters or supple micro-catheters allow hyperselective catheterization of very small arteries. These improvements have allowed development of endovascular therapeutics such as transcatheter embolization which now constitutes an alternative or a complementary procedure to surgery.

Techniques of vascular exploration

Global opacification is presently a routine vascular examination

The puncture site is selected in relation to the region of interest;
the size of the needle used depends on the age and morphology of the patients and should allow the flow of a sufficient amount of contrast media;
a 16 G needle is commonly used in adult vascular exploration;
the injection should be performed with an electric injector.
The techniques of global opacifications are resumed in Table 1.

Selective catheterization

Selective catheterization is often required after global opacification procedures. The thin, flexible catheters currently available, facilitate this procedure

Table 1

	Puncture site	Flow	Volume
Upper limb shoulder	(retrograde) humeral	$15\,cm^3/sec$	$35\,cm^3$
Lower limb	(retrograde) femoral	$15\,cm^3/sec$	$40\,cm^3$
Cervical spine	(simultaneous) bi-humeral	$15\,cm^3/sec$	$40\,cm^3$
Lombosacral spine	(simultaneous) bi-femoral retrograde	$15\,cm^3/sec$	$40\,cm^3$
Thoracic spine	selective catheterization		

even in the exploration of atheromatous patients. The femoral approach is most commonly used. A 16 G needle allows introduction of a 5 F catheter which is positioned using Seldinger's technique. The use of a sheath catheter prevents repeated friction against the wall of the artery during catheter manipulation. It also allows fast replacement of the catheter during the investigations. The catheters are either pre-shaped, or shaped manually with steam prior to the examination. The form of the catheter is adapted to conform with anatomical curves.

The metallic guides are used to rigidify the catheter or modify the curve. Great care should be taken when using metallic guides.

Precautions

Catheters should permanently contain contrast material so as to avoid stagnation of the blood, which could be the future source of clotting. Before injection, reflux of blood flow in the syringe should be checked to guarantee the absence of clotting and air bubbles. To avoid spasm which can occur during catheterization of small arteries, the catheter should be handled with great care, without force while using the guide with parcimoni. For extremely small vessels, a system of progressively smaller coaxial catheters sliding one over the other should be used. This type of exploration is rarely used for strictly diagnostic means and is generally reserved for embolization techniques only [7, 8].

Special cases

Although the femoral approach is frequently used, catheterization can be performed at other sites.

The humeral artery should be avoided if possible due to the major risk of spasm or thrombosis.

The axillary approach is a better alternative. Compression should be carried out carefully in order to avoid compression of the brachial plexus by hematoma.

Finally, the carotid approach is frequently used for head and neck exploration.

Embolization

Endovascular therapy results from these major developments. Endovascular therapy comprises: angioplasty (remodeling of the arteries using a balloon catheter), selective arterial perfusion (arterial chemotherapy, or fibrinolysis in situ) and, finally, embolization – i.e. using catheters to position neutral particles or chemotherapeutic embolization. The choice of embolization material depends on the angiographic structure of the lesion (vascular malformation, tumor), the size and number of the relating pedicles and the size of arteriovenous shunts. Healthy arteries or anastomoses arising from the pedicle to be embolized or adjacent pedicles are first carefully located. Catheterization must be stable and avoid spasm. New catheters are undoubtedly better tolerated [2, 3, 4, 5]. Several embolization products are currently available. The choice depends on the type of the lesion and the purpose of embolization (preoperative or embolization treatment alone).

For a very proximal occlusion of a single large pedicle, a detachable balloon system can be used. The latex balloon, attached to a microcatheter with a ligature, is positioned using a coaxial system. The balloon is inflated with contrast media. The result of occlusion is controlled using a second catheter. The balloon is then detached. The contrast media contained in the balloon allows verification of its position with plain films.

Gelfoam is resorbed in just a few days. It is recommended to use this particulate agent in preoperative investigation. The date for embolization should thus be chosen with respect to the date of surgery.

Liquid alcohol emboli (used in renal pathology) and biological adhesive agents (Isobutyl-2-Cyanoacrylate) (IBC) are efficient, permanent materials requiring very careful handling.

In cases of a lesion with multiple vascular pedicles or when the nidus of a malformation should be directly occluded, either particles or liquid emboli can be used. The particles are fragments of lyophilized dura mater, or Gelfoam trimmed to an appropriate size by the operator. Pre-calibrated polyvinyl alcohol (Ivalon) (100, 200, 300 microns) can be used without preparation. Particles are mixed with serum and injected slowly. Frequent testing with contrast media injections is carried out to monitor correct catheter placement and effective embolization. The progressive disappearance of the shunts confirms the correct placement of emboli. Embolization should be stopped once reflux of contrast occurs at the level of the catheter. Washout before catheterization of another pedicle or catheter withdrawal should be carefully performed if there is even the slightest doubt of remaining particles or clot in the catheter.

IBC is the most commonly used liquid product. It polymerizes on contact with the blood and should be injected between 2 columns of glucose serum. Its viscosity becomes variable by adding Lipiodol and the mixture becomes opaque with tantalum powder.

More recently, microspheres containing a cytotoxic agent have been developed. The microcapsule consists of the cytotoxic agent (50–80%) as the core and Ethylcellulose as the shell [6].

Following embolization, liberation of the drug takes place progressively over a few hours or a few days. Chemoembolization allows a combination of the effects of embolization (hypoxia) and an increase in the contact time between the tumor and the cytotoxic agent. The general complications of chemotherapy are also diminished due to the reduction of the systemic concentration.

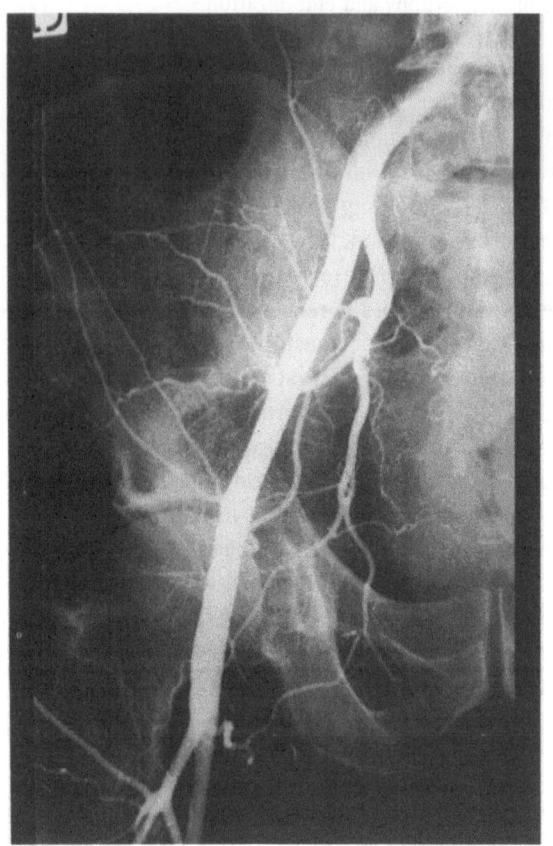

a

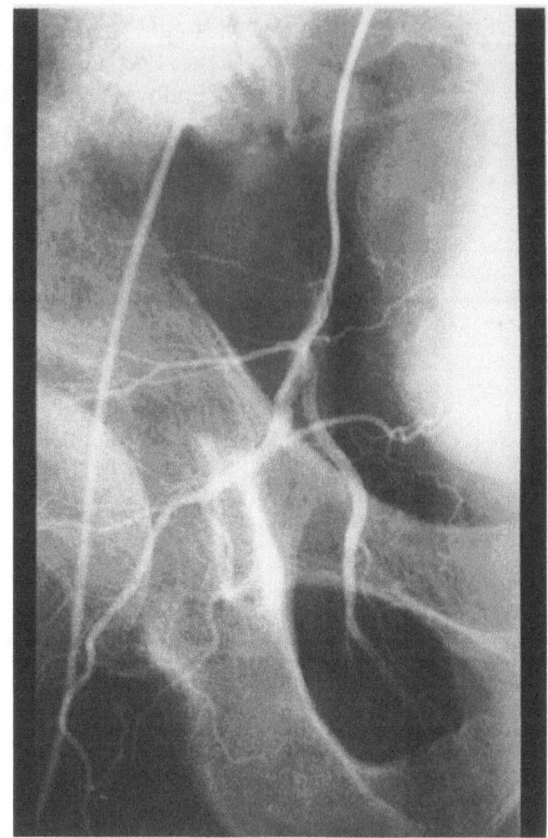

b

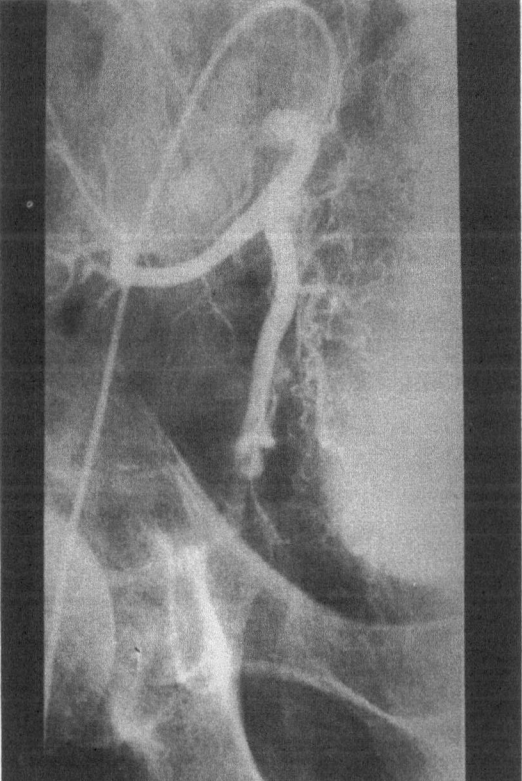

c

Fig. 1. Embolization accident. Patient with hyperalgic lytic metastasis (arrow) of pelvis. Antalgic embolization. **a** Angiography before embolization. **b** Selective opacification of the inferior gluteal and obturator arteries. **c** After embolization with absolute ethanol and gelfoam. After embolization the patient presented a complete and definitive paralysis of the left lower limb due to the occlusion of the feeding arteries of nerves

Accidents –
embolization side effects (Fig. 1)

Accidents or side effects occurring during procedures are rare for a well-experienced interventional team. Two types of accidents have to be considered: accidents related to hyperselective catheterization (thrombosis, dissection spasm), and those related to embolization (vascular rupture by an over-inflated balloon, obliteration of normal vessels ...). However, some anatomical regions are more fragile than others. This is particularly the case for the spine (medullary risk) or the pelvis (risk of ischemia of the nerve at hypogastric embolization) (Fig. 1).

References

1. Berenstein A, Kricheff I (1979) Catheter and material selection for transarterial embolization. Radiology 132: 619–630
2. Cromwel JD, Kerber CW (1979) Modification of cyanoacrylate for therapeutic embolization. AJR 132: 799–801
3. Debrun G, Lacou P, Caron JP (1978) Detachable balloon and calibrated leak balloons techniques. J Neurosurg 49: 635–649
4. Ellman BA, Green GE, Eigenbrodt E (1980) Renal infarction with absolute ethanol. Invest Radiol 15: 318–322
5. Jack CR, Forbes G, Dewanjee MK (1985) Polyvinyl alcohol sponge for embolotherapy: particle size and morphology. AJNR 6: 595–597
6. Kato T, Nemoto R, Mori H, Takahashi M, Harada M (1981) Arterial chemoembolization with mitomycin C microcapsules: an approach to selective cancer chemotherapy with sustained effects. JAMA 245: 1123–1127
7. Melki JP, Riche MC, Reizine D, Assouline E, Aymard A, Merland JJ (1986) Arteriographie bifemorale simultanée sous pression. J Neurol 13: 62–70
8. Rufenacht D, Merland JJ (1986) A new and original microcatheter system for the hyperselective catheterization and endovascular treatment avoiding the risk of arterial rupture. J Neuroradiol 13: 85–87
9. Rufenacht D, Merland JJ, Guimaraens L, Reizine D (1986) A simple propulsion chamber system for the 16 gauge approach. Neuroradiology 28: 355–358

Interventional vascular radiology in musculo-skeletal trauma

A. Aymard, S. Marciano, and J.-J. Merland

Department of Neuroradiology and Therapeutic Angiography, Hôpital Lariboisière,
Paris, France

Introduction

Vascular injuries are frequently observed in cases of the pelvic trauma and femoral diaphysis. Hemostasis is often spontaneous. However, vascular injuries may cause severe internal hemorrhage which can quickly compromise vital prognosis. In these cases, arteriography provides localization of the origin of bleeding in order to activate hemostasis through an endovascular approach. This examination must be performed as quickly as possible.

In vascular wounds after pelvic surgery, post-traumatic arteriovenous fistulae are generally revealed some time after surgery by local or general hemodynamic perturbations. Arteriography plays an important role in the diagnosis and treatment of these fistulae.

Patient management and care

The management of these patients should be multidisciplinary. Trauma patients are usually immobile and present hemodynamic conditions requiring intensive care initiated before the endovascular procedure. Angiographic exploration should be performed as early as possible with the support of a well-experienced team. The multiple transfusions required to maintain hemodynamics can in themselves provoke serious visceral lesions. Angiographic procedures should be performed within 12–18 hours following the start of intensive care.

The arterial approach

The femoral approach is the most frequently used. The side chosen depends on the suspected site of bleeding. If this approach is impossible or fails due to hematoma the upper humeral or left axillary approach should be used.

Catheterization material

Depending on the arterial anatomy, flexible 5 F catheters should be used. They provide distal catheterization which is usually sufficient. A global angiographic catheter allows opacification of the abdominal aorta, its visceral branches and the iliopelvic arterial territory.

Embolization material [1, 5, 6, 8]

Gelfoam and autologous clot are the most appropriate material for embolization indicated for hemostasis. It is easy to use and can be fragmented and trimmed according to the size of the arteries to be embolized and the proximity of the tip of the catheter to the lesion. The efficiency of embolization is confirmed by a significant diminution of the arterial flow with disappearance of all extravasation. Total arterial devascularization should be avoided due to the risk of severe ischemia. Very small emboli fragments may give rise to ischemia of healthy areas, thus complicating the embolization procedure.

The effectiveness of embolization can only be appreciated clinically on the amount of blood transfusion necessary to maintain a stable hemodynamic condition. Embolization material for arteriovenous fistulae will be described in detail later.

Indications
Hemorrhage with pelvic trauma [1, 2, 3, 4, 7]

Retroperitoneal hemorrhage with fracture of the pelvic girdle is frequent especially after car accidents. The mortality rate in such cases is as high as 5 to 25%. In isolated fracture of the pelvic girdle, retroperitoneal hemor-

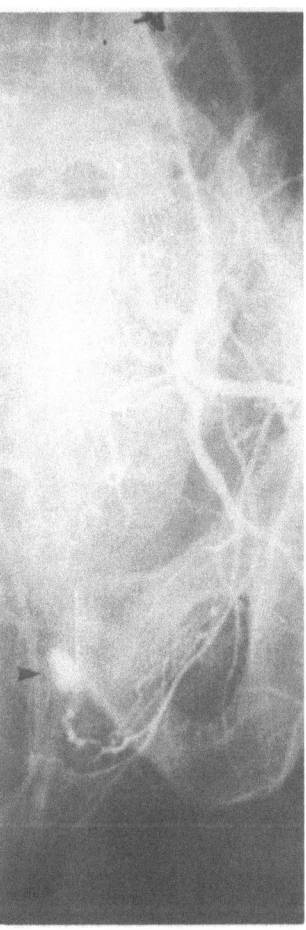 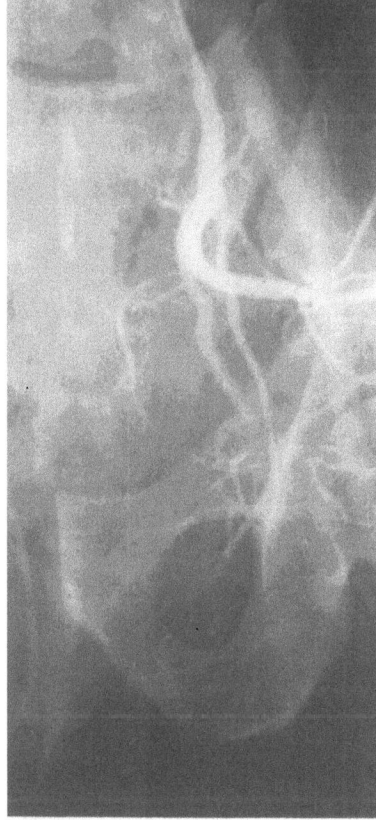

Fig. 1. Traffic accident. Spillage of contrast medium (arrow) from the left internal pudendal artery

Fig. 2. Following clot embolization

Fig. 1   Fig. 2

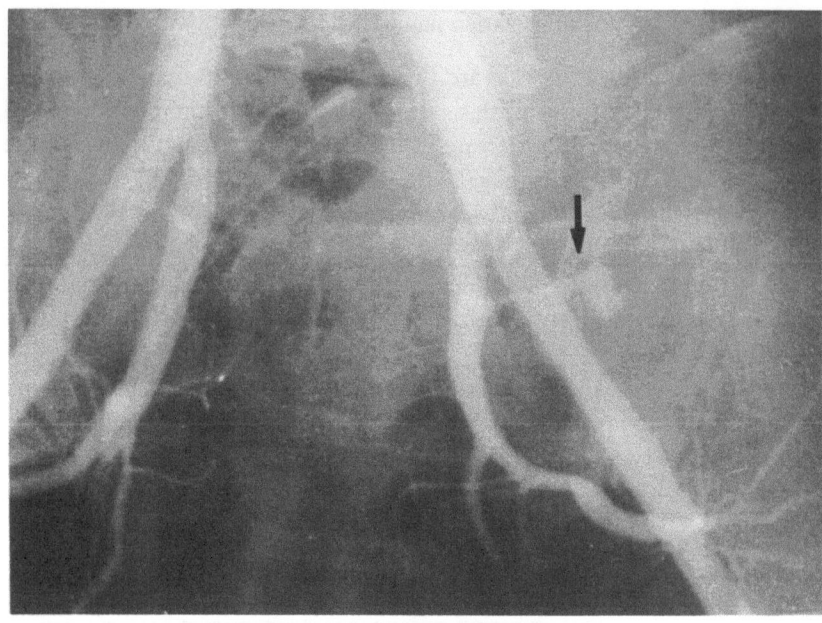

Fig. 3. Global arteriography after insertion of a catheter by the axillary route shows spillage from the left iliolumbar artery

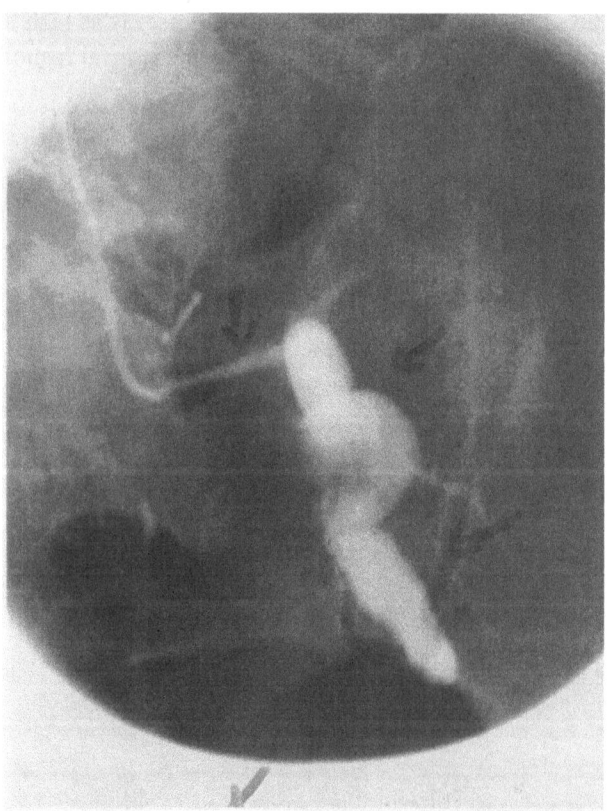

Fig. 4. Selective injection into the iliolumbar artery shows major arterial blood loss. Selective embolization of this artery allowed discontinuation of blood replacement

rhagic complications are responsible for two thirds of mortalities. Angiographic exploration allows localization of bleeding sites and implementation of hemostasis using the endovascular approach.

Abdominal aortography

The examination includes global abdominal and pelvic aortography. The catheter is placed at the level of the coeliac artery for mesenteric and renal

hepatosplenic evaluation when one of these organs is clinically suspect. In isolated fracture of the pelvic girdle, only global pelvic angiographic investigation should be performed.

Pelvic arteriography

The tip of the catheter is positioned at the aortic bifurcation. Iliac crests and the pelvic symphysis should be included in the X-ray field. A flow rate of 10 to 15 ml per second for 3 to 5 seconds with serial exposures extending

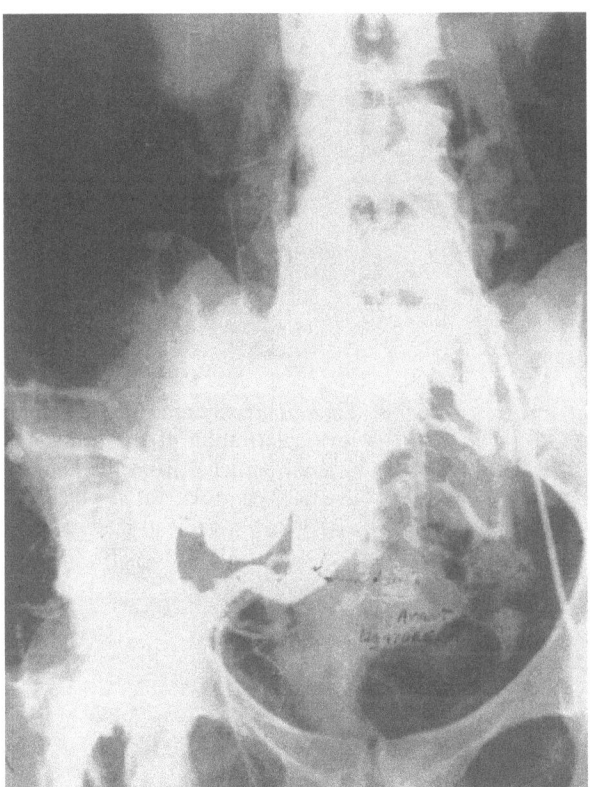
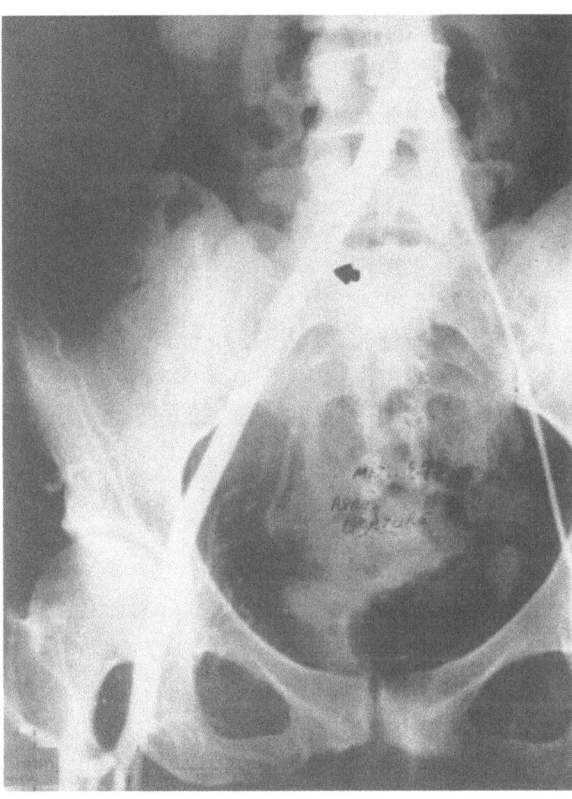

Fig. 5   Fig. 6

Fig. 5. Arteriovenous fistula following pelvic injury
Fig. 6. Postoperative appearance one month after surgical ligation of the origin of the hypogastric artery

over a 25-second period are programmed. Digital angiography is particularly suited for this procedure since smaller quantities of contrast product are sufficient for excellent visualization of small extravasations. Bleeding is visualized as a pool of contrast material or a false arterial aneurysm. Arterial interruptions or irregularities may correspond to a vascular wound masked by a temporary spasm. Absence of abnormalities on global opacification imperatively leads to selective injections of the two hypogastric arteries and the last lumbar arteries for depiction of small areas of extravasation. Once the vascular lesions has been identified, its arterial pedicle is selectively catheterized, and embolized. Embolization material selected should be resorbable (gelfoam or autologous clot) to allow secondary repermeabilization of the artery. If the arteriography does not reveal any bleeding, proximal embolization of the two hypogastric arteries by autologous clot is often effective.

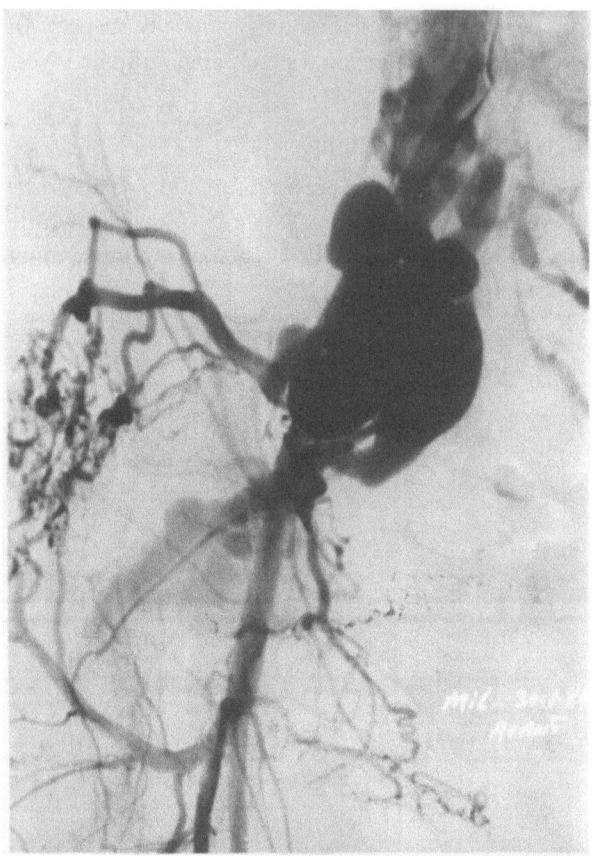

Fig. 7. Three months after ligation: revascularization of the fistula by arterial branches from the lumbar arteries, controlateral hypogastric artery and superficial iliac circumflex artery

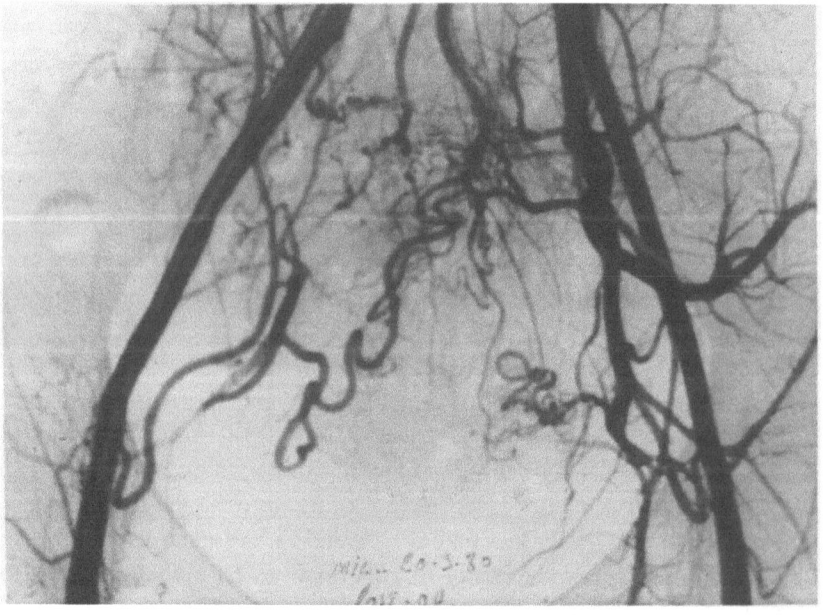

Fig. 8. Control two months after embolization with complete removal of the fistula

Hemorrhage of the lower limb associated with bone and
soft tissue trauma [1]

For femoral fractures, endovascular access can only be controlateral due to the presence of hematoma on the injured side. Arteriography determines the bleeding site (most frequently the circumflex arteries and branches of the deep femoral artery) and permits selective embolization at the same time.

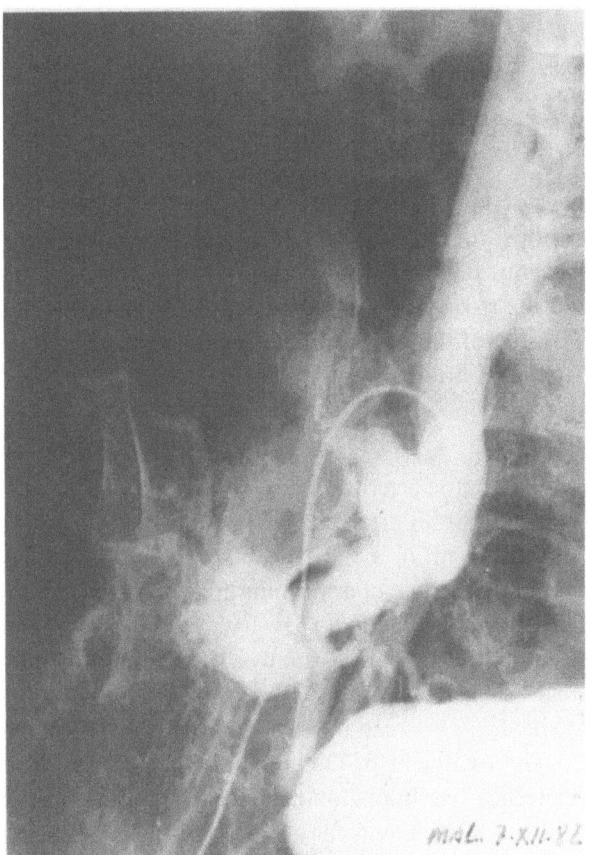

Fig. 9. Arteriovenous fistula discovered 10 years after an orthopedic surgical procedure on the hip

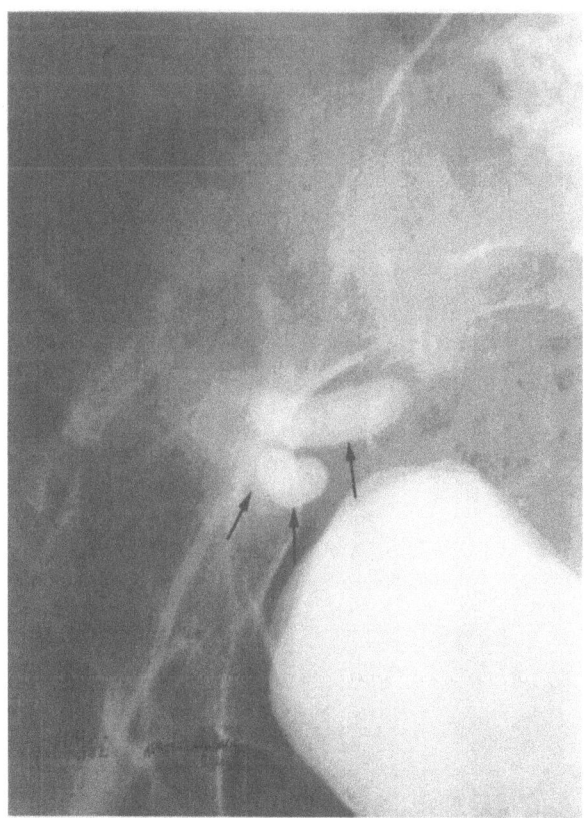

Fig. 10. After insertion of three balloons: two in the venous flow, one in the arterial flow (in the gluteal artery) (arrows)

For trauma of the leg, a homolateral approach with antegrade femoral puncture is preferable. This allows optimal visualization of arterial lesions and the distal arterial circulation and careful selective embolization of the injured muscular branches using resorbable material.

Postoperative hemorrhage [1, 7, 8]

Postoperative hemorrhage following hip replacement or after synthesis of a fracture of the pelvis is somewhat rare. Subsequent surgery is often difficult and the risk of sepsis is significantly increased. Arteriography allows identification of the origin of bleeding, as well as simultaneous embolotherapy. Circonflex arteries are frequently involved.

Post-traumatic arteriovenous fistulae [1, 7]

Post-traumatic arteriovenous fistulae often occur in the case of incomplete disruption of an artery and its satellite vein, or in the case of rupture of a false aneurysm in a vein. They are generally revealed by signs of venous pressure, presence of a murmur at auscultation and in some cases, general hemodynamic echoing. They have the arteriographic appearance of true malformations. There is usually a single shunt but multiple arterial branches are involved and converge towards this zone. These aspects can be explained by the delay between the constitution of the fistula and its revelation.

The treatment of these complex fistulae is difficult. To preclude any recurrence, the shunt should be totally occluded or excised surgically. Surgery is often difficult and may be complicated by hemorrhage. The isolated ligature of the hypogastric artery at its origin or of an artery located above the fistula is ineffective. Due to the persistence of the shunt, neighboring arteries are progressively involved and later treatment is increasingly complex.

We advocate therefore that endovascular therapy should be the first step treatment. Occlusion material should be of sufficient size by large, so

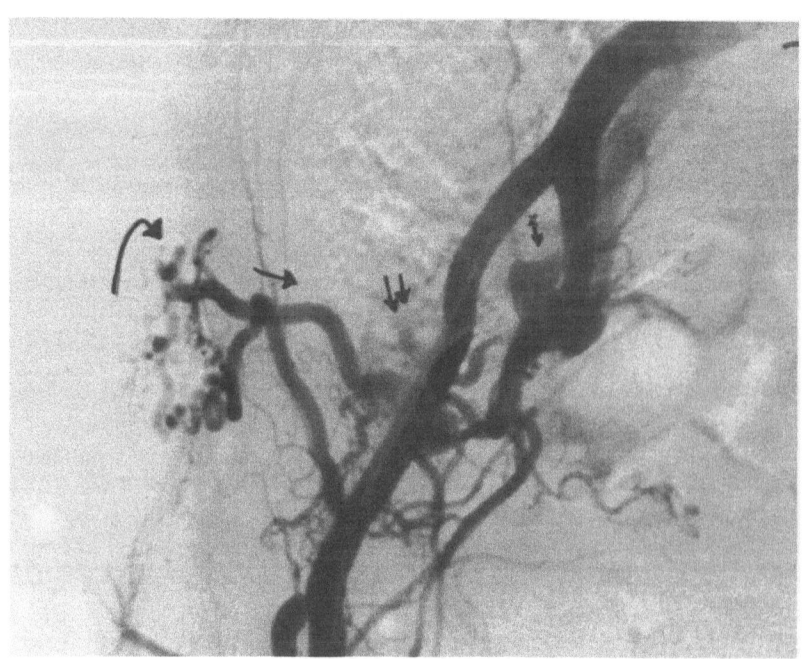

Fig. 11. Angiographic control demonstrated the significance of a small flow through the iliac circumflex artery into the distal segment of the gluteal artery with countercurrent opacification (arrow) (double arrow: site of shunt, aneurysm)

as not to pass into the general venous, then pulmonary circulation. It should be capable of completely blocking the site of the fistula. Finally, it should not be resorbable. At present, detachable balloon embolotherapy is the technique of choice in the treatment of fistulae. Using the arterial approach, this method allows accurate localization of the shunt by yielding remarkable morphologic study of the fistula thanks to slow-flow opacification. The balloon is inflated to the exact volume of the shunt, positioned on the arterial wall of the fistula, and then gently detached from the catheter. As an additional safety precaution, a second balloon can be detached in the artery under study to protect the first one. Other embolization materials can be used (Gianturco coils, Isobutylcyanoacrylate). The procedure can be facilitated by compression of the vena cava or the use of an additional balloon to reduce the blood flow in the iliac artery. Post-embolization surgery is not systematic and depends on the quality of the endovascular treatment. In any case, surgical conditions are considerably improved by a previous embolization. These complex pelvic fistuale should be treated by teams of skilled, well-equipped radiologists and surgeons. Poorly conducted initial treatment can lead to an almost insoluble situation.

Conclusion

Endovascular treatment has become an efficient alternative to surgery in bleeding associated with bone and soft tissue trauma. However, this technique has to be carried out by radiologists well-experienced in vascular interventional radiology.

References

1. Athanasoulis CA, Green BE, Pfister RL, Roberson GH (1982) Interventional radiology. Saunders, Philadelphia, pp 174–195
2. Ayella R, Dupriest R, Khanejas (1978) Transcatheter embolization of autologous clot in the management of bleeding associated with fractures of the pelvis. Surg Gynecol Obstet 117: 849–852
3. Baylis SM, Lansing EH, Glas UW (1962) Traumatic retroperitoneal hematoma. Am J Surg 103: 477–480
4. Fleming WH, Bowen JC (1973) Control of hemorrhage in pelvic crush injuries. J Trauma 13: 567–570
5. Gatterson FP, Morton KS (1973) The cause of death in fractures of the pelvis. J Traum 13: 849–853
6. Matalon T, Athanasoulis CA, Margoles MN (1979) Pelvic fractures with hemorrhage: efficacity of transcatheter embolization. Am J Roentgenol 133: 859–865
7. Paster S, van Houton F, Adams D (1974) Percutaneous balloon catheterization. A technique for the control of arterial hemorrhage caused by pelvic trauma. JAMA 230: 573–575
8. Reizine D, Merland JJ, Birkui P, Chauffour J (1984) Apports de l'angiographie et de l'embolisation dans le traitement des complications vasculaires des traumatismes pelviens. Ann Radiol 27: 457–463
9. Stock JR, Athanasoulis CA, Harris WH (1980) Transcatheter embolization for the control of wound hemorrhage following hip surgery. J Bone Joint Surg [Am] 62: 1000–1006

Indications of embolization in bone and soft tissue tumors

D. Reizine, Eva Assouline, Françoise Gelbert, and J.-J. Merland

Department of Neuroradiology and Therapeutic Angiography, Hôpital Lariboisière,
Paris, France

Despite recent developments in noninvasive imaging modalities such as computed tomography or magnetic resonance, angiography maintains a firm hold in pretherapeutic work-up and treatment of certain bone tumors. Angiography is vital for determining arterial structure of certain areas especially in the spine. Furthermore, it allows embolization and intra-arterial chemotherapy.

Angiography

Hyperselective angiography is a key factor in pretherapeutic work-up of vertebral tumors. It determines the level of origin of anterior radiculomedullary arteries. It should be remembered that the anterior spinal artery feeds the anterior two thirds of the spinal cord and is supplied by only a few segmental arteries (3 to 5 for the whole spinal cord). The ligature of a medullary artery can provoke acute irreversible medullary ischemia. Selective medullary arteriography must be performed to determine the level of origin of the radiculo-medullary arteries and their relation with the vertebral tumor. The results of the arteriographic examination will thus provide guidance in the surgical approach. Emergence of the Adamkiewicz artery at the level of the tumor was considered a contraindication to radical excision (Fig. 1).

More recently, it has been proposed to first determine the functional value of a radiculo-medullary artery by clamping of the artery followed by investigation of the subadjacent artery for checking continuity of the anterior spinal axis.

Embolization

Embolization consists of using different materials to occlude the arterial pedicles or the capillary tumor bed.

Hemostatic embolization

Embolization has considerable importance in hemostasis in postoperative as well as tumoral bleeding. Angiography pinpoints the area of bleeding which is visualized as a pool of contrast material. Hemostasis is then ensured using nonresorbable particle emboli. Hemostasis can be guaranteed even in extremely dramatic cases such as intravascular coagulopathy.

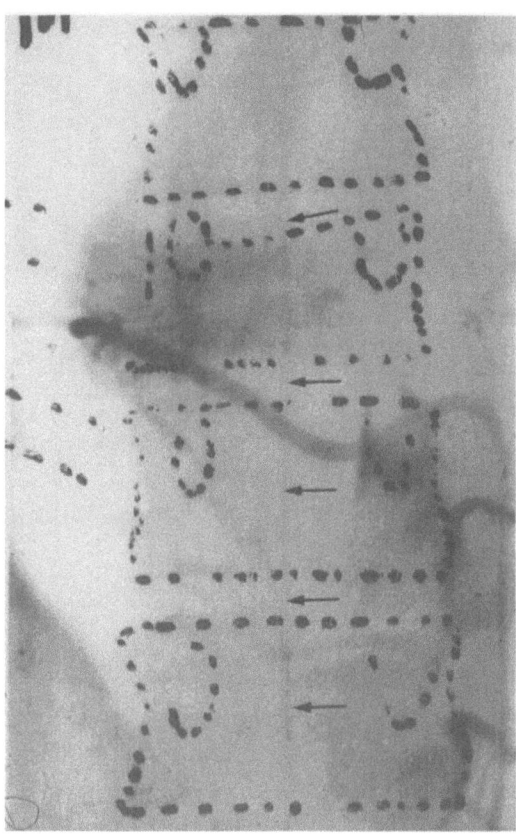

Fig. 1. Plasmocytoma of T 8. Adam-kiewicz artery (arrow) originating at the same level: this case is a classical contraindication for radical excision

Preoperative embolization

Preoperative embolization (Fig. 2) of bone tumors is the other major indication of hemostatic embolization. It is recommended for all hypervascular tumors especially in regions such as the spine and girdles where peroperative hemostasis is difficult to obtain. The main tumors included in this category are primary malignant tumors (osteosarcoma), metastasis of thyroid or renal carcinoma and certain benign tumors such as osteoblastoma or aneurysmal cyst. To be really efficient, embolization needs to be complete. All feeding pedicles should be catheterized and embolized. Inactive, non-resorbable particles should be used. Gelfoam fragments can be used provided that surgery is performed in the 3 to 5 days following embolization. Embolization is continued until complete disappearance of blush is observed while reflux of particles is carefully avoided. The goal of preoperative embolization is to reduce blood loss while ensuring an easier approach to the lesion and a more radical excision. Bowers reports that the average blood loss in embolized patients operated for metastatic cancer of the kidney was 500 ml as compared to a 5-liter loss in patients who had not undergone embolization or in whom embolization was incomplete [2].

Embolization aimed at reducing the tumorous mass [1–3] (Fig. 3)

When very small particles (50 to 100 microns) or liquid emboli are used, embolization provokes tumor necrosis. Catheterization should be particularly selective so as to preclude necrosis of neighboring tissues. Obtaining tumor necrosis by embolization has several advantages: whether associated

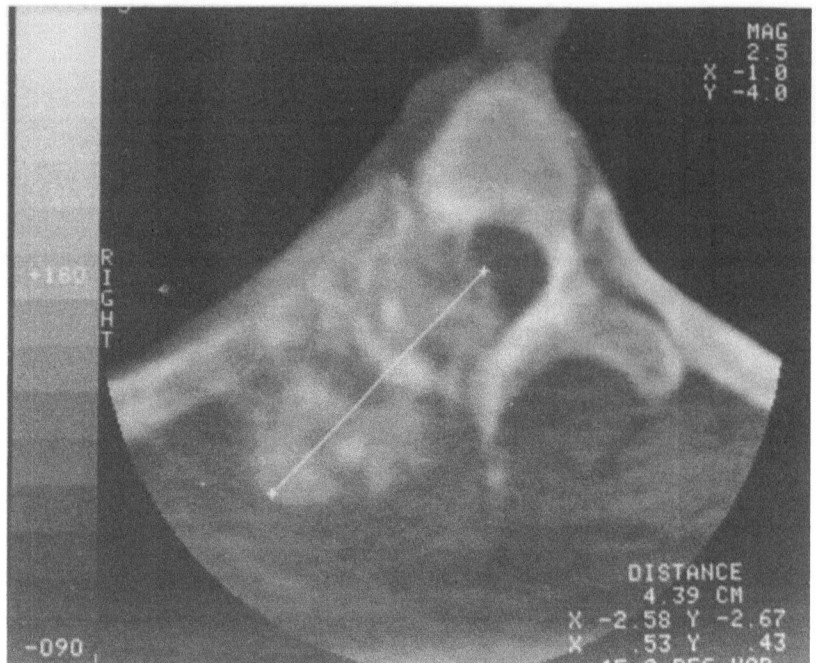

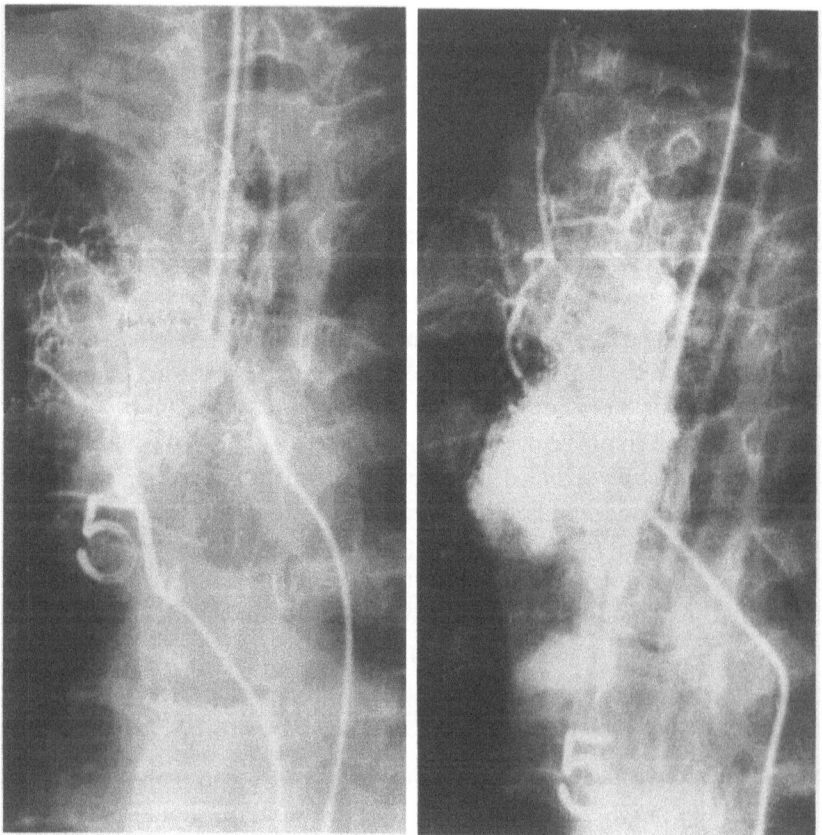

Fig. 2. Preoperative embolization of a vertebral osteosarcoma of T 5. **a** CT showing the tumor extension. **b** Selective angiography before embolization. **c** Selective angiography after embolization. Radical surgery was possible with minimal blood loss

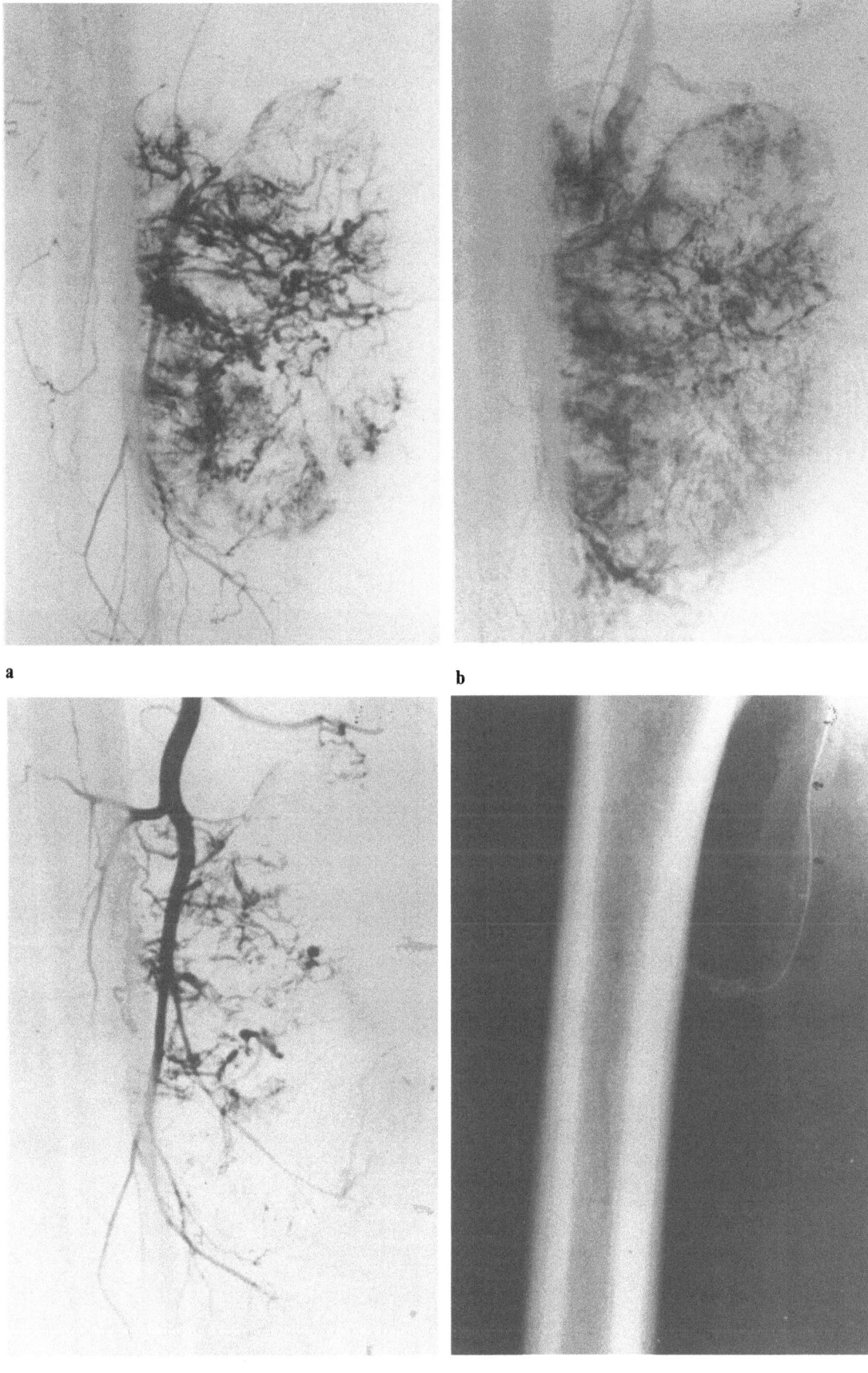

a

b

c

d

Fig. 3. Embolization combined with intraarterial chemotherapy. Rhabdomyosarcoma of the left tigh: conservative surgery was not possible firstly. **a, b** Angiography before embolization. **c** Angiography after embolization. **d** A microcatheter was left for continuous chemotherapy (21 days). Surgical conservative treatment was then possible

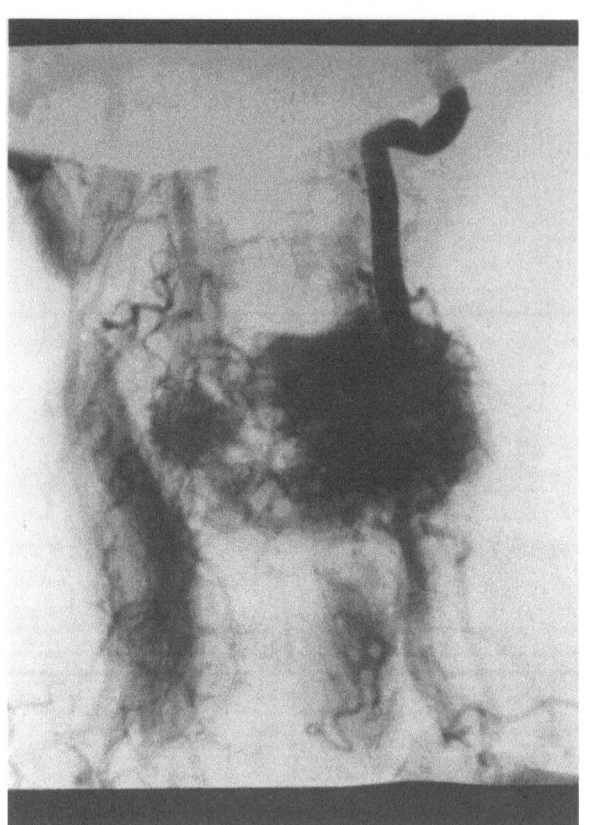

a

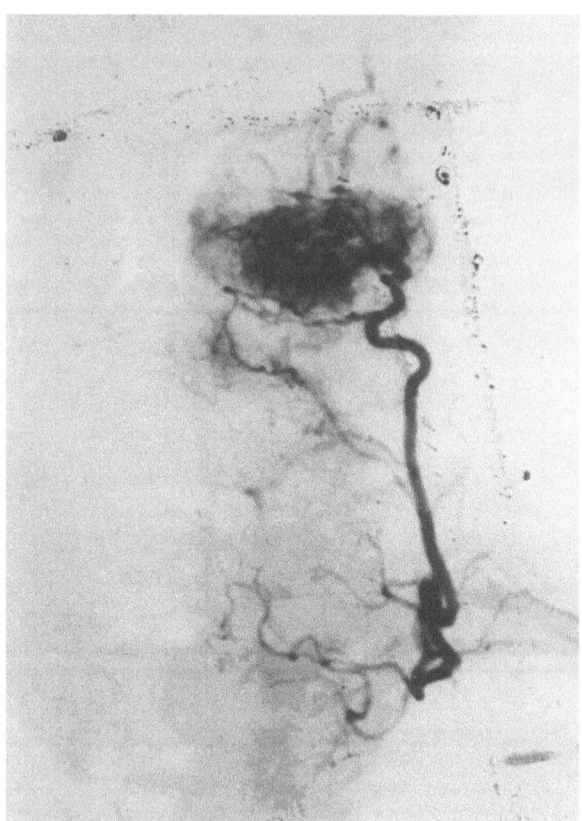

b

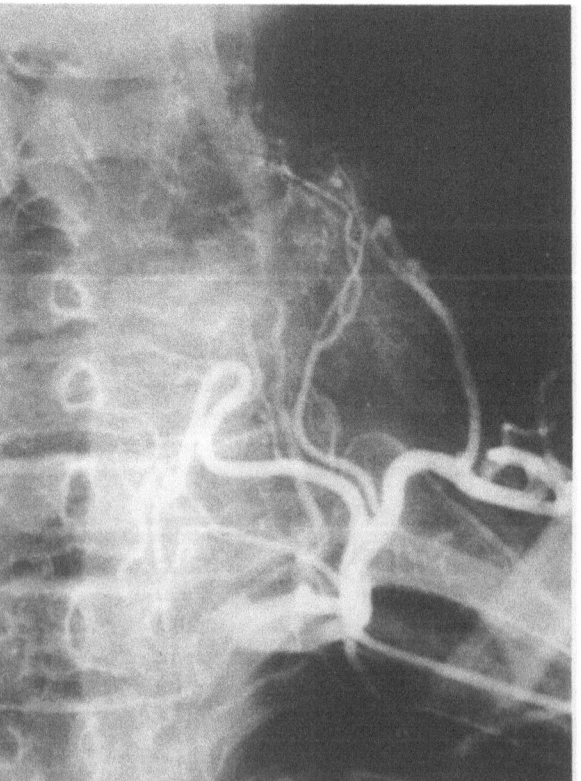

c

d

Fig. 4. Antalgic embolization of a cervical kidney metastasis. Tetraplegia and acute pain. **a** Global angiography shows the hypervascularization. All pedicles were successively embolized: **b** Cervical ascending artery. **c, d** Thyro-bicervical artery. **e, f** And a pedicle originating from the right vertebral artery (before and after embolization). Embolization allowed relief of pain

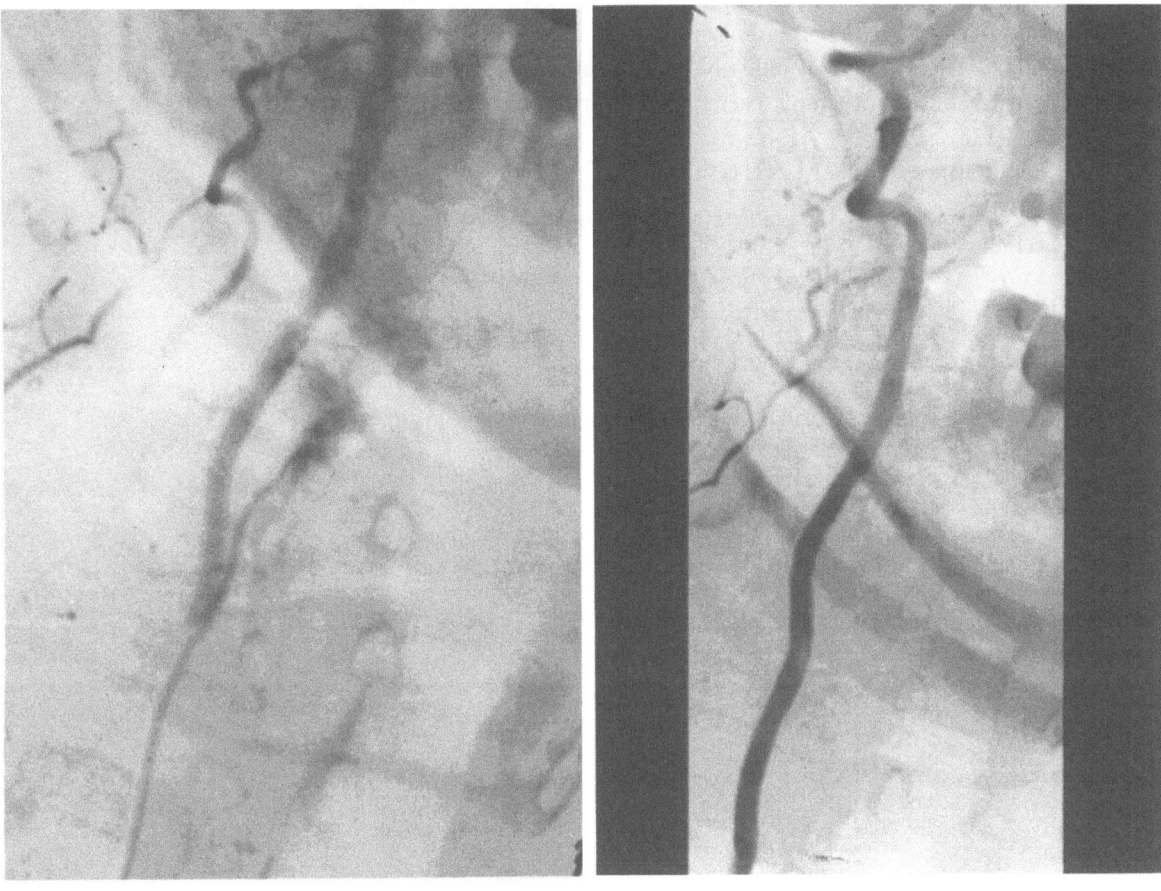

e   f

or not with chemotherapy, embolization can reduce the volume of a tumor judged as inoperable thus allowing conservative excision surgery (Fig. 3). Certain metastatic tumors are accompanied by hormonal secretion. This is the case in spinal metastases of malignant pheochromotytoma [5]. These metastases are accompanied by medically uncontrolled hypertension. Embolization using a necrosing agent (absolute ethanol) allows control of arterial hypertension. In our experience, patients had normal blood pressure readings without medical treatment six months after the embolization. The control of tumor hormonal secretion by secreting bone metastases is similar to that observed after embolization of hepatic carcinoid metastatic tumors [7]. Embolotherapy has also proved effective in the treatment of multiple metastases of thyroid cavernomas. In these highly vascular tumors, surgery is difficult and treatment by radioactive iodine is rarely effective when the tumoral mass is of considerable size. A possible course of therapy associates a preoperative embolization in case of surgically accessible tumoral masses, necrosing embolization for residual tumorous masses and finally, a course of radioactive iodine treatment.

Antalgic embolization (Fig. 4)

Embolization alone, or in addition to radiotherapy also has antalgic effects [1]. These effects are all the more pronounced in cases where the bone tumor is vascularized. The antalgic effect is probably secondary to diminished hypervascularization. It is variable in its intensity and duration from one case to another. In certain patients, embolization can be repeated to produce an antalgic effect. This antalgic or palliative embolization is recommended when other treatment (radiotherapy, chemotherapy) has failed.

References

1. Chuang VP, Wallace S (1981) Transcatheter management of neoplasms. JAMA 245: 1151–1152
2. Chuang VP, Wallace S, Swanson DA (1979) Arterial occlusion in the management of pain from metastatic renal cell carcinoma. Radiology 133: 611–617
3. Bowers T, Murray J, Charnsangavej C (1982) Bone metastasis from renal cell carcinoma: the preoperative use of transcatheter arterial occlusion. J Bone Joint Surg 64: 749–754
4. Soo CS, Chuang VP, Wallace S (1980) Interventional angiography in the treatment of metastasis. Radiology 135: 295–300
5. Soo CS, Wallace S, Chuang VP (1982) Lumbar artery embolization in cancer patients. Radiology 145: 655–660
6. Hortow SA, Hrabousky E, Klingberg WG (1983) Therapeutic embolization of a hyperfunctioning pheochromocytoma. AJR 140: 987–989
7. Mitty HA, Warner RRP, Newman LH (1985) Control of carcinoid syndrome with hepatic artery embolization. Radiology 155: 623–625

Radiologic Management of Vertebral hemangiomas

J.-D. Laredo[1], D. Reizine[2], J.-J. Merland[2], and M. Bard[1]

Departments of [1] Bone and Joint Radiology and of [2] Neuroradiology and
Therapeutic Angiography Hôpital Lariboisière, Paris, France

Vertebral hemangiomas (VH) include a range of clinical entities from the frequent, incidentally noted, asymptomatic lesion to the rare expansile tumor that may compress the spinal cord.

Asymptomatic vertebral hemangiomas (AVH)

AVH are very frequent. AVHs are seen at autopsy in 10–11% of cases [8]. Only a small percentage of these AVHs are detected on plain radiographs. Incidental finding of AVH is more frequent at CT study. In most cases, AVH are located at the lumbar spine and thoraco-lumbar junction. They usually have the characteristic vertically striated radiographic appearance described by Perman [12]. In addition, AVHs have the radiologic features of inactive (regular striations and normal cortex), localized (incomplete involvement of the vertebral body, normal neural arch and soft tissue) angiomatous dystrophy. Selective arteriography, when performed, shows few or no abnormal vessels. These radiologic features are different from those of VH responsible for cord compression [8] (Figs. 1–3).

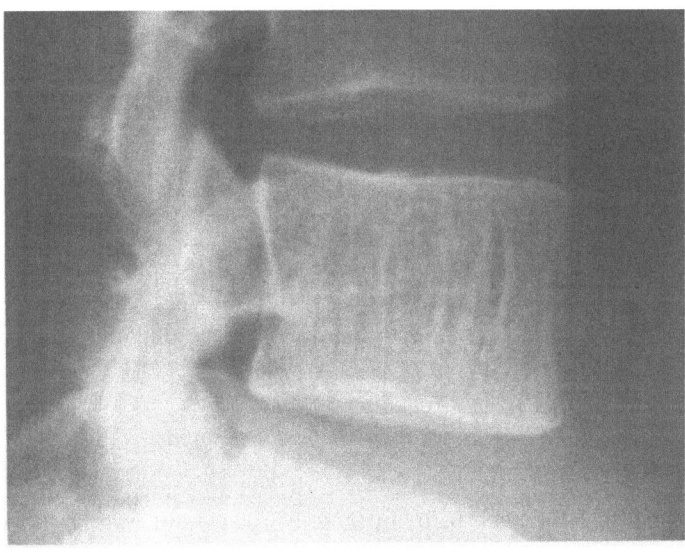

Fig. 1. Asymptomatic VH of L4 with regular sticated appearance. Note normal cortex and sparing of the pedicles

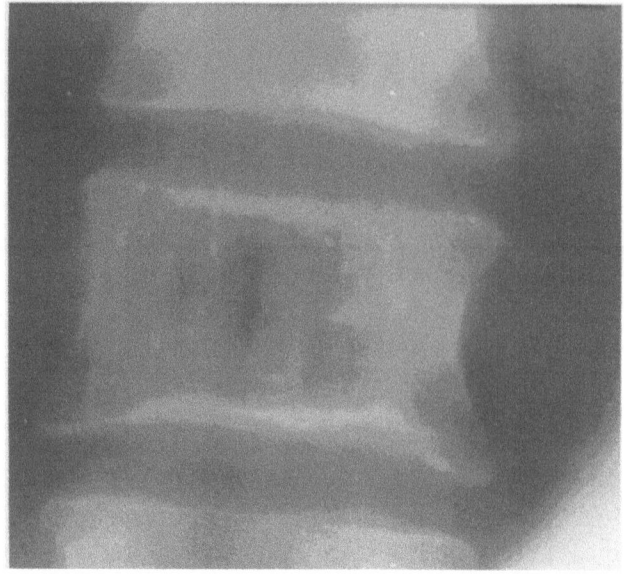

Fig. 2. AP tomogram of an asymptomatic VH of L 2. Note incomplete involvement of the vertebral body, normal cortex and regular striation

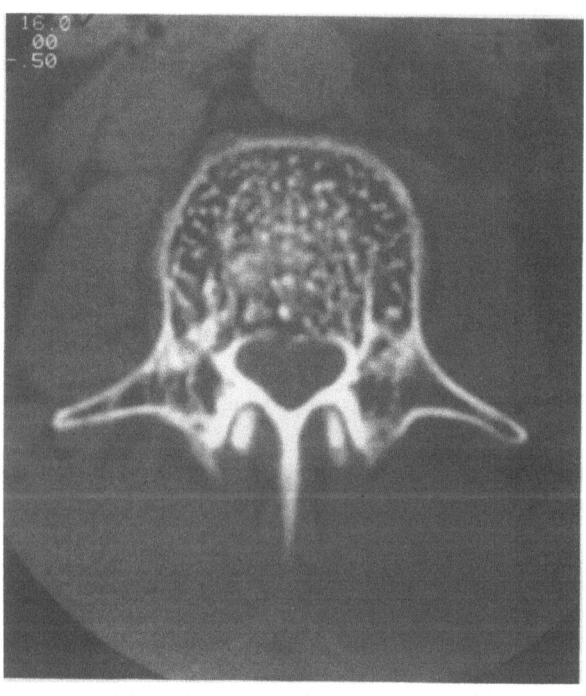

Fig. 3. CT scan of an asymptomatic VH of L 1: note regular trabeculation, sparing of the neural arch and normal soft tissue. As usual in asymptomatic VH the cortex appears almost normal. The minute cortical discontinuity of the mid-posterior vertebral body corresponds to normal venous drainage of the vertebral body

Compressive vertebral hemangiomas (CVH)

CVH occur more rarely than AVHs. The 14 cases of CVH that we previously reported were seen over a 20-year period through selective referrals [8]. The role of vascular radiologic procedures in the diagnosis and treatment of CVH is discussed below.

Symptomatic vertebral hemangiomas (SVH)

A radiographic image suggestive of VH is sometimes found in patients consulting for back pain. In these cases, careful radiologic evaluation as detailed below, may help to determine appropriate management.

CVH

Rare in the first decade of life, CVH occurs at any age, with a peak frequency in the young adult. CVH preferentially occurs in the thoracic spine. More specifically 72% to 93% [8] of cases are located from T-3 to T-9, that is, on seven thoracic vertebrae.

CVH outside the thoracic spine are rare. Only a few cases of cervical CVH are cited in the literature, and these are often cases of regional skeletal angiomatosis [8]. The lumbar and sacral spine are infrequently involved by CVH. CVH must be distinguished from extradural, foraminal, or paravertebral arteriovenous malformations, in which involvement of bone is secondary. In most cases, CVHs are responsible for slow, progressive compression of the spinal cord. Months or even years of back or nerve-root pain may precede spinal cord compression. Frequently, pain is absent, and sensory or motor deficiency, often of insidious onset, is the presenting complaint. As a result, the patient is frequently seen at a late stage of the condition. A unique form of onset is the development of paraplegia in women in their third trimester of pregnancy. In these cases, onset of impairment is often sudden and complete. The spontaneous evolution of neurologic signs is most often unfavorable but unpredictable, and partial or even complete remissions of varying duration have been reported [8]. Palpation of local tumefaction is rare and more suggestive of an adjacent angiomatosis in the soft tissue.

The radiographic appearance of CVH is usually characteristic of VH with a vertical striated appearance. However, the radiographic appearance of CVH can be misleading. In some cases, demineralization of the vertebral body and nonvisualization of the pedicles suggest metastasis [4]. In others, the dominant radiographic feature may be an expanded cortex [10] or a paravertebral soft-tissue swelling [10]. Even in those lesions with radiologic characteristics of VH in general, CVH shows additional features of active (irregular trabeculae with lytic zones, poorly defined and expanded cortex) and extensive (involvement of the entire vertebral body, neural arch and soft-tissue extensions) vascular tumors (Fig. 4 a–c) [8].

Selective arteriography of the spinal cord, first developed by Djindjian et al. [3], is an obligatory procedure in cases of CVH (Fig. 4 d–g). The arteriographic appearance is usually characteristic: dilatation of arterioles of the vertebral body, multiple blood pools in the capillary phase, and, finally, intense opacification extending beyond the normal hemivertebral territory throughout the entire vertebral body. Opacification beyond normal cortical limits corresponds to extension of the angioma into the paravertebral soft tissues and the spinal canal (best seen on lateral radiographic views). The absence of early venous draining distinguishes CVH from high-flow arteriovenous malformations. The origin of the anterior spinal artery supplying the cord is visualized and can thus be avoided during embolization and/or the surgical approach. Embolization of one or more feeding arteries greatly reduces bleeding and facilitates surgery [1, 9]. In the absence of preoperative embolization, bleeding during surgery may be extremely profuse. Surgical findings confirm that compression of the spinal cord is more often due to extradural extension of the angioma than to bone expansion.

The treatment of VH accompanied by neurologic signs has been much debated. Decompressive laminectomy is the usual treatment. Radiation therapy is sometimes recommended as an adjunct to decompressive laminectomy.

214   J.-D. Laredo et al.

Others have proposed radiation therapy as the sole treatment for CVH [8] because of the high rate of surgical morbidity and failures. Current treatment of CVH associates preoperative embolization and decompressive laminectomy [1, 2, 7, 9]. Inert, non-resorbable particles are generally used. All feedings pedicles must be embolized with respect to the anterior spinal artery. In certain cases, embolization alone allowed complete and durable neurologic recovery. Hekster [6, 7] reported a case report concerning a patient with paraplegia due to spinal compression by a vertebral hemangioma which was treated in 1972 by embolization and radiotherapy [7]. Fifteen years later the patient was still in excellent condition and without motor impairment [6].

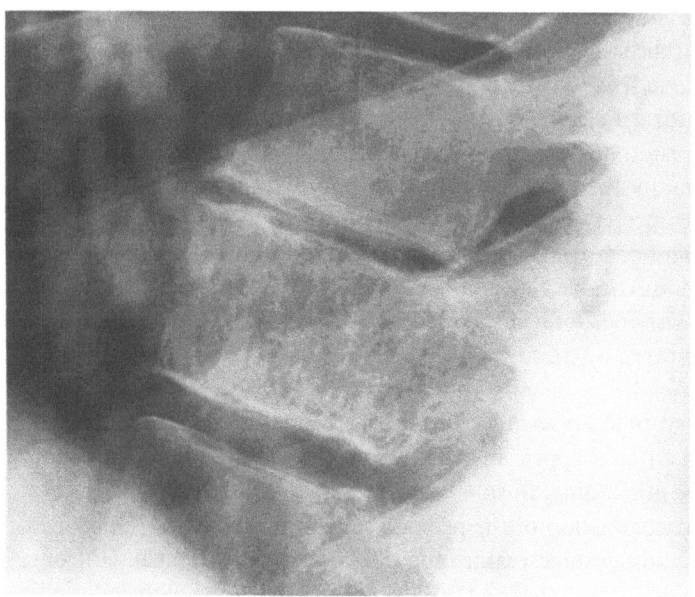

a

Fig. 4a, b. Plain films of a compressive vertebral hemangioma. **c** CT of the same patient: irregular trabeculation, extension in the paravertebral soft tissue. **d, e** Angiography shows hypervascularisation. **f** Lateral view of the left 10th intercostal artery. **g** Anterior spinal artery arises from the left 9th intercostal artery: embolization can be performed

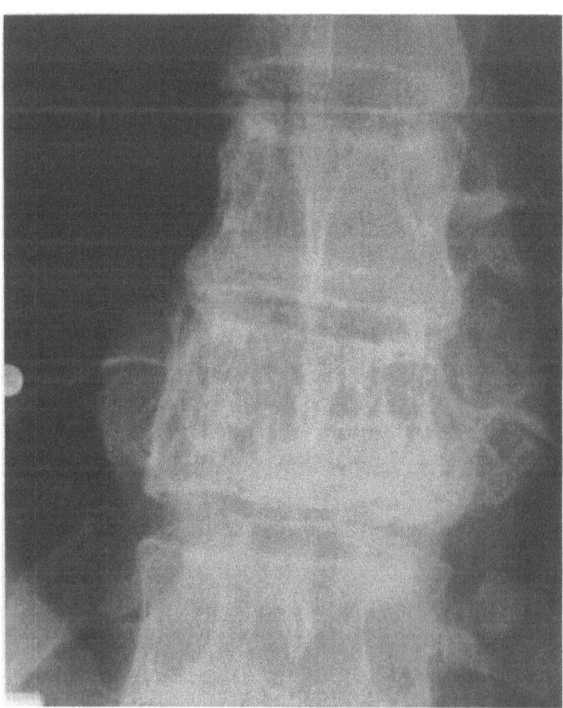

b

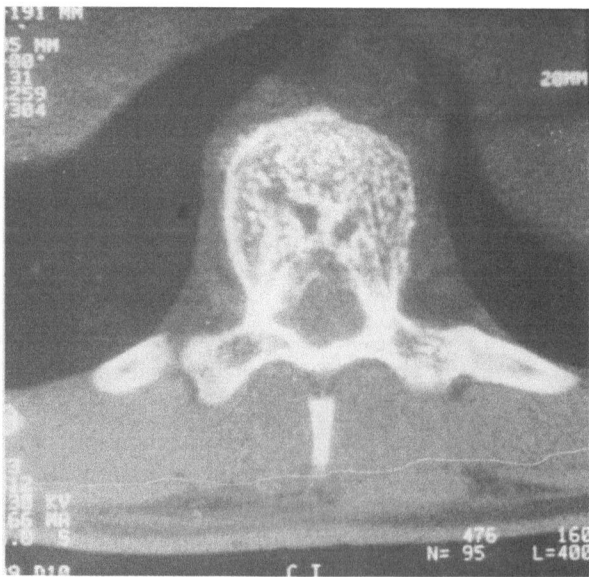

c

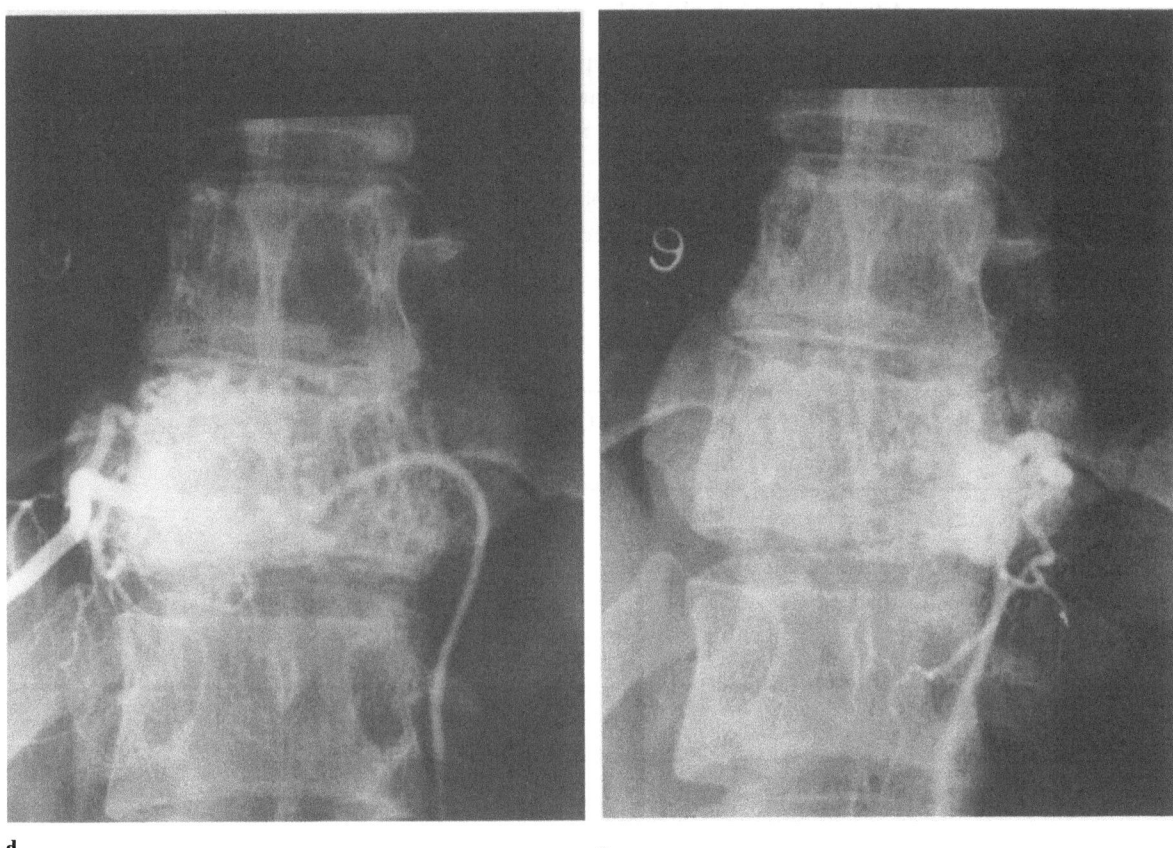

d

e

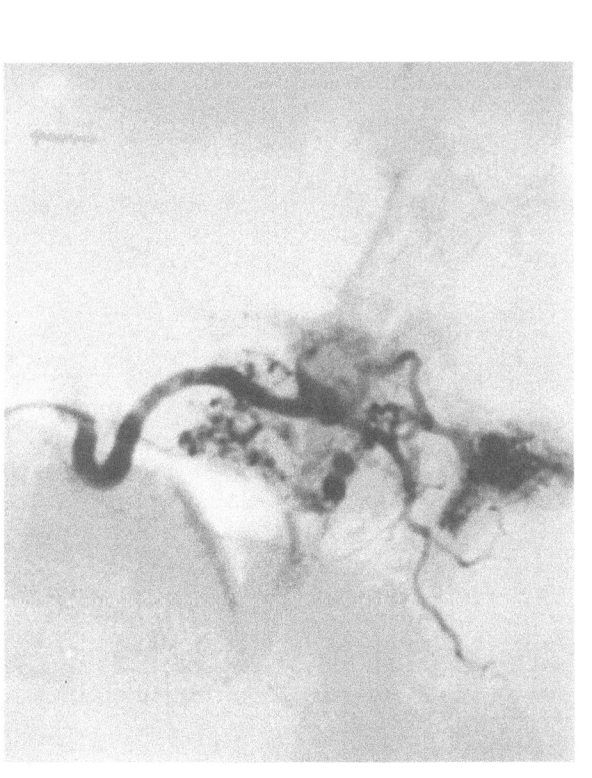

f

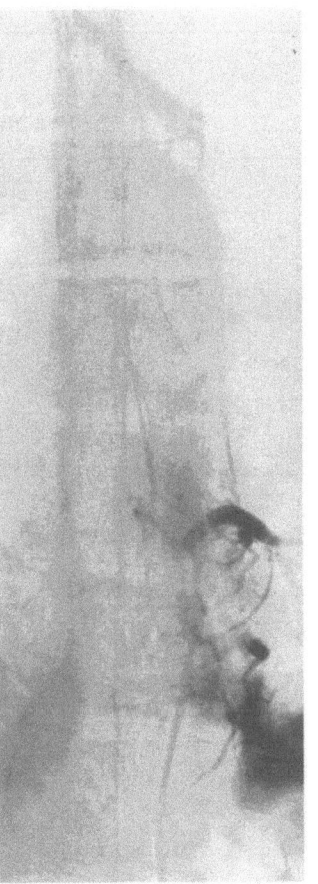

g

Clinical approach of SVH

Patients with VH in whom the only complaint is local pain (with no objective evidence of spinal cord compression) must be distinguished from those with asymptomatic and compressive VH. In these cases it is difficult to select adaquate management.

Since CVHs and AVHs have different radiographic and arteriographic features (Table 1), it has been stated that, in individual cases, the radiographic

Table 1. Radiologic signs of CVH versus AVH [8]

Signs	CVH (n = 14)	AVH (n = 11)
T-3 to T-9 location	13	1
Involvement of entire vertebral body	13	4
Involvement of neural arch (pedicles)	13	3
Irregular trabeculation	10	0
Expanded and poorly defined cortex	13	1
Soft tissue swelling	10	0

Table 2. Management of VH using a scoring system (plain films and CT)

Signs	Score Value	
	Sign present	Sign absent
1 T-3 to T-9 location	1	0
2 Involvement of entire vertrebral body	1	0
3 Involvement of neural arch (pedicles)	1	0
4 Irregular trabeculation (lytic zone)	1	0
5 Expanded and poorly defined cortex	1	0
6 Soft tissue swelling	1	0
Final score	6	0

* Score 0–2 with or without local pain:

  → Probable inactive VH

  ⇨ Routine clinical and radiological follow-up

* Score 3 and/or nerve root pain:

  → Possible aggressive VH

  ⇨ Selective arteriography in specialized center

appearance may help to evaluate the aggressiveness of a VH. In a previous paper [8], we postulated that those SVH with radiologic features of CVH may correspond to aggressive lesions while those those with the features of AVH are likely to be inactive. We have proposed a scoring system based on 6 radiologic criteria evaluated on plain films and CT (Table 2). This scoring system helps to place an individual lesion in the group of aggressive or inactive VH. For each criteria, one point in given if that feature is present and no points if it is absent. A score ranging from 0 to 6 is thus obtained for each lesion. With that scoring system, CVH have a score of 4 or more while AVH have a score of 2 or less [8]. Therefore, in the evaluation of an individual case of SVH, a score from 0 to 2 suggests inactive, localized, angiomatous dystrophy, necessitating only routine clinical and radiologic follow-up (Table 2). By contrast, a score of 3 or more and/or the presence of nerve-root pain in the territory of the lesion seen on the radiographs raises the possibility of an active VH and requires the use of selective arteriography followed by embolization of the feeding arteries if hypervascularization is noted. The score is also useful as an element of comparison for follow-up radiologic examinations.

References

1. Benati A, da Pian R, Mazza C et al (1974) Preoperative embolization of a vertebral haemangioma compressing the spinal cord. Neuroradiology 7: 181–183
2. Bouchez B, Gozet G, Lecoutour X, Kassiotis P, Arnott G, Delecour M (1984) Compression médullaire par angiome vertébral au cours de la grossesse: un cas traité par embolisation. Presse Med 13: 1696–1697
3. Djindjian R, Cophignon J, Merland J-J, Theron J, Houdard R (1973) Embolization in vertebro-medullary pathology. Neuroradiology 6: 132–142
4. Gaston A, Nguyen JP, Djindjian M et al (1985) Vertebral haemangioma: CT and arteriographic features in three cases. J Neuroradiol 12: 21–33
5. Gross CE, Hodge CHJ, Binet EF, Kricheff H (1976) Relief of spinal block during embolization of a vertebral body hemangioma. J Neurosurg 45: 327–330
6. Hekster REM, Endtz LJ (1987) Spinal cord compression caused by vertebral hemangioma relieved by percutaneous catheter embolization 15 years later. Neuroradiology 29: 101
7. Hekster REM, Luyendijk N, Tan TI (1972) Spinal cord compression caused by vertebra hemangioma relieved by percutaneous catheter embolization. Neuroradiology 3: 160–164
8. Laredo JD, Reizine D, Bard M, Merland J-J (1987) Vertebral hemangioma, radiologic evaluation. Radiology 161: 186–189
9. Lepoire J, Montaut J, Picard L, Heppner H, Masingue M, Arnould G (1973) Embolisation préalable à l'exérèse d'un hémangiome du rachis dorsal. Neurochirurgie 19: 173–181
10. McAllister VL, Kendall BE, Bull JWD (1975) Symptomatic vertebral haemangiomas. Brain 98: 71–80
11. Mohan V, Gupta SK, Tuli SM, Sanyal B (1980) Symptomatic vertebral haemangiomas. Clin Radiol 31: 575–579
12. Perman F (1926) On hemangiomata in the spinal column. Acta Chir Scand 61: 91–105

Intra-arterial treatment of primary bone tumors

S. Wallace[1], C. H. Carrasco[1], C. Charnsangavej[1], W. Richli[1], N. Jaffe[2],
J. Murray[3], A. Ayala[4], A. K. Raymond[4], S. P. Chawla[5],
and R. S. Benjamin[5]

Departments of [1] Diagnostic Radiology, of [2] Pediatrics, of [3] General Surgery,
of [4] Anatomical Pathology, and of [5] Medical Oncology, The University of Texas
System Cancer Center M. D. Anderson Hospital and Tumor Institute, Houston,
Texas, U.S.A.

Transcatheter intra-arterial infusion of chemotherapeutic agents for osteosarcomas and embolization of inoperable giant cell tumors of the axial
skeleton have become essential components in patient management [1–13].

Giant cell tumor of bone (GCT)

Giant cell tumor of bone is a destructive neoplasm that apparently arises
from non-bone-forming, supportive connective tissue of the marrow. The
tumor is composed of a vascularized network of spindle-shaped or oval
stromal cells regularly and rather heavily insterspersed with multinuclear
cells [14]. GCT are relatively rare and occur more often in women. Of the
neoplasma, 85% occur in patients beyond the age of 19 years with a peak
incidence in the third decade [14]. The tumor is generally benign, but rarely
may give rise to distant metastases. Its biologic behavior cannot be predicted
on the basis of the histologic appearance.

Management

The accepted treatment for giant cell tumor is removal by curettage with
subsequent bone grafting. However, this therapy is generally associated with
a 50% incidence of local recurrence. Total excision of the tumor is also
performed when small bones are involved. Radiation therapy is not as well
accepted because these tumors are relatively radioresistant and because of
the potential risk of inducing malignant transformation, 10% [14, 15].

Giant cell tumors involving the axial skeleton are rare and occur more
frequently in the sacrum. Surgical excision of tumors located in the spine
is often incomplete; therefore alternative modes of therapy including irradiation, chemotherapy, and embolization have been employed.

Technical considerations

Angiography is performed to determine the vascular supply to the neoplasm;
GCT is usually hypervascular. For tumors located in the thoracolumbar
area, angiography of the intercostal or lumbar arteries supplying the segment

above and below the lesion, in addition to the arteries at the level of the neoplasm, is performed (Fig. 1). In patients who have undergone previous surgery, ligation of the pertinent vessels may have been done and it will be necessary to identify the collateral blood supply. The origin of the anterior spinal artery (anywhere from T_8 to L_2 usually on the left) must be delineated, the potential occlusion of which contraindicates embolization. To decrease the risk to the spinal cord, at least non-neurotoxic contrast medium should be employed. In cases of sacral GCT both internal iliac arteries, the middle sacral artery, both 4th and 5th (when present) lumbar arteries and, at times, the inferior mesenteric artery must be studied.

The purpose of embolization is to create ischemia of the neoplasm, to arrest tumor growth, or hopefully to induce regression [6, 7, 12, 16]. Interruption of a vessel at its origin (central occlusion) is similar to surgical ligation. Collateral circulation is available immediately and abundantly, the more central the occlusion. The closer the occlusion is to the tumor (peripheral embolization), the smaller is the opportunity for collateral circulation. The size and nature of the particles will determine the site of vascular interruption. At M. D. Anderson Hospital and Tumor Institute, polyvinyl alcohol foam particles (Ivalon, Unipoint Laboratories, High Point, NC) 150–500 µm, and surgical gelatin (Gelfoam, Upjohn, Kalamazoo, MI) 1–3 mm are most commonly used for peripheral embolization, whereas stainless steel coils (Cook Inc., Bloomington, IN) or Gelfoam segments, $3 \times 3 \times 20$ mm, are used for central occlusion. Dehydrated absolute ethanol, a thrombosing or sclerosing agent, theoretically diffuses throughout the

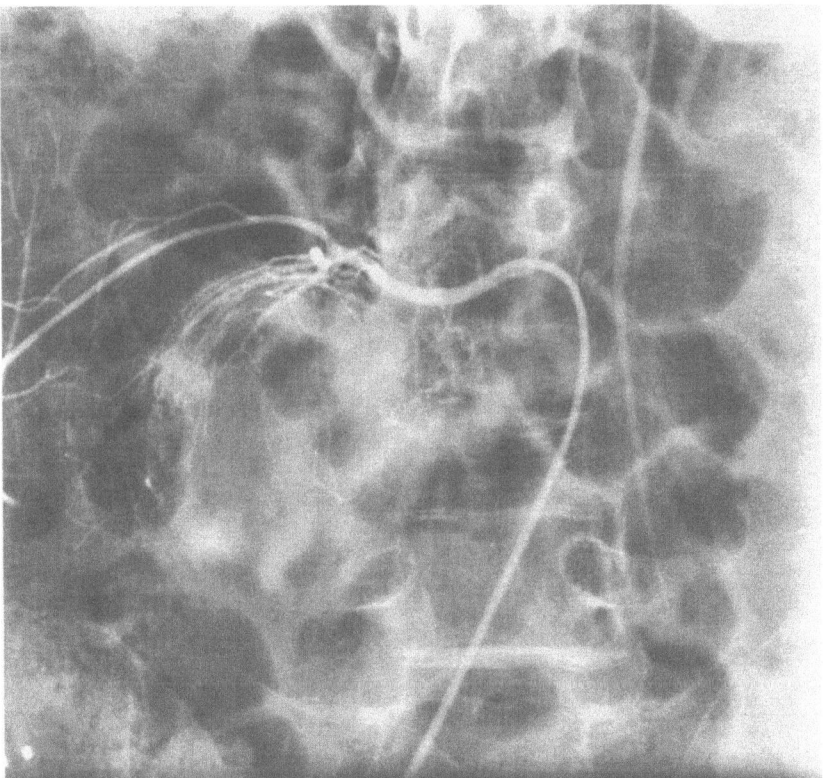

Fig. 1 **a**

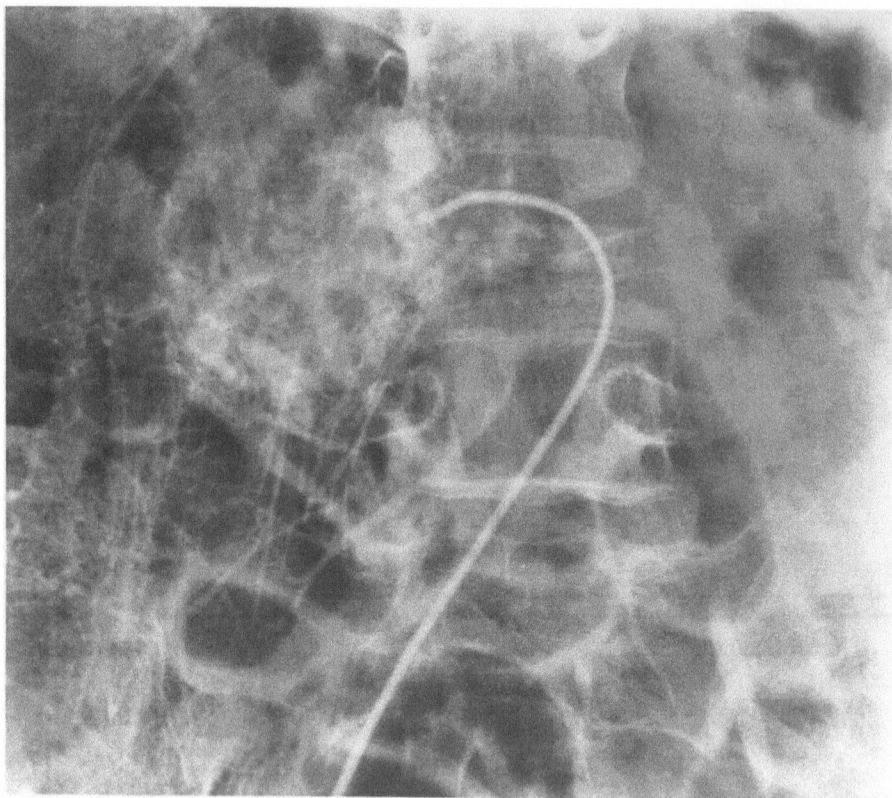

b

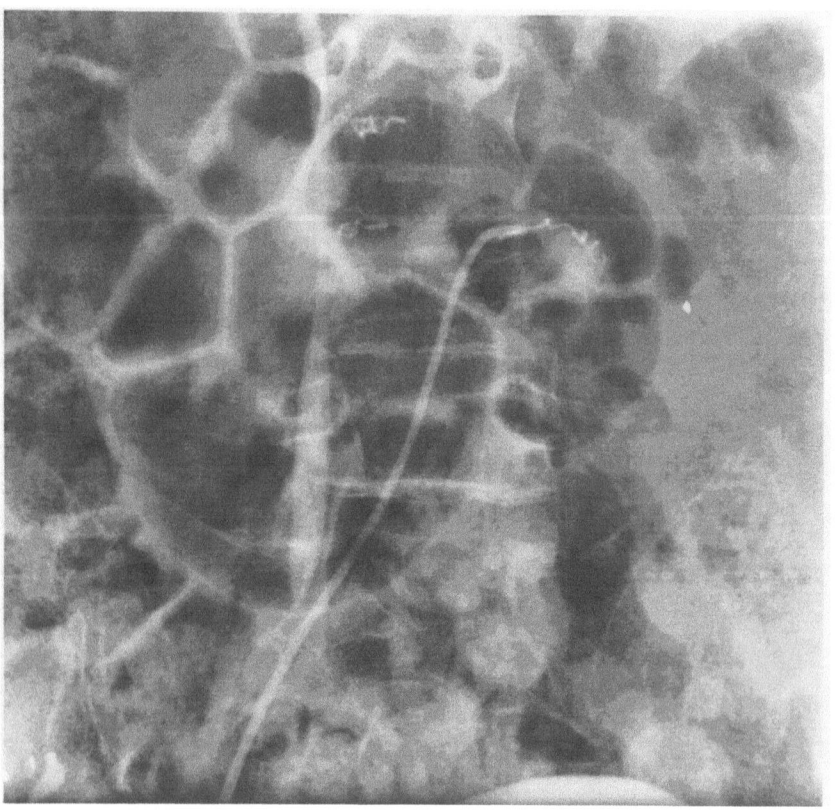

c

Fig. 1. Giant cell tumor (GCT) of L 4. **a** A hypervascular GCT of L 4 was demonstrated by the catheterization and opacification of the right third lumbar artery and bilateral fourth lumbar arteries in this 12-year-old female. **b** The capillary phase better defined the extent of the soft tissue mass on the right. There is compression of the fourth lumbar vertebral body. **c** The arteries were occluded with Gelfoam particles and stainless steel coils. **d** GCT of L 4. Anteroposterior tomography. **e** GCT of L 4. Lateral tomography. **f, g** Follow-up examinations have been stable from 2 years postembolization to the present, 9 years later. Anteroposterior and lateral conventional radiographs illustrate healing of the GCT

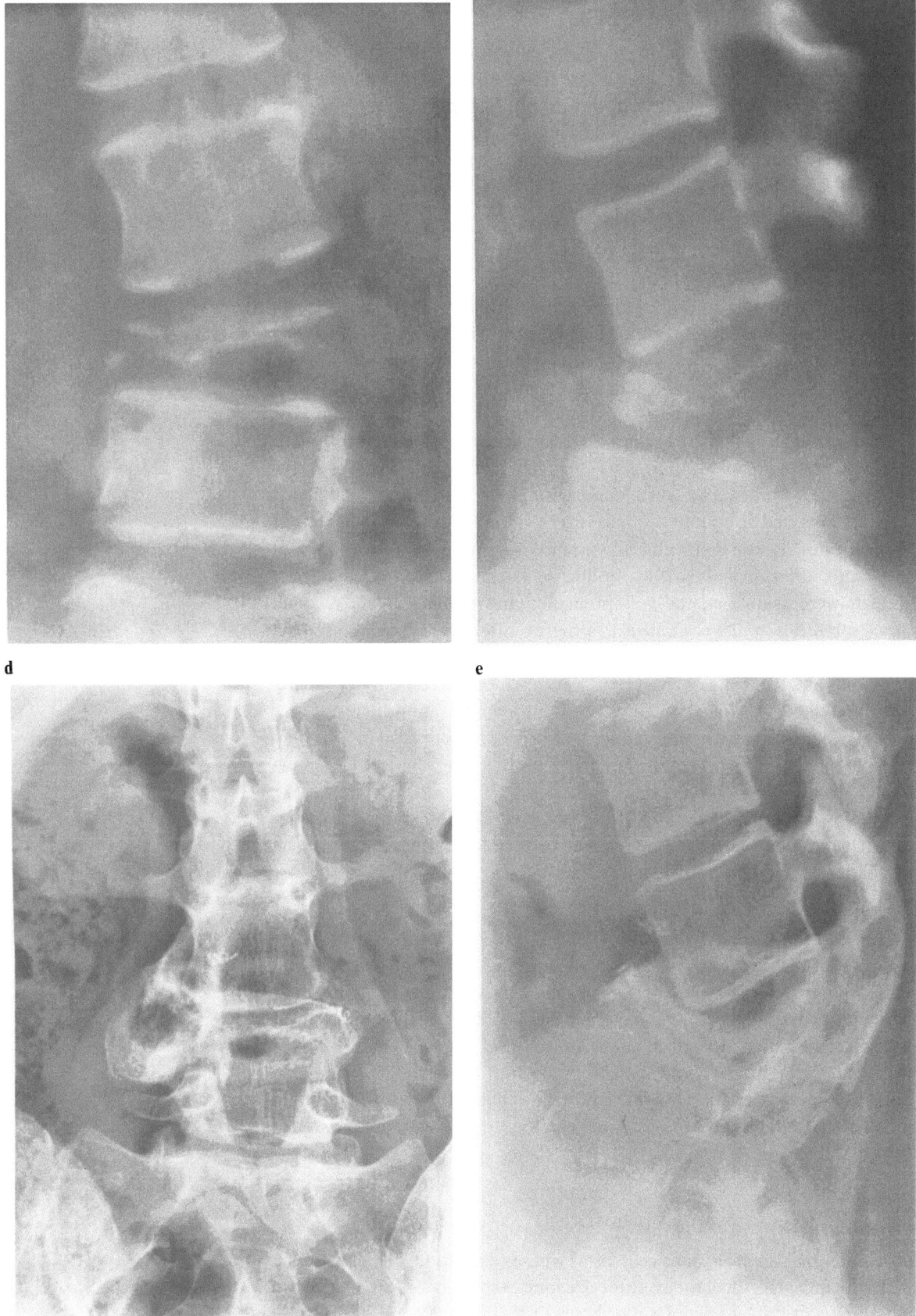

d

e

f

g

entire vascular bed and is used alone or with polyvinyl alcohol foam particles [8, 9, 11].

The vessels supplying the neoplasm are embolized with particulate material, preferably Ivalon. Reconstitution of the arterial blood flow to the neoplasm by collateral vessels occur invariably. The procedure is repeated at monthly intervals until symptomatic improvement is experienced and radiographic changes of healing are observed. The embolic episodes are then spaced at longer intervals depending on the clinical course and should be repeated when symptoms recur.

Results

Twenty-one patients with unresectable giant cell tumors were treated with arterial embolization during a period of 9 years. The tumors were located in the sacrum in 9, the thoracolumbar spine in 3, the ilium alone in 3, and involved both the sacrum and ilium in 6 patients. Eighteen patients had received prior therapy consisting of chemotherapy, irradiation, or surgery.

Ten patients (48%) had complete disappearance of their symptoms. Radiographic signs of healing consisting of reactive calcification at the periphery and in the center of the tumor were noted in all of these patients (Fig. 2). The median follow-up time was 25 months ranging from 1 to 7 years after the last embolization. Recurrence of symptoms at an average of 45 months occurred in 4 patients. Following re-embolization three of these patients were again rendered asymptomatic. One patient did not respond to re-embolization and continued to progress after more than three years of an excellent response. Four (19%) patients had partial relief of their symptoms and two patients relapsed at 5 and 30 months, respectively. Therefore 14 of the 21 patients (67%) responded to embolization therapy. Five of the 21 patients failed embolization, four of whom were dead at a median of 16 months after the initial procedure. Two patients were lost to follow-up.

Following embolization, 7 patients received chemotherapy or irradiation. One patient underwent subsequent surgery for an iliac tumor which responded to embolization [8, 9, 11, 17].

Complications

Apart from the side effects of the embolization consisting of nausea, vomiting, low grade fever, and transient pain, there were few complications. Ischemic neuropathy resulted in foot drop and foot numbness in three patients. One patient developed mild signs of rectal ischemia which resolved without sequelae after embolization of the superior hemorrhoidal artery and both internal iliac arteries. One patient with a moderate but brief response died suddenly 10 days after re-embolization for a relapse. An autopsy failed to reveal the cause of death [8, 9, 11, 17].

Osteosarcoma

Osteosarcomas are malignant neoplasms whose cells produce tumor osteoid matrix in at least a small focus: After myeloma, these tumors are the second most frequent primary neoplasm of bone. Osteosarcoma comprises several subtypes that vary in their degree of aggressiveness: the high grade intra-

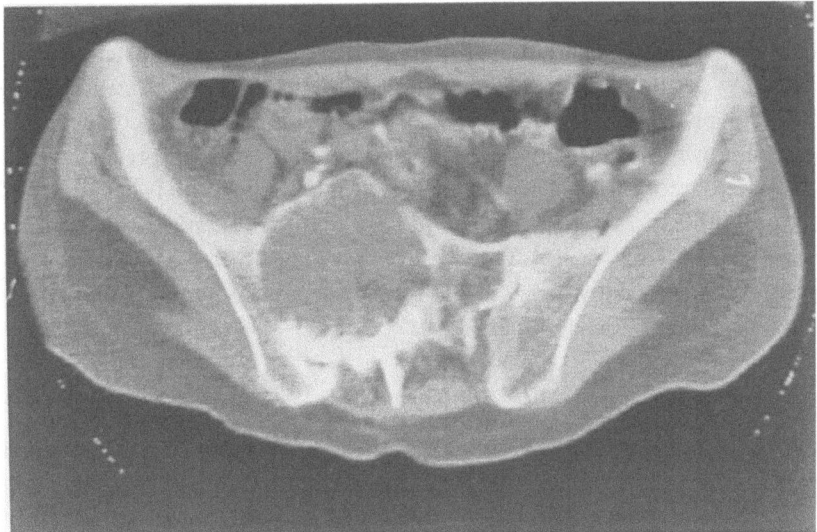

a

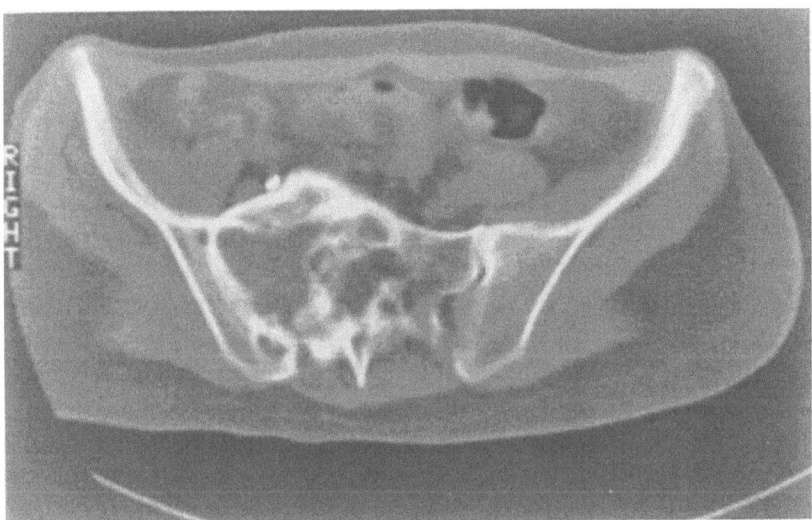

b

Fig. 2. Giant cell tumor of sacrum (right). **a** Computed tomography of the pelvis of a 31-year-old female with a GCT of the right sacrum extending into the right ilium. **b** Two years following embolization with Gelfoam and coils there was sclerosis of the margin of the tumor and calcification within. **c** Four years following embolization there was further healing of the GCT without any additional treatment. The patient is now asymptomatic $8^{1}/_{2}$ years after the embolization

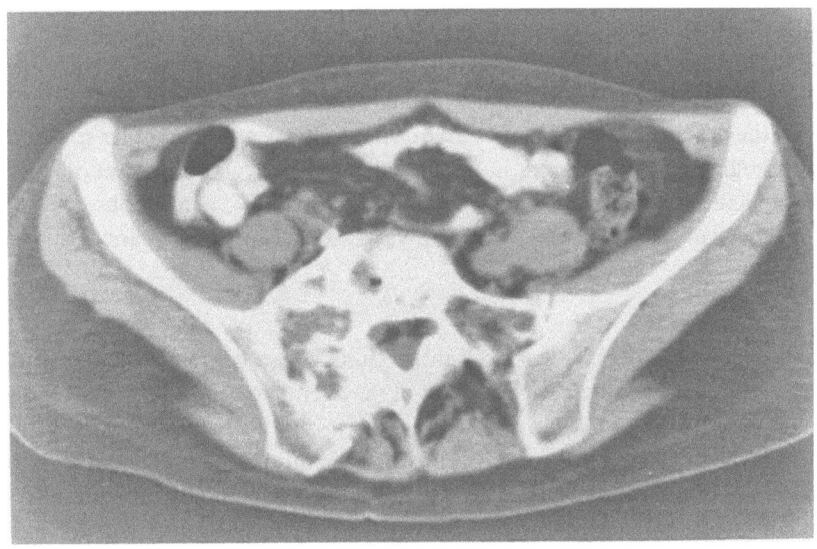

c

medullary conventional (fibroblastic, chondroblastic, and osteoblastic) the small cell, and telangiectatic. In addition, special types of osteosarcoma including periosteal, parosteal, cortical, and extraskeletal generally have a different prognosis. The etiology of the majority of osteosarcomas is unknown however, Paget's disease and irradiation are known precursors.

Osteosarcomas may occur in any age group but the peak is in the second decade of life; they are rare before the age of 10 years, and 70–80% occur before age 30 years. Males are twice as commonly affected as females. They occur most frequently in the metaphyses of the distal femur, the proximal tibia, and the proximal humerus.

Management

Radical surgery had been the principal therapy for primary osteosarcomas with an overall survival rate of 20% [18]. Radiation therapy used for local control alone or in combination with surgery did not yield better results [19, 20, 21]. Radiologic evidence of pulmonary metastases was seen at a median of 8.5 months following potentially curative surgery and patients usually died within 6 months after pulmonary metastases were detected [22, 23].

Chemotherapeutic agents were also utilized in an attempt to improve the poor results achieved by surgery alone [24, 27]. Response rates of 35%–40% have been obtained using methotrexate [28, 29], Adriamycin [30], Cisplatin [31], and Cytoxan [32]. Administration of these agents, alone or in combination, led to eradication of established metastases, destruction of the primary tumor, and prolongation of the disease-free survival time. Because osteosarcoma is believed to be microscopically disseminated at the time of diagnosis, adjuvant chemotherapy is administered following surgery [33, 34].

Advances achieved with chemotherapy led to the search for alternative methods, limb salvage, to treat the primary tumor short of amputation [35, 36, 37]. Preoperative chemotherapy was also used initially in an attempt to contain the primary tumor while awaiting the production of a customized endoprosthesis for limb salvage surgery [38, 39]. Subsequently, preoperative chemotherapy and delayed surgery were employed with the intent to treat the primary tumor and identify an effective chemotherapeutic regime for adjuvant therapy based on the degree of tumor necrosis [1, 36, 40, 41, 42, 43]. The most convincing argument in favor of adjuvant chemotherapy comes from the studies of Rosen and his colleagues who demonstrated that patients with the best response to preoperative chemotherapy with a complex multiple drug regimen (high dose methotrexate, bleomycin, cyclophosphamide, dactinomycin, and adriamycin) had the best postoperative continuous disease-free survival (92% for a median of 2 years in a group of 79 patients) [44].

Chemotherapy

Theoretically, any drug delivered intravenously can be administered intra-arterially as long as the concentration of the agent is tolerated by the arterial endothelium and local normal tissues. Intra-arterial infusion exposes the neoplasm to a higher concentration of the chemotherapeutic agent than is achieved by its intravenous delivery, attaining a greater therapeutic effect in a limited anatomic area. The increase in local concentration also depends upon the blood flow through the tumor and the "first pass" effect. The cytotoxic effect is not only concentration dependent but varies with the amount of uptake by the tumor, the metabolic activity of the tumor and the drug in the tumor, the sensitivity of the tumor to the drug, the local tumor environment (PH, pCO_2, and pO_2) and the total body clearance of the agent. Jaffe et al. described consistently elevated cisplatin levels in the local draining veins as compared to peripheral veins during arterial infusion of patients with osteosarcomas of the lower extremity. The concentration of platinum in the neoplasm and the degree of tumor destruction were directly related to the number of arterial infusions and the total dose of platinum [45].

Vigorous hydration with intravenous fluids is started on the night prior to the administration of cisplatin and is continued for 24 hours afterwards

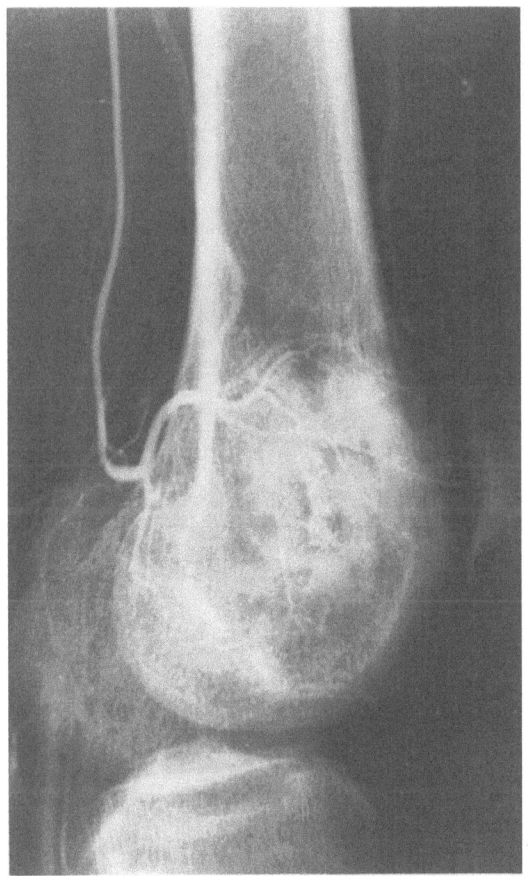 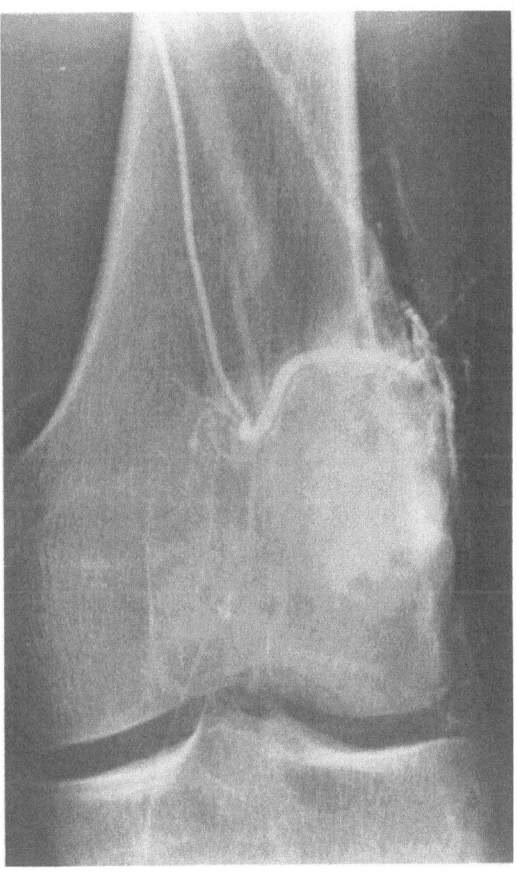

a   b

Fig. 3. Osteosarcoma. **a** Selective catheterization and opacification of the middle geniculate branch of the superficial femoral artery demonstrated the hypervascular osteosarcoma in an 18-year-old male – lateral projection. **b** Anteroposterior projection. **c** Computed tomography (CT) delineated the osteosarcoma of the lateral aspect of the distal left femur. **d** CT was performed following 5 courses of intraarterial cisplatin and intravenous Adriamycin. Note the healing of the neoplasm. **e** Magnetic resonance image of the distal femur defines a bright area (increase signals) within the treated zone. **f** Local resection and graft placement was performed because the patient refused more extensive surgery and a prosthesis. No viable tumor was found. The patient is living and well without recurrence or metastases two years after treatment

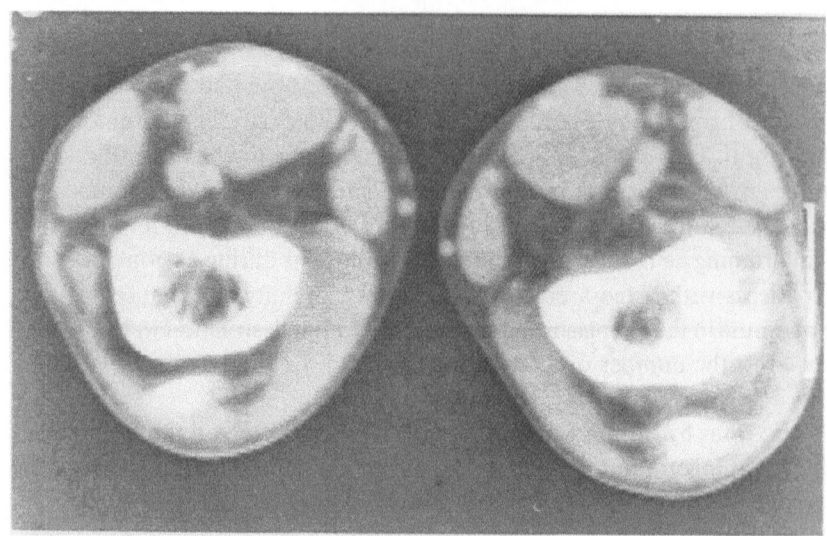

Fig. 3 c

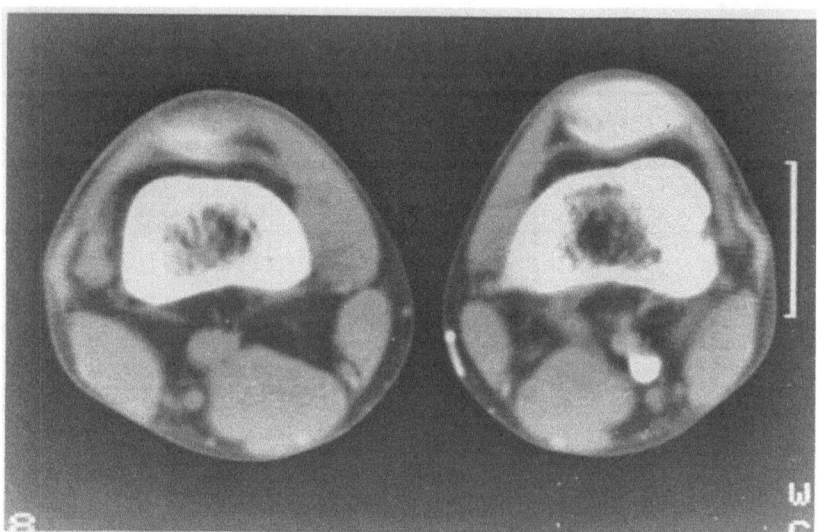

Fig. 3 d

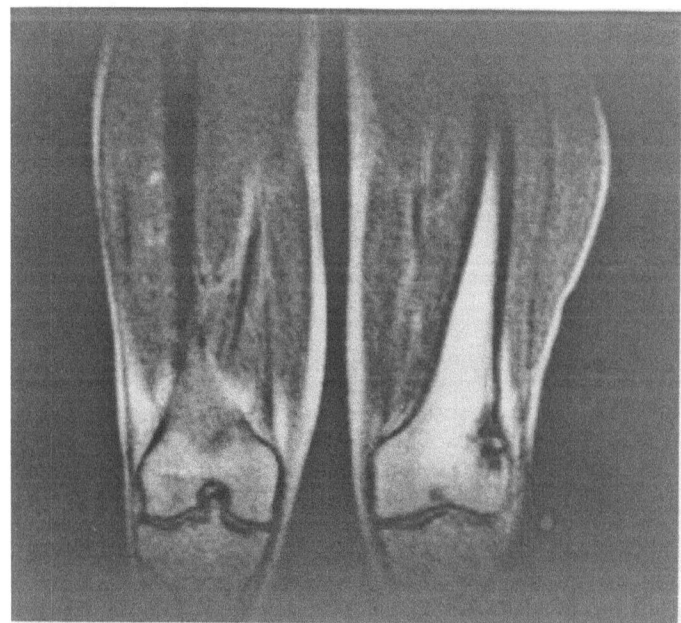

Fig. 3 e

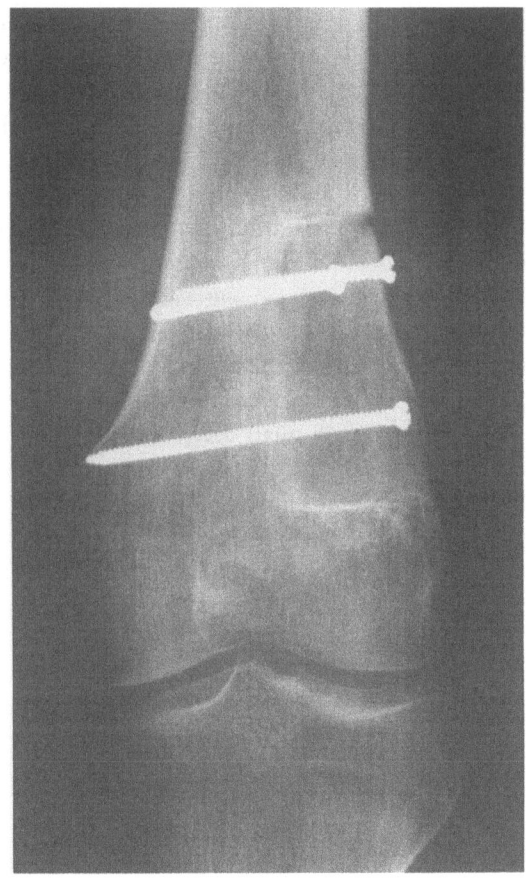

Fig. 3 f

[46–50]. Mannitol diuresis is obtained throughout the course of the infusion. Cisplatin at a dose of 120 to 200 mg/m^2 is diluted in 300 ml of 3% saline solution and administered intra-arterially over 2–24 hours [2, 3, 4].

Patients younger than 16 years of age receive 7 courses of intra-arterial cisplatin (150 mg/m^2) alone delivered every two weeks. Older patients receive intravenous adriamycin (90 mg/m^2) infused over 96 hours prior to each of 3–6 courses of intra-arterial cisplatin [3, 5, 51, 52, 53]. The primary tumor is downstaged so that a larger number of skeletally mature patients can undergo limb salvage rather than amputation [5].

Technical considerations

In pediatric patients, the procedure is performed under general anesthesia. In older patients, mild sedation and local anesthesia suffice. The usual access route is through the femoral artery by the Seldinger technique. The contralateral femoral artery is used for insertion of the catheter in patients with tumors in the lower extremities, whereas, either femoral artery is adequate to approach all other tumors [17].

As soon as the catheter is placed in the arterial system, patients are anticoagulated with aqueous heparin, 45 units/kg, and an equal dose is administered during the course of the 2-hour infusion of cisplatin. For higher doses of cisplatin, 160–200 mg/m^2 infused over 24 hours to minimize toxicity of nausea and vomiting, anticoagulation is continued throughout maintaining the clotting parameters at 1.5–2.0 times normal.

Thrombotic complications are also minimized by the use of 3.7–5 French catheters. In children the use of straight guide wires will decrease the likelihood of producing vascular spasm which occurs more frequently when curved guide wires are employed. The risk of producing subintimal dissections with the straight guide wires is minimal because of the absence of atherosclerosis in this age group.

Straight catheters are used to decrease the possibility of chemotherapy induced endothelial injury which more frequently occurs when the tip of curved catheters rests on the vessel wall. A deflector wire is employed to advance the catheter over the aortic bifurcation into the contralateral common iliac artery. For the upper extremity neoplasms, the catheter's tip is preshaped in a gentle curve to engage the brachiocephalic vessels [17].

Rarely, a single small vessel will supply most of the blood flow to the osteosarcoma, e.g. geniculate or circumflex branches [54]. At least one of the courses of preoperative cisplatin should be delivered selectively into the predominant vessel (Fig. 3). Usually, the catheter tip is placed proximal to the branches supplying the tumor. Because of the slow rate of the infusion (50–150 ml/h) laminar flow or streaming prevents optimal mixing of the

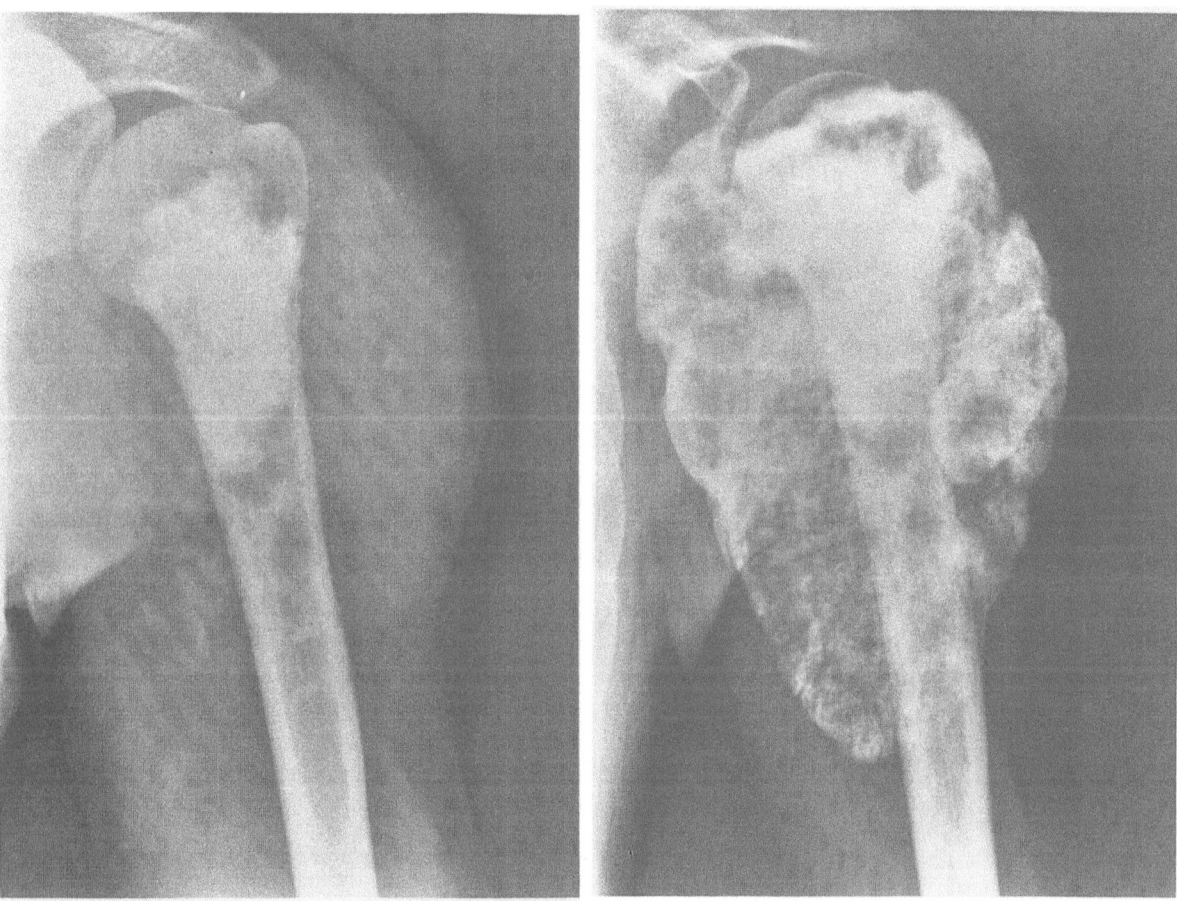

a   b

Fig. 4. Healing osteosarcoma. **a** Pretreatment osteosarcoma of the proximal left humerus. **b** Following intraarterial cisplatin and intravenous Adriamycin there is further calcification of the tumor. The margins are distinct

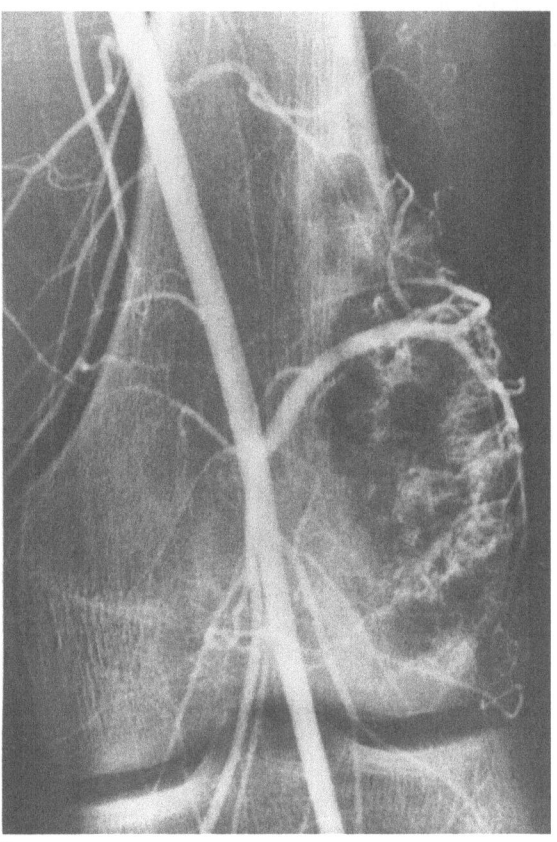

a

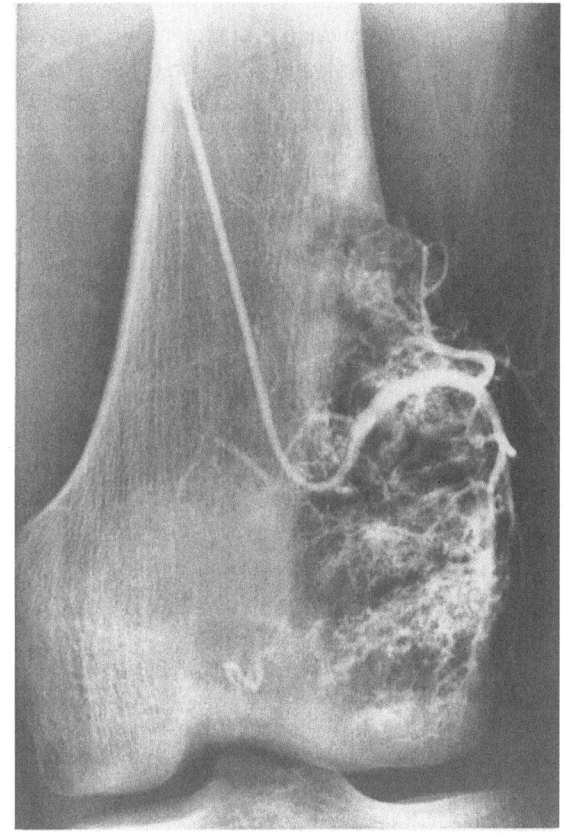

b

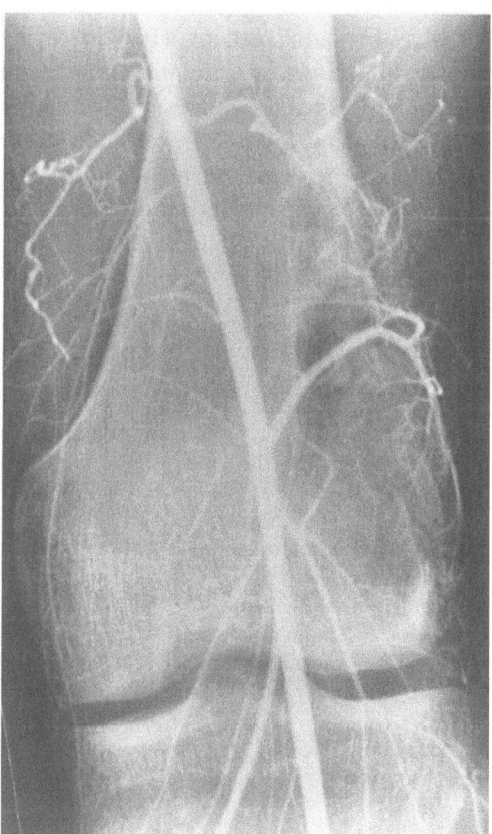

c

Fig. 5. Healing osteosarcoma. **a** This hypervascular osteosarcoma in an 18-year-old male is supplied by branches of the superficial femoral artery. **b** The dominated supply originated from the middle geneculate branch through which two (cisplatin intraarterially) of the 5 courses of chemotherapy were given. **c** Following 5 courses of therapy, cisplatin intraarterially and Adriamycin intravenously were given, there is a marked reduction of tumor vascularity. This correlated well with a least 90% destruction of the neoplasm. The tumor was resected and a knee prosthesis put in place

chemotherapeutic agent. This at times causes a high concentration of cis-platin in a musculocutaneous branch resulting in a severe local inflammatory reaction. The use of a pulsatile pump (Pulser, Cook Inc., Bloomington, IN) at 70 pulses/minute creates turbulence, better mixing and a more favorable distribution [55].

Criteria for response

A cytotoxic effect, i.e. healing of the osteosarcoma, will appear on conventional radiographs as an increase in the reactive calcifications and a decrease in the soft tissue mass. The outline of the neoplasm will become better defined with varying degrees of remodeling of the cortex (Fig. 4) [5, 56].

Computed tomography and especially magnetic resonance imaging will better delineate the extent of disease but are not accurately predictive of the degree of tumor necrosis that is to be observed histologically (Fig. 3 c, d).

Angiography with subtraction is performed in the AP and lateral projections at the time of catheterization to assess anatomic arterial supply as well as the degree of vascularity. Osteosarcomas are most frequently hypervascular. The total disappearance of the tumor vascularity best correlates with over 90% histologic tumor necrosis (Fig. 5) [5, 54].

Results

Pain relief occurs in most patients within days of the initial administration of cisplatin. Limb salvage surgery was possible in 24 of the initial 40 patients treated, whereas, prior to preoperative chemotherapy, only six patients were considered candidates for the procedure. Currently, approximately 80% of our skeletally mature patients undergo limb salvage procedures.

Sixty-five adult patients (16 years of age and over) treated with preoperative intra-arterial cisplatin and systemic adriamycin since 1980 were evaluated. Surgery was followed with adjuvant chemotherapy of adriamycin and dacarbazine. Since 1983 patients with less than 90% tumor necrosis in the resected area received high dose methotrexate and bleomycin, cyclophosphamide and actinomycin D. The overall disease-free survival is 65%. The 28 patients treated since 1983 have a 75% disease-free survival at 2 years compared with 62% for those treated before 1983 [5].

References

1. Akahoshi Y, Takeuchi S, Chen S et al (1976) The results of surgical treatment combined with intra-arterial infusion of anti-cancer agents in osteosarcoma. Clin Orthop 120: 103–109
2. Calvo DB, Patt YZ, Wallace S, Chuang VP, Benjamin RS, Pritchard JD, Hersh EM, Bodey GP, Mavligit GM (1980) Phase I trial of percutaneous intra-arterial (IA) cis-diamminedichloride-platinum II (CDDP) for regionally confined malignancies. Cancer 45: 1278–1283
3. Benjamin RS, Chuang VP, Wallace S et al (1982) Preoperative chemotherapy for osteosarcoma, abstract C-675. Am Soc Clin Oncol 1: 174
4. Jaffe N, Knapp J, Chuang VP et al (1983) Osteosarcoma: Intra-arterial treatment of the primary tumor with cis-diammine-dichloroplatinum II (CDP). Cancer 51: 402–407
5. Benjamin RS, Murray JA, Wallace S et al (1984) Intra-arterial preoperative chemotherapy for osteosarcoma – a judicious approach to limb salvage. Cancer Bull 36: 32–36
6. Feldman F, Casarella WJ, Dick HM, Hollander BA (1975) Selective intraarterial embolization of bone tumors. Am J Roentgenol 123: 130–139

7. Dick HM, Bigliani LU, Michelsen WJ et al (1979) Adjuvant arterial embolization in the treatment of benign primary bone tumors in children. Clin Orthop 139: 133–141

8. Wallace S, Granmayeh M, de Santos LA et al (1979) Arterial occlusion of pelvic bone tumors. Cancer 43: 322–328

9. Chuang VP, Soo CS, Wallace S, Benjamin RS (1981) Arterial occlusion: management of giant cell tumor and aneurysmal bone cyst. AJR 136: 1127–1130

10. Murphy WA, Strecker WB, Schoenecker PL (1982) Transcatheter embolization therapy of an ischial aneurysmal bone cyst. J Bone Joint Surg [Br] 64: 166–168

11. Eftekhari F, Wallace S, Chuang VP, Soo CS, Cangir A, Benjamin RS, Murray JA (1982) Intraarterial management of giant cell tumors of the spine in children. Pediatr Radiol 12: 289–293

12. Channon GM, Williams LA (1982) Giant cell tumor of the ischium treated by embolization and resection: a case report. J Bone Joint Surg [Br] 64: 164–165

13. Keller FS, Rosch J, Bird CB (1983) Percutaneous embolization of bony pelvic neoplasms with tissue adhesive. Radiology 147: 21–27

14. Lichtenstein L (1972) Giant cell tumor of bone (osteoblastoma). In: Lichtenstein L (ed) Bone tumors. Mosby, St. Louis, pp 135–165

15. Dahlin DC (1978) Giant cell tumor (osteoclastoma). In: Dahlin DC (ed) Bone tumors. Thomas, Springfield, pp 99–115

16. Hilal SK, Michelsen JW (1975) Therapeutic percutaneous embolization for extra-axial vascular lesions of the head, neck and spine. J Neurosurg 43: 275–287

17. Wallace S, Carrasco CH, Charnsangavej C, Lee YY, Wright K, Gianturco C (1984) Percutaneous transcatheter infusion and infarction in the treatment of human cancer: part II. Curr Probl Cancer 8: 5–76

18. Friedman MA, Carter SK (1972) The therapy of osteogenic sarcoma: current status and thoughts for the future. J Surg Oncol 4: 482–510

19. Cade S (1955) Osteogenic sarcoma: a study based on 113 patients. J R Coll Surg Edinb 1: 79–111

20. Lee ES, Mackenzie DH (1964) Osteosarcoma: a study of the value of preoperative megavoltage radiotherapy. Br J Surg 51: 252–274

21. Jenkin RDT, Allt WEC, Fritzpatrick PJ (1972) Osteosarcoma: an assessment of management with particular reference to primary irradiation and selective delayed amputation. Cancer 30: 393–400

22. Marcove RC, Mike V, Hajek JV et al (1971) Osteogenic sarcoma in childhood. NY State J Med 71: 855–859

23. Jeffree CM, Price CHG, Sessons HA (1975) The metastatic patterns of osteosarcoma. Br J Cancer 32: 87–107

24. Greesbeck HP, Cudmore JTP (1963) Evaluation of 5-fluorouracil (5-FU) in surgical practice. Am Surg 29: 638–641

25. Sullivan MP, Sutow WW, Taylor G (1963) L-phenylalanine mustard as treatment for osteogenic sarcoma in children. J Pediatr 63: 227–237

26. Finkelstein J, Hittle RE, Hammond UD (1969) Evaluation of high dose cyclophosphamide regimen in childhood tumors. Cancer 23: 1239–1244

27. Sutow WW (1976) Evaluation of dosage schedules of mitomycin C (NSC-26980) in children. Cancer Chemother Rep 55: 285–289

28. Jaffe N (1976) Osteogenic sarcoma: state of the art with high-dose methotrexate treatment. Clin Orthop 120: 95–102

29. Jaffe N, Link M, Traggis D et al (1981) The role of high-dose methotrexate in osteogenic sarcoma: sarcomas of soft tissue and bone in childhood. Natl Cancer Inst Monogr 56: 2101–2106

30. Cortes EP, Holland JF, Wang JJ et al (1974) Amputation and adriamycin in primary osteosarcoma. N Engl J Med 291: 998–1000

31. Ochs JJ, Freeman AI, Douglass HO et al (1978) Cis-dichloro-diammineplatinum (II) in advanced osteogenic sarcoma. Cancer Treat Rep 62: 239–245

32. Pinkel D (1969) Cyclophosphamide in children with cancer. Cancer 15: 42–49

33. Cortes EP, Necheles TF, Holland JF et al (1979) Adrimycin (ADR) alone versus ADR and high dose methotrexate-citrovorum factor rescue (HDM-CFR) as adjuvant to operable primary osteosarcoma: a randomized study by Cancer and Leukemia Group B (CALGB). Proc Am Assoc Cancer Res 20: 412

34. Sutow WW, Sullivan MP, Wilbur JR et al (1975) A study of adjuvant chemotherapy in osteogenic sarcoma. J Clin Pharmacol 7: 530–533

35. Jaffe N, Watts N, Fellows KE et al (1978) Local en bloc resection for limb preservation. Cancer Treat Rep 62: 217–223

36. Rosen G, Marcove RC, Caparros B et al (1979) Primary osteogenic sarcoma: The rationale for preoperative chemotherapy and delayed surgery. Cancer 43: 2163–2177

37. Morton DL, Eilber FR, Townsend CN Jr et al (1976) Limb salvage from a multidisciplinary treatment approach for skeletal and soft tissue sarcomas of the extremity. Ann Surg 184: 268–278

38. Rosen G, Murphy ML, Huvos AG et al (1976) Chemotherapy, en bloc resection and prosthetic bone replacement in the treatment of osteogenic sarcoma. Cancer 37: 1–11

39. Marcove RC (1977) En bloc resection for osteogenic sarcoma. Can J Surg 20: 521–528

40. Huvos AG, Rosen G, Marcove RC (1977) Primary osteogenic sarcoma: Pathologic aspects in 20 patients after treatment with chemotherapy, en bloc resection and prosthetic bone replacement. Arch Pathol Lab Med 101: 14–18

41. Rosen G, Caparros B, Huvos A et al (1982) Preoperative chemotherapy for osteogenic osteosarcoma: Selection of postoperative adjuvant chemotherapy based on the response of the primary tumor to preoperative chemotherapy. Cancer 49: 1221–1230

42. Eilber FR, Grant T, Morton DL (1978) Adjuvant therapy for osteosarcoma: Pre-operative treatment. Cancer Treat Rep 62: 213–216

43. Jaffe N, Prudich J, Knapp J et al (1981) Osteosarcoma: treatment of the primary tumor with intra-arterial high-dose methotrexate (MTX-CF): pharmcokinetic, clinical, radiographic and pathologic studies. Abstract C-409. Proc Am Assoc Cancer Res 22: 195

44. Rosen G, Marcove RC, Huvos AG et al (1983) Primary osteogenic sarcoma: Eight-year experience with adjuvant chemotherapy. J Cancer Res Clin Oncol 106 [Suppl]: 55–67

45. Jaffe N, Bowman R, Wang Y-M et al (1984) Chemotherapy for primary osteosarcoma by intra-arterial infusion: review of the literature and comparison with results achieved by the intravenous route. Cancer Bull 36: 37–42

46. Nitsche R, Starling KA, Vats T et al (1978) Cis-diammine-dichloroplatinum (NSC-119875) in childhood malignancies: a Southwest Oncology Group study. Med Pediatr Oncol 4: 127–132

47. Baum F, Greenberg L, Gaynon P et al (1978) Use of cis-platinum diammine dichloride (CPDD) in osteogenic sarcoma (OS) in children. Proc Am Assoc Cancer Res 19: 385

48. Freeman AI, Ettinger LJ, Brecher ML (1979) Cis-dichloro-diammineplatinum II in childhood cancer. Cancer Treat Rep 63: 1615–1620

49. Pratt CB, Hayes A, Green AA et al (1981) Pharmacokinetic evaluation of cis-platin in children with malignant solid tumors: a phase II study. Cancer Treat Rep 65: 1021–1026

50. Rosen G, Nirenberg H, Caparros B et al (1980) Cisplatin in metastatic osteogenic sarcoma. In: Prestayko AW, Crooke ST, Carter SK (eds) Cisplatin: current status and new developments. Academic Press, New York, pp 465–475

51. Ettinger LJ, Douglass HO Jr, Higby IJ et al (1981) Adjuvant adriamycin and cis-diammine-dichloroplatinum (cis-platinum) im primary osteosarcoma. Cancer 47: 248–254

52. Baum ES, Gaynon P, Greenberg L et al (1981) Phase II trial of cisplatin in refractory childhood cancer: Children's Cancer Study Group report. Cancer Treat Rep 65: 815–822

53. Rosen G, Caparros B, Nirenberg A et al (1982) Cisplatinum (DDP)-adriamycin (ADR) combination chemotherapy (CT) in evaluable osteogenic sarcoma (OS), abstract C-672. Am Soc Clin Oncol 1: 173

54. Carrasco CH, Charnsangavej C, Raymond AK, Richli WR, Wallace S, Chawla SP, Ayala AG, Murray JA, Benjamin RS (1987) Angiographic assessment of response to preoperative chemotherapy in osteosarcoma. Radiology (in press)

55. Wright KC, Wallace S, Kim EE, Haynie T, Charnsangavej C, Carrasco C, Chuang VP, Gianturco C (1986) Pulsed arterial infusions. Chemotherapeutic implications. Cancer 57: 1952–1956
56. Chuang VP, Benjamin RS, Jaffe N, Wallace S, Ayala AG, Murray J, Charnsangavej C, Soo CS (1982) Radiographic and angiographic changes in osteosarcoma after intraarterial chemotherapy. AJR 139: 1065–1069
57. Ayala A, Raymond AK, Jaffe N (1984) The pathologist's role in the diagnosis and treatment of osteosarcoma in children. Human Pathol 15: 258–266

Direct steroid injection in the treatment of tumor-like lesions

Direct steroid injections in the treatment of single bone cyst

M. Campanacci and R. Capanna

Orthopaedic Clinic I, Istituto Ortopedico Rizzoli, Bologna, Italy

In the past, conventional surgical procedures for treatment of single bone cysts (curettage and bone grafting through a limited cortical window) were not satisfactory, showing a recurrence rate around 30–40% [3, 6, 19, 25].

Only a more aggressive surgical approach (hemi or total subperiosteal resection of the cystic wall) allowed a decreased recurrence rate to 5–8% [16].

Regardless of the performed procedure, surgery presents several disadvantages (prolonged immobilization and hospitalization; surgical scars) and may have major complications (growth plate injury – infection).

In 1974 Scaglietti [21] introduced local injections of methylprednisolone acetate (MPA) as treatment of simple bone cysts. Many authors [1, 7, 8, 9, 12, 17, 20, 22, 23, 24] with this simple, not invasive procedure reported clinical and radiographical results comparable, or even better, than those obtained with surgery.

The healing mechanism of the cyst after MPA injections is still unknown. Corticosteroids may cause destruction of the cystic wall connective tissue membrane [24]: their inhibitory effect on fibroblast growth was in vitro demonstrated [4]. Corticosteroids may prevent formation or induce resorption of the cystic fluid, whose chemical nature is similar to a tissue transudate [14, 24, 26]. The microcrystalline suspension of MPA being relatively insoluble, shows prolonged pharmacological effects. Other corticosteroids with topical action had less effective results than MPA in clinical application [24]. The injection related injuries to the cyst (hole drilling; fluid aspiration; ex vacuo hemorrhage; hydrodynamical changes of the intracystic pressure) may play an important role in the healing of the cyst [13, 18].

Direct cortisone injection was also used as treatment of eosinophilic granuloma of bone and usually complete healing of the lesion was achieved [11, 15, 24].

Technique of MPA injections

All patients with a simple bone cyst (either recurrent or virgin) are candidates to MPA injection. The diagnosis of bone cyst is based on clinical data (age of the patient, location of the cyst) and the rather typical radiographic features of the cyst. Diagnosis is definitely confirmed at the time of treatment when the characteristic serous fluid is aspirated from the cyst cavity. An immediate and copious blood spilling from the needle without evidence of the typical serous fluid, is, on the contrary, suggestive of an aneurysmal bone cyst. If any doubt exists, a needle biopsy may be performed as proposed by some authors [8, 20].

If a pathological fracture is present at the onset, we conservatively treat the fracture and then we start the MPA treatment 40 days later, after fracture healing.

The injection is done in the operative room under sterile conditions either under general or local anesthesia: usually we prefer Ketamine anesthesia. The entire procedure takes from 5 to 10 minutes (from 15 to 25 if a contrast roentgenographic examination of the cyst and measure of the internal pressure are performed). The treatment requires 1 day of hospitalization if general anesthesia is used, otherwise it may be done on an outpatient basis.

Special instruments particularly fit for perforation of bony walls with different hardness (cannulated needles, trocar, and manual drills) were developed by Bartolozzi [2]. We usually perform the injections using two standard large gauge needles with stylets and a mallet to penetrate the cortex. If the cortex cannot be penetrated, a bone punch is used to produce a defect in the bone through which the needle can then be passed.

An image intensifier is used to assist the placement of the needles. The procedure consists in the following surgical steps:

(1) A single needle is introduced percutaneously into the lower end of the cystic cavity through a lateral approach and the trocar is removed.

(2) The cannula may be connected by a semirigid polithene tube filled with heparinized saline to a pressure transducer and recording system and the internal pressure recorded until a stable pressure is obtained. This surgical step is performed only for research purposes, and it is not mandatory in clinical practice.

(3) A second needle is then inserted through a lateral approach into the upper end of the cavity and its trocar is removed to permit the spontaneous escape of fluid from the cyst. Forceful aspiration through a single needle is likely to be followed by venous hemorrhage.

The lateral approach facilitates the escape of the fluid and a more complete emptying of the cyst, both needles being horizontally placed at the level of the cyst. To obtain the same effect with an anterior approach and vertically placed needles, aspiration is required and secondary bleeding may occur. To avoid secondary bleeding some authors [24] use, whenever possible, a tourniquet placed above the level of the injection. In practice the tourniquet is rarely used, most cysts being located in the proximal part of the humerus or femur. We do not believe that secondary bleeding could negatively affect the outcome of the treatment, altering the dilution of the corticosteroid on the cyst. On the contrary, venous hemorrhage may contribute in promoting the healing of the cyst. For this reason we do not pay particular attention to avoid secondary bleeding except when a contrast examination of the cyst is planned: in this case a venous hemorrhage may impair the injection and the distribution of radiopaque medium. If a contrast examination is not performed, any approach may be used for placement of the needles.

(4) Contrast material (Pielographin) is then injected through one needle into the lesion to outline any solid components or intracystic fibrous septa. Filling of the cavity with the radiopaque medium is observed fluoroscopically

and serial X-rays may be obtained. An increasing quantity of the medium is injected into the cavity until escape of the contrast material through venous drainage occurs (phlebography) with no further filling of the cavity. Final X-rays in two planes are then obtained. Preliminary contrast study of the cyst is useful to differentiate unilocular (Fig. 1 a) from multilocular cysts. Multilocular cysts are more frequent than unilocular ones (3:1 ratio). In multilocular cysts, the intracystic fibrous septa may form loculations without intercommunication between adjacent areas of the cyst (Fig. 1 b). It is important to recognize these "separate" areas to avoid an unequal distribution of MPA [10, 20]. After the roentgenographic examination the contrast medium is aspirated.

(5) If the cyst is unilocular, one needle is removed and MPA is injected into the cyst. If the cyst is multilocular, a preliminary forceful irrigation of the cavity with normal saline instilled through both needles is recommended in an attempt to break the fibrous septa. The MPA is then injected through both needles. Separate areas may require selective injections using adjunctive needles.

(6) The amount of injected MPA (40 mg/ml) is empirically established. Usually 80–200 mg (2–5 ml) of MPA are injected into the cyst: the amount is determined by the age and size of the patient and by the size of the cyst.

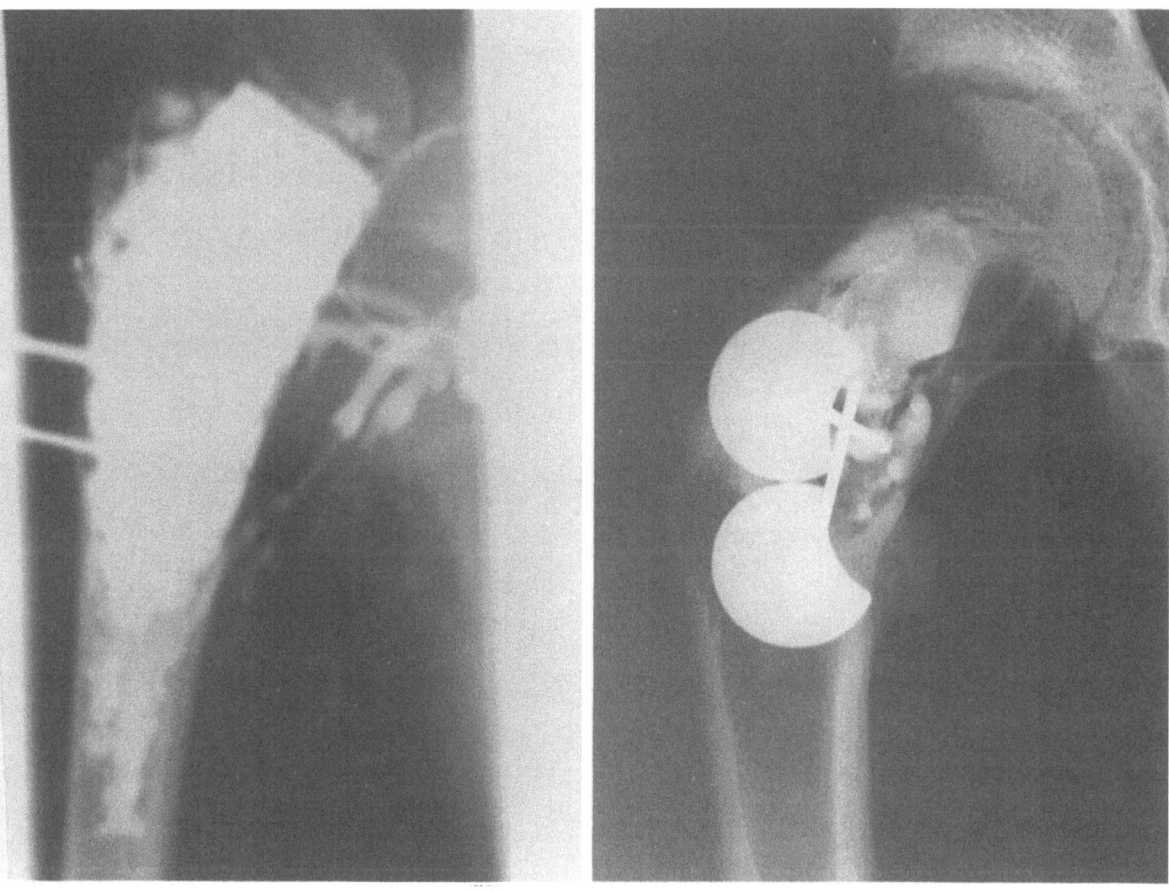

b

Fig. 1. a Complete filling of the cavity with radiopaque medium in a cyst of the proximal humerus. **b** Multilocular cyst of the proximal femur; a fibrous septa forms loculations without intercommunication between adjacent areas of the cyst

(7) The involved extremity is immobilized only in the presence of pathologic fracture. The injections are given routinely at two months intervals
until advanced healing of the cyst is evident on X-rays (phase 2: frosted
glass opacification). In most cases, from 2 to 4 injections suffice: the maximal
number of injections given, is 7 in our experiment. Every time the injection
is repeated, the needles have to be placed in a different area of the cyst from
those previously injected.

(8) In eosinophilic granuloma of bone, the contrast examination is ob-
viously not performed; only one needle is used and placed in the epicenter
of the tumor; 1–4 ml of MPA are injected, depending on the sizes of the
radiolucent defect and the soft tissue mass; most of the lesions heal with
only 1 or 2 injections; a 3 month interval between injections is recommended
[11, 15, 24].

Radiographic evolution – outcome and prognostic factors

Four responses to MPA treatment may be observed.

Healing

After the first injections of MPA, cystic expansion ceases and the cortex
may thicken (phase 1). After 6 months (range 2–14 months) from the onset

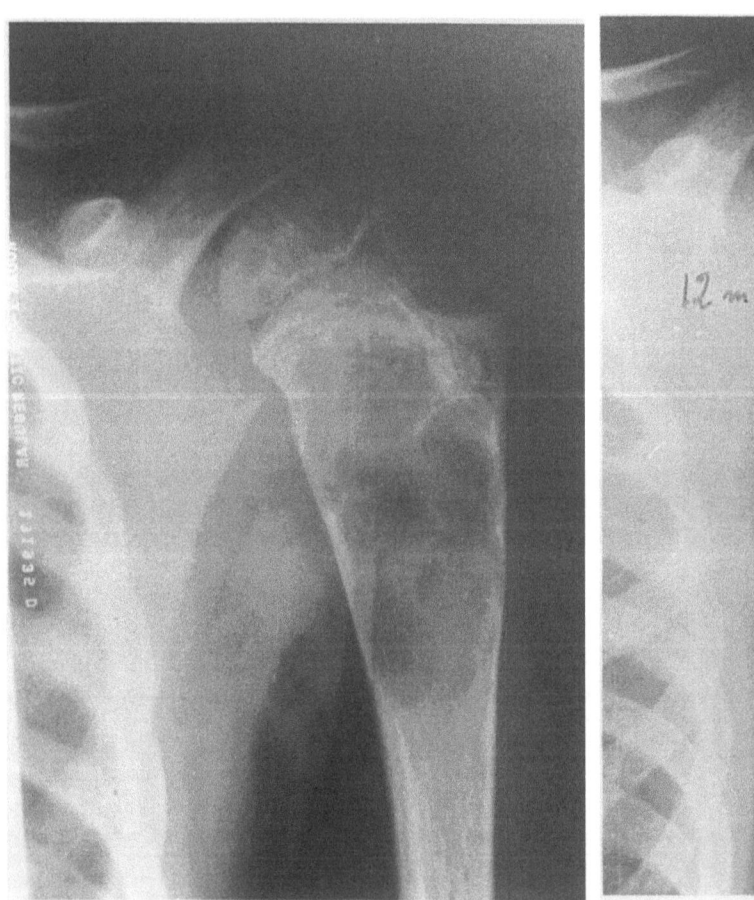 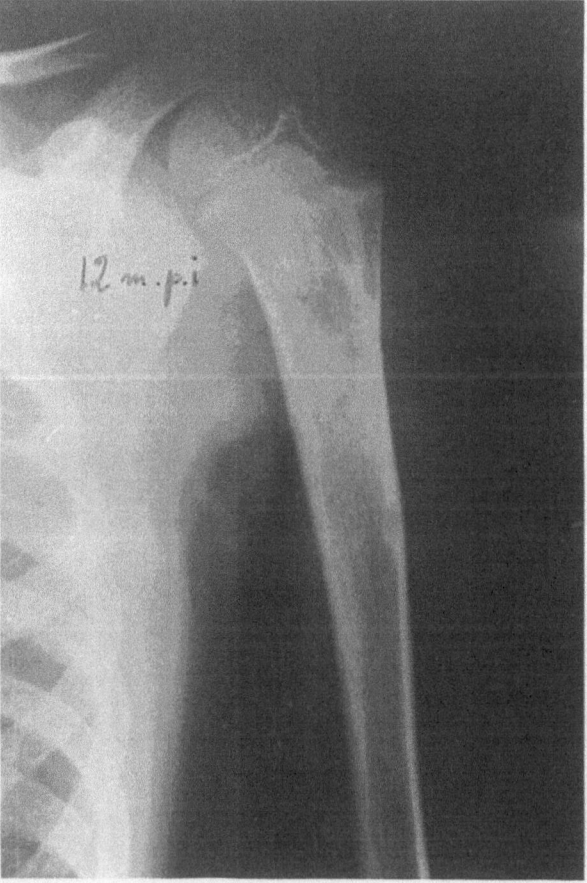

a   b

Fig. 2 a, b. Complete healing of a simple bone cyst of the proximal humerus after
MPA injection

of the treatment, the cyst has a "frosted glass" type of opacification (phase 2). After 12 months (range 4–26 months), the cyst becomes consolidated with dense sclerotic bone (phase 3). The temporal progression from phase 2 to phase 3 does not necessarily occur simultaneously in all areas of the cyst, especially in multiloculated cysts. Although a delayed consolidation in certain subcysts is not a poor prognostic sign (because most of the cavities will eventually fill with bone using repeated selective injections), healing of the cyst cannot be definitely stated until the cyst completely fills in with bone (Fig. 2). Only rarely, mainly in diaphyseal cysts, and after several years, the bone segment will recover its normal morphology (phase 4). No recurrences were observed when a complete healing of the cyst was achieved and a close radiographic surveillance of these patients is unnecessary.

Incomplete healing

New bone formation fills the area previously occupied by the cyst and cortical margins thicken; however, small sites of osteolysis may be seen within the boundaries of the cyst. An incomplete healing may be represented by:

    a residual peripheral rim of osteolysis;

    a small residual osteolytic area;

    multiple osteolytic areas between thickened bony septa;

    irregular bone formation inside the cyst.

An incomplete healing may be stable in time; however, from 30 to 50% of the cysts with an incomplete healing will have a local recurrence. Radiographic surveillance (every 4–6 months) is advisable in these patients for at least 3 years after treatment: if there is any evidence of cyst recurrence, additional MPA injections may be given into the persistent osteolytic area.

Recurrence

Initially, the cyst is partially filled with bone; however, large areas of osteolysis and cortical thinning subsequently returned. The recurrence may be late (within 1 year in 37% of patients; between 1 and 3 years in 44%; more than 3 years in 19%). Most of the recurrences do not need treatment if the patient age is over 14–16 (the cyst tends to heal spontaneously when skeletal maturity is achieved). If treatment is required, most of the cysts (90%) still respond to a new cycle of MPA injections [12].

No response

The cyst shows no evidence of response to MPA treatment. In this case, a new cycle of MPA injections is usually without effect.

In a recent multicentric study of 234 patients treated with MPA injections [12], 26% of the patients had complete healing, 43% an incomplete healing, 20% had a local recurrence and 11% showed no response. In a comparable group of 209 patients surgically treated (curettage and bone grafting) there was 41% complete healing, 22% incomplete healing, and 37% recurrences.

Three prognostic factors appeared to have a positive effect on the outcome of the cyst after MPA treatment: the unilocular aspect (vs multilocular), the small size (vs medium and large sizes) and the diaphyseal location (vs metaphyseal and metadiaphyseal) (Fig. 3).

The sex and age of the patient, the site of the cyst (humerus or femur),

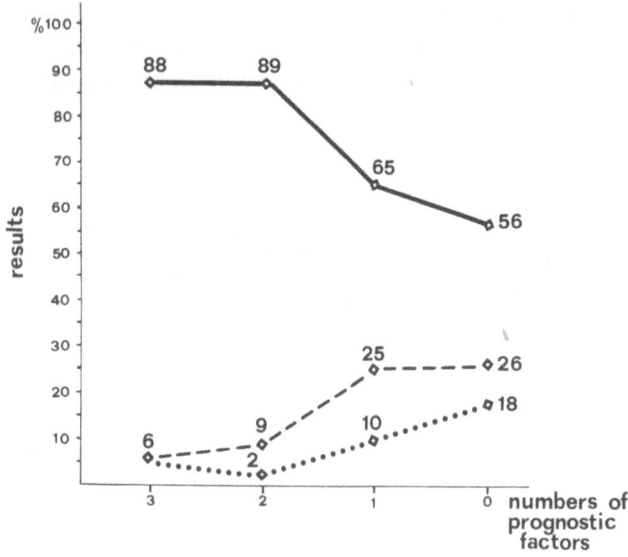

Fig. 3. Outcome of simple bone cyst treated with MPA, having no or some or all the three positive prognostic factors (unilocular cyst, small size, diaphyseal location). ●———● % Complete and incomplete healing; ● - - - - ● % recurrence; ● · · · · · ● % no response

the number of previous fractures, or a previous surgery had apparently no bearing on the outcome of the MPA treatment.

Complications

Pathological fracture during the course of the treatment was rare (2%). It is difficult to determine if this is a complication of the treatment. Although the holes made in the cyst are small, they potentially could act as stress risers and cause the pathologic fracture.

Pathological fractures were also uncommon (16%) in patients who had a recurrent cyst or no response after MPA injections: in most instances, the treatment stopped the expansion of the cyst and produced a sufficient increase in the mechanical strength of the affected bone. Among 234 consecutive cases, all initially treated with MPA, only 9 (3%) underwent a surgical treatment.

Limb length discrepancy was rare (5%) in patients treated only with MPA injections, while it was more frequent (13%) in surgically treated patients: moreover the severity of the limb length discrepancy was usually minor (less than 2 cm in 85% of the cases) in the first group, while major in the latter (more than 3 cm in 60% of the cases).

Laboratory investigations showed no significant changes in biohumoral parameters, ensuing MPA injections [5].

No infections were reported. Among 234 cases, only one major complication was observed: a patient aged 6 who developed avascular necrosis of the proximal femoral epiphysis during MPA treatment. A case of vascular necrosis was also found in the surgical group.

Partially supported by grant N. 86.02679.44 Italian National Research Council, special project "Oncology".

References

1. D'Astous J (1984) Unicameral bone cysts – steroid injections. Round Table Discussion. In: Uhthoff HK (ed) Current concepts of diagnosis and treatment of bone and soft tissue tumors. Springer, Berlin Heidelberg New York Tokyo, p 297–304

2. Bartolozzi F (1976) Su due aghi cannula per il trattamento topico di lesioni scheletriche. Arch Putti Chir Organi Mov 27: 251–254

3. Bensahel N, Aigrain Y, Desgrippes Y (1982) Bilan du traitement du kyste essential des os de l'enfant. J Chir 119: 319–323

4. Berliner DL, Nabors CJ (1968) Effects of corticosteroids on fibroblast functions. Perlman D (ed) Topics in pharmaceutical sciences, vol 1

5. Brillante C, Manduchi R, Guachardo E (1985) Cisti ossee infantili: controllo delle modificazioni bioumorali indotte dal trattamento topico steroideo. La Clinica 40: 131–136

6. Campanacci M, de Sessa L, Bellando Randone P (1975) Cisti ossea (Revisione di 275 osservazioni. Risultati della cura chirurgica e primi risultati della cura incruenta con metilprednisolone acetato). Chir Organi Mov 62: 471–482

7. Campanacci M, Capanna R, Ricci P (1986) Unicameral and aneurysmal bone cysts. Clin Orthop 204: 25–36

8. Campos OP (1982) Treatment of bone cysts by intracavity injection of methylprednisolone acetate. Clin Orthop 165: 43–48

9. Capanna R, Dal Monte A, Gitelis S, Campanacci M (1982) The natural history of unicameral bone cyst after steroid injection. Clin Orthop 166: 204–211

10. Capanna R, Albisinni U, Caroli GC, Campanacci M (1984) Contrast examination as a prognostic factor in the treatment of solitary bone cyst by cortisone injection. Skeletal Radiol 12: 97–102

11. Capanna R, Springfield DS, Ruggieri P, Biagini R, Picci P, Bacci G, Giunti A, Lorenzi E, Campanacci M (1985) Direct cortisone injection in eosinophilic granuloma of bone: a preliminary report on 11 patients. J Pediatr Orthop 5: 339–342

12. Capanna R, Bettelli G, dal Monte A, de Sanctis M, de Christofaro R, Gagliardi S, Ippolito E, Manes E, Martelli C, Mastrogostino S (1987) Cisti ossee: trattamento chirurgico (parte I). Cisti ossee: trattamento infiltrativo (parte II). G Ital Ortop Pediatr (in press)

13. Chigira M, Shimizu T, Arita S, Watanabe H, Heshiki A (1986) Radiological evidence of healing of a simple bone cyst after hole drilling. Arch Orthop Trauma Surg 105: 150–153

14. Cohen L (1960) Simple bone cysts: studies of cyst fluid in six cases with a theory of pathogenesis. J Bone Joint Surg [Am] 42: 609–616

15. Cohen M, Zornoza Y, Cangir A, Murray JA, Wallace S (1980) Direct injection of MPA in the treatment of solitary eosinophilic granuloma of bone. Radiology 136: 289–293

16. Fahey JJ, O'Brien ET (1973) Subtotal resection and grafting in selected cases of solitary unicameral bone cyst. J Bone Joint Surg [Am] 55: 59–68

17. Kohler R (1982) Traitement de kystes essentiels des os par injections de corticoides. Lyon Chirurgical 78: 158–161

18. Kuboyama K, Shidou T, Harada A, Yokoe S (1981) Therapy of solitary unicameral bone cyst with percutaneous trepanation. Rinsho Seikei Geka 16: 288

19. Mastragostino S, Sanguinetti C (1960) I trapianti auto-omoed eteroplatici nel trattamento delle cisti ossee. Arch Putti Chir Organi Mov 13: 96–111

20. Oppenheim WL, Galleno H (1984) Operative treatment versus steroid injections in the management of unicameral bone cysts. J Pediatr Orthop 4: 1–7

21. Scaglietti O (1974) L'azione osteogenetica dell'acetato di metilprednisolone. Bull Sci Med Bologna 146: 159–160

22. Scaglietti O, Marchetti PG, Bartolozzi P (1976) Sull'azione topica del corticosteroidi in microcristalli in alcume lesioni dello scheletro. Arch Putti Chir Organi Mov 27: 9–31

23. Scaglietti O, Marchetti PG, Bartolozzi P (1979) The effect of methylprednisolone acetate in the treatment of bone cysts. Results of three years follow-up. J Bone Joint Surg [Br] 61: 200–204

24. Scaglietti O, Marchetti PG, Bartolozzi P (1982) Final results obtained in the treatment of bone cysts with methylprednisone acetate (Depo-Medrol) and a discussion of results achieved in other bone lesions. Clin Orthop 165: 33–42

25. Spence KF, Sell KW, Brown RH (1969) Solitary bone cyst: treatment with freeze-dried cancellous bone allograft: a study of one hundred seventy-seven cases. J Bone Joint Surg [Am] 51: 87–96

26. Villani E (1956) Studio clinico ed enzimologico sulle cisti solitarie. Congresso SICOT, pp 387–390

Eosinophilic granuloma of bone: direct steroid injection

S. Wallace, C. H. Carrasco, C. Charnsangavej, and M. Cohen*

Department of Diagnostic Radiology, The University of Texas System Cancer Center
M. D. Anderson Hospital and Tumor Institute, Houston, Texas, U.S.A.

Eosinophilic granuloma of bone is a benign condition that was described as a distinct clinical and pathological entity by Lichtenstein and Jaffe [1] and Otani and Ehrlich [2] in 1940. One year later, Farber suggested a pathological gradation including eosinophilic granuloma, Hand-Schuller-Christian disease, and Letterer-Siwe disease [3]. In 1953, Lichtenstein consolidated these three entities under the term "histiocytosis X", thereby emphasizing the inflammatory proliferative reaction common to all three forms of the disease as well as their indefinite etiologies [4]. Eosinophilic granuloma is generally accepted to represent the benign localized form of the disease, while Hand-Schuller-Christian and Letterer-Siwe diseases are thought to be disseminated types; however, the validity of this unifying concept is controversial and other authors stress their individual nature inspite of certain histological similarities [5, 6, 7].

In eosinophilic granuloma, which comprises approximately 50–60% of all cases of histiocytosis X, one or occasionally multiple bones demonstrates a histiocytic and eosinophilic leukocyte infiltrate of unknown etiology. The highest incidence occurs during the first three decades of life, with children between 5 and 10 affected more than other children or adults and males approximately 50% more often than females. The lesions most frequently involve the skull and femur in patients under 20 and the ribs and mandible in patients over 20.

Symptoms are usually limited to pain and tenderness in the area of the lesion.

Management

It is generally accepted that solitary eosinophilic granuloma of bone is a benign process that tends to heal spontaneously, though how long it takes has not been well documented. Den Herder [8] reported a large eosinophilic granuloma of the ilium for which no therapy was administered; it remained unchanged for 5 months, after which it regressed to nearly complete healing by 18 months. Though this is a single case, a period of several months seems

* Present address: Department of Radiology, Grossmont Hospital, 5565 Grossmont Center Drive, La Mesa, CA 92041, U.S.A.

to reflect a reasonable rate of natural bone repair. Circumstances which would suggest the need for therapeutic intervention in an effort to speed healing and allay complications include (a) unremitting pain, (b) significant restriction of motion and limitation of activity, (c) radiographic documentation of an extremely aggressive lesion, (d) threatened involvement of the adjacent growth plate, and (e) the likelihood of a pathological fracture based on the size and/or position of the lesion.

Systemic administration of corticosteroids has been shown to be effective in patients with histiocytosis X. Thompson et al. reported rapid clearing of pathologically proved eosinophilic granuloma of the lung within 3 weeks after starting systemic methylprednisolone [9]. Local injection of hydrocortisone produced marked involution of skin lesions in a case report by Carrie [10]. Similarly, Starling indicates that steroids are particularly useful when large areas of the skin are involved and in cases of pulmonary involvement [11]. At present the trend is toward combining steroids with antineoplastic drugs. Isolated lesions have been treated successfully with curettage or moderate doses of radiation therapy. Intralesional injection of methylprednisolone constitutes an alternative mode of treatment [12].

Direct injection of steroids into bone lesions is not without precedent. In 1974, Scaglietti et al. began a trial of methylprednisolone injections into bone cysts [13]. A 2-year experience with 72 cases followed up for more than 18 months revealed that 69 (96%) showed clearly positive results, with healing in 60% and/or more or less complete bone repair in 36%; pain promptly abated following injection. Since that time, others have obtained similar healing responses in bone cysts treated with one or more injections of methylprednisolone [14–17].

Technical considerations

Following identification of the lesion suspected of representing an eosinophilic granuloma based on the clinical and roentgenographic findings, a percutaneous needle biopsy is performed under fluoroscopic guidance. General anesthesia is required in children, whereas local anesthesia suffices in the older patients. If the overlying cortex is intact, it is perforated with a small drill bit and a spinal type needle (18–20 gauge) is introduced through the orifice. Tissue is aspirated for cytological analysis and once the diagnosis is confirmed, methylprednisolone sodium succinate, 125–150 mg is infiltrated into the lesion (Figs. 1 and 2).

Results

Twelve patients with 14 lesions in a series of 50 patients with eosinophilic granuloma were treated with intralesional injections of methylprednisolone sodium succinate at The University of Texas M. D. Anderson Hospital and Tumor Institute [18]. Two patients required two injections. All patients treated experienced relief of pain within the first two weeks following injection and did not require any additional therapy for pain. Follow-up in nine of the patients ranged from 25 to 48 months.

Radiographic features of healing occurred in all patients and were usually apparent three months after intralesional injection of methylprednisolone. These changes consisted of solidification of the periosteal reaction followed by a progressive decrease in the cortical thickening. The lytic area then

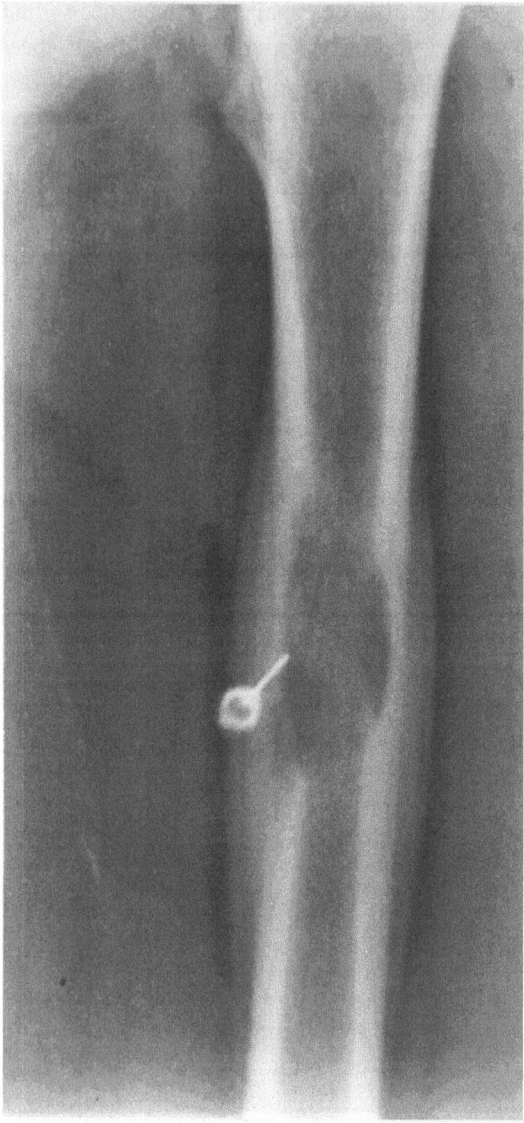
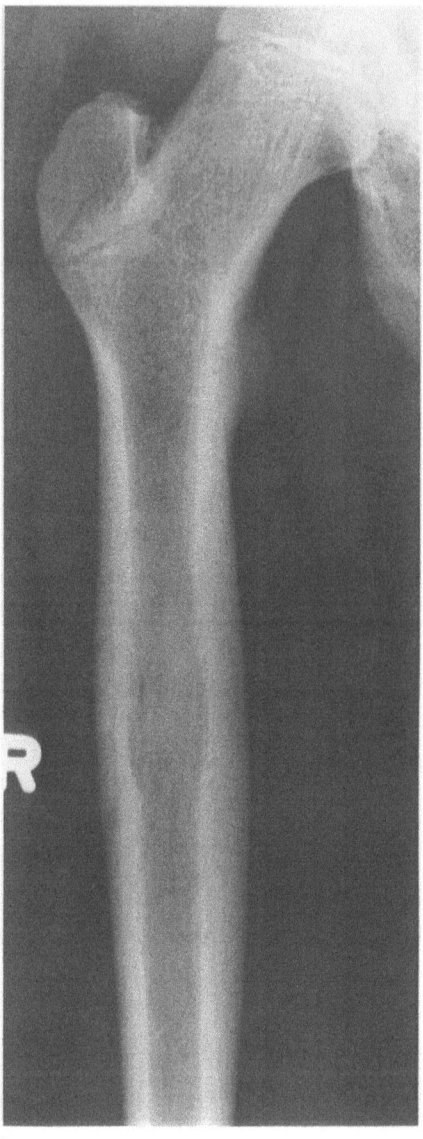

a   b

Fig. 1. Eosinophilic granuloma of bone. **a** 11-year-old male with a lytic lesion of the femoral shaft and an associated benign periosteal reaction. The diagnosis was established by percutaneous needle biopsy. **b** Thirteen months following the injection of methylprednisolone

gradually filled in with trabeculated bone attaining at times a completely normal architecture. Minimal cortical thickening may be the only residual evidence of the treated lesion (Figs. 1 and 2) [12, 18].

While it is not possible to document a cause and effect relationship between steroids and healing, especially in view of the tendency of the lesion to heal spontaneously, the rapidity of radiologically demonstrated response and relief of pain obtained from the injection strongly supports the efficacy of the medication. Finally, because other proved and accepted modes of therapy exist, it is impossible to set up satisfactory control studies. Possibly, though improbably, percutaneous biopsy and/or the introduction of the needle may trigger a healing response. Our limited results, coupled with the

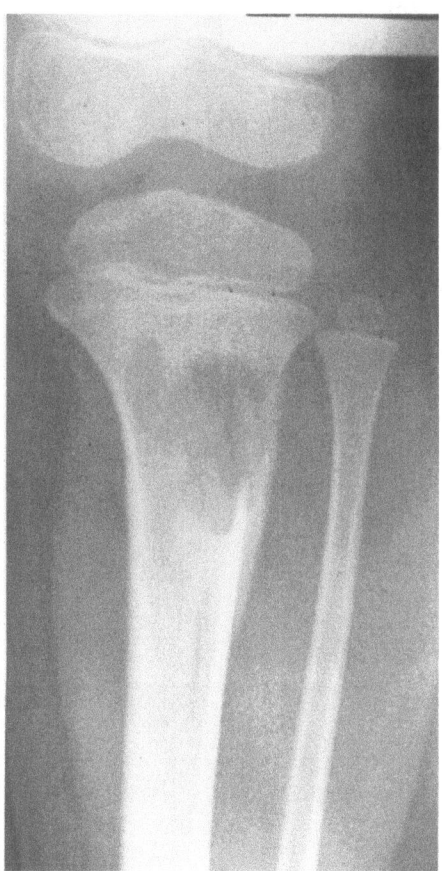

a

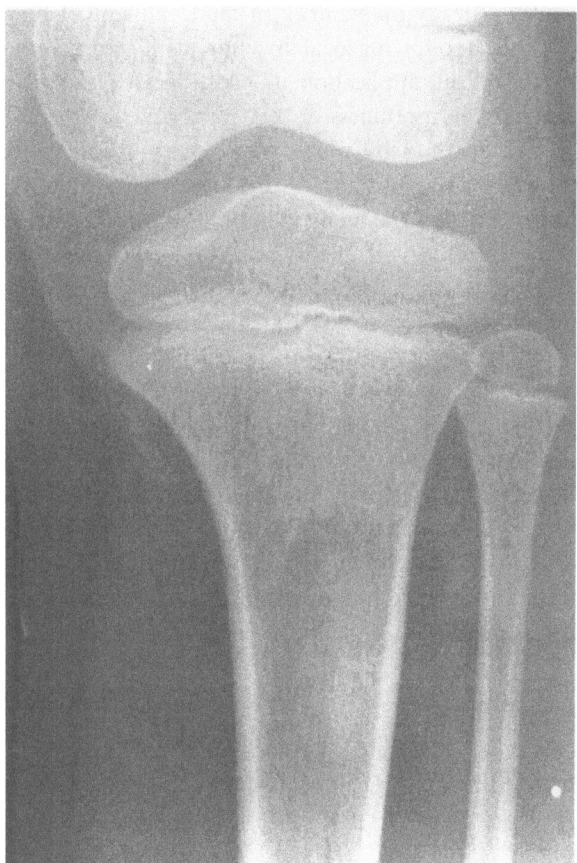

c

Fig. 2. Eosinophilic granuloma of bone. **a** 3-year-old female with a lytic lesion in the proximal tibia and an associated benign periosteal reaction. **b** The diagnosis was established by percutaneous needle biopsy. Iodinated contrast material was injected into the lesion to determine distribution of the methylprednisolone which was then injected. **c** Eight months after injection there was considerable healing

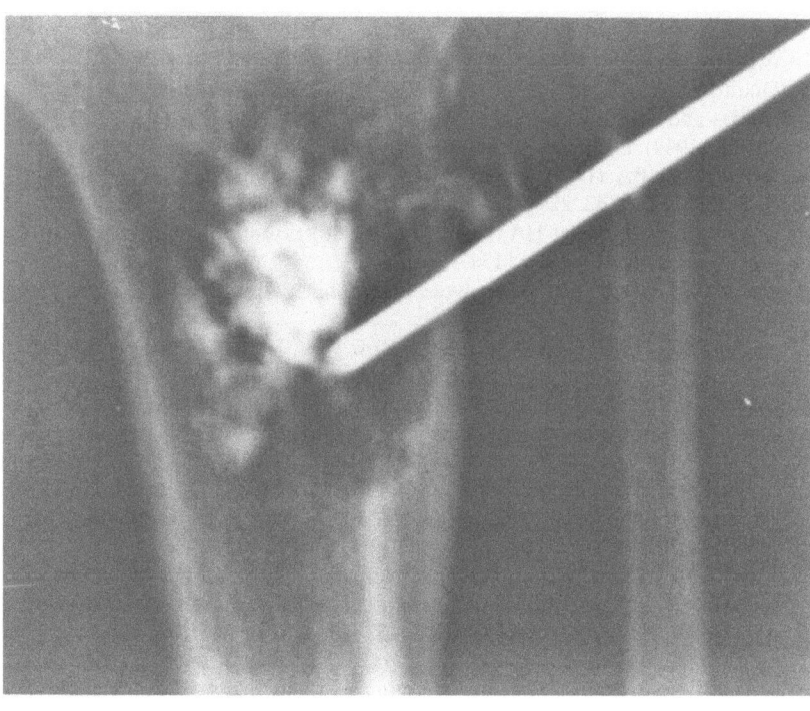

b

proved efficacy of the drug in the treatment of bone cysts and its lack of side effects following local injection justify continued use. Long-term follow-up as well as application in additional cases will be necessary to firmly establish the usefulness of this approach.

The method by which methylprednisolone affects bone repair is unknown. With respect to bone cysts, Scaglietti et al. [13] postulate that steroids destroy or alter the lining membrane, allowing osteogenesis. They dissected a cyst injected with methylprednisolone 2 months earlier and found the cavity filled with edematous fibroblastic connective tissue and active proliferation of trabeculae or reticular and osteoid bone on the wall of the cyst. This is in direct contrast to the usual histological findings in bone cysts: a thin fibrous, poorly visualized lining with little active proliferation and a few giant cells. Eosinophilic granuloma, however, has no definite lining membrane, and one is left to postulate some direct suppressive effect upon the cells of the lesion or antigens contained within it if one is to parallel the bone cyst hypothesis. The absence of an animal model, coupled with the lack of post-therapy tissue, limits any real investigation into the area of healing. Any valid hypothesis must await further clarification of the etiology of this disease.

References

1. Lichtenstein L, Jaffe HL (1940) Eosinophilic granuloma of bone. With report of a case. Am J Pathol 16: 595–604
2. Otani S, Ehrlich JC (1940) Solitary granuloma of bone. Simulating primary neoplasm. Am J Pathol 16: 479–490
3. Farber S (1941) The nature of "solitary or eosinophilic granuloma" of bone (abst). Am J Pathol 17: 625–626
4. Lichtenstein L (1953) Histiocytosis X. Integration of eosinophilic granuloma of bone. "Letter-Siwe disease", and "Schuller-Christian disease" as related manifestations of a single nosologic entity. Arch Pathol 56: 84–102
5. McGavran MH, Spady HA (1960) Eosinophilic granuloma of bone. A study of twenty-eight cases. J Bone Joint Surg [Am] 42: 979–992
6. Otani S (1957) A discussion on eosinophilic granuloma of bone, Letter-Siwe disease and Schuller-Christian disease. J Mt Sinai Hosp NY 24: 1079–1092
7. Siwe S (1949) The reticulo-endothelioses in children. Adv Pediatr 4: 117–143
8. Den Herder BA (1973) Changing views on eosinophilic granuloma of bone. Radiol Clin Biol 42: 218–221
9. Thompson J, Buechner HA, Fishman R (1958) Eosinophilic granuloma of the lung. Ann Intern Med 48: 1134–1145
10. Carrie C (1958) Zur Behandlung des eosinophilen Granuloms. Hautarzt 9: 38–40
11. Starling KA (1977) Histiocytosis. In: Sutow WW, Vietti TJ, Fernbach DJ (eds) Clinical pediatric oncology, 2nd edn. Mosby, St. Louis, pp 467–486
12. Cohen M, Zornoza J, Cangir A, Murray JA, Wallace S (1980) Direct injection of methylprednisolone sodium succinate in the treatment of solitary eosinophilic granuloma of bone. A report of 9 cases. Radiology 136: 289–293
13. Scaglietti O, Marchetti PG, Barolozzi P (1979) The effects of methylprednisolone acetate in the treatment of bone cysts. Results of three years follow-up. J Bone Joint Surg [Br] 61: 200–204
14. Campanacci M, de Sessa L, Bellando Randone P (1975) Cisti ossea. (Revisione de 275 osservazioni. Risultati della cura chirurgica e primi risultati della cura incruenta con metilprednisolone acetato). Chir Organi Mov 62: 471–482
15. Gualtieri I, Gualtieri G, Montefusco E (1976) Risultati ottenuti nel trattamento delle cisti ossee mediante infiltrazione con acetato di metilprednisolone. Osp Ital Chir 29: 155–160

16. Campanacci M, de Sessa L, Trentani C (1977) Cura incruenta della cisti ossea con iniezioni locali di metilprednisolone acetata secondo Scaglietti. Ital J Orthop Traumatol 3: 27–36

17. Savastano AA (1979) The treatment of bone cysts with intracyst injection of steroids. Injection of steroids will largely replace surgery in the treatment of benign bone cysts. Rhode Island Med J 62: 93–95

18. Nauert T, Zornoza J, Ayala A, Harle TS (1983) Eosinophilic granuloma of bone: diagnosis and management. Skeletal Radiol 10: 227–235

Miscellaneous

Direct intra-foraminal injection of corticosteroids in the treatment of cervico-brachial pain

G. Morvan, D. Monpoint, M. Bard, and G. Levi-Valensin

Department of Bone and Joint Radiology, Hôpital Lariboisière, Paris, France

Most cases of cervico-brachial pain (CBP) are relieved by medical treatment. However, in a few cases, pain is resistant to all medical treatments and surgery may be required. In such cases, some authors propose local intrathecal injection of corticosteroids as the last step of conservative treatment prior to surgery [2, 5]. Following the same concept, we have been performing foraminal injection of corticosteroids (FIC) in selected patients with CBP since 1977.

Patients and methods

Seventy-three FIC procedures were performed in 51 patients (37 men and 14 women) 25 to 69 years old (mean 46 years) over a nine-year period, from 1977 to 1986. Twenty-three cases of CBP involved the right side, and 28 involved the left side. In 36 patients, signs and symptoms concerned a single nerve root, while three had signs involving two different nerves. In 12 cases, the neurological territory was imprecise. The levels injected are indicated in Table 1. The mean duration of pain prior to injection was 4.4 months (15

Table 1. Cervical levels injected

C 4–C 5	4
C 5–C 6	22
C 6–C 7	43
C 7–D 1	4
Total	73

days to 16 months). Five patients whose pain had lasted 3, 4, 5, 9 and 20 years were excluded from this mean. The onset of CBP was sudden in 21 cases, progressive in 28, and unknown in 2. Pain was severe in 20 cases, significant in 23, moderate in 2, and of unknown intensity in 6. Pain was sufficiently intense, long-lasting and resistant to all medical treatment to require hospitalization. Signs of nerve root involvement consisting of motor or reflex impairment were present in 40 patients and absent in 11. None had signs of spinal cord compression. Radiographic and, in some cases, CT features at relevant cervical levels [1, 3, 4, 7] were compared with results of FIC.

Technique

The method used derives from the technique of cervical discography [6]. The radiographic table is placed vertically. The patients is sitting on a revolving stool with his head well propped up. Under fluoroscopic control, the patient is turned in an oblique position with the painful side forward

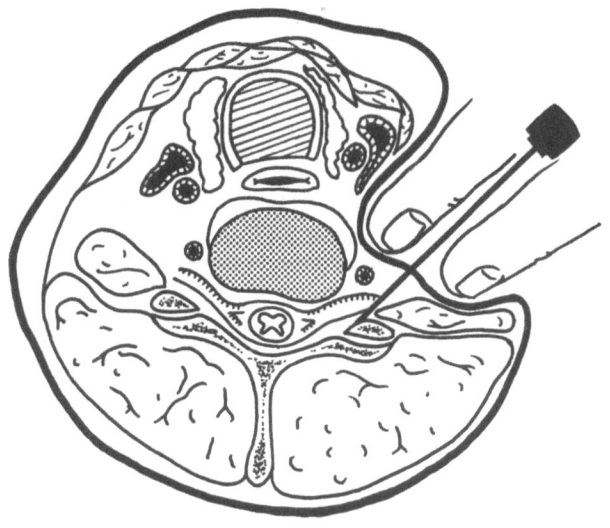

Fig. 1. Technique of approach

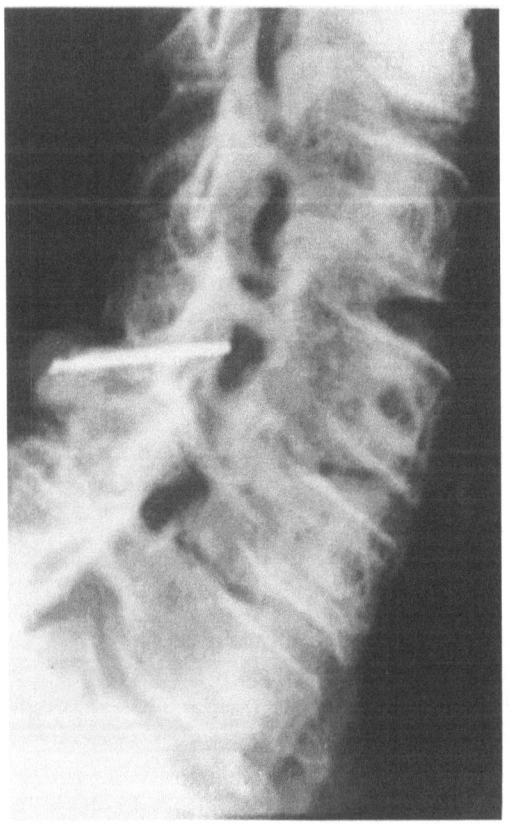

Fig. 2. Oblique view of the cervical spine. The needle tip is placed at the posterior aspect of the foramen, against the superior articular process

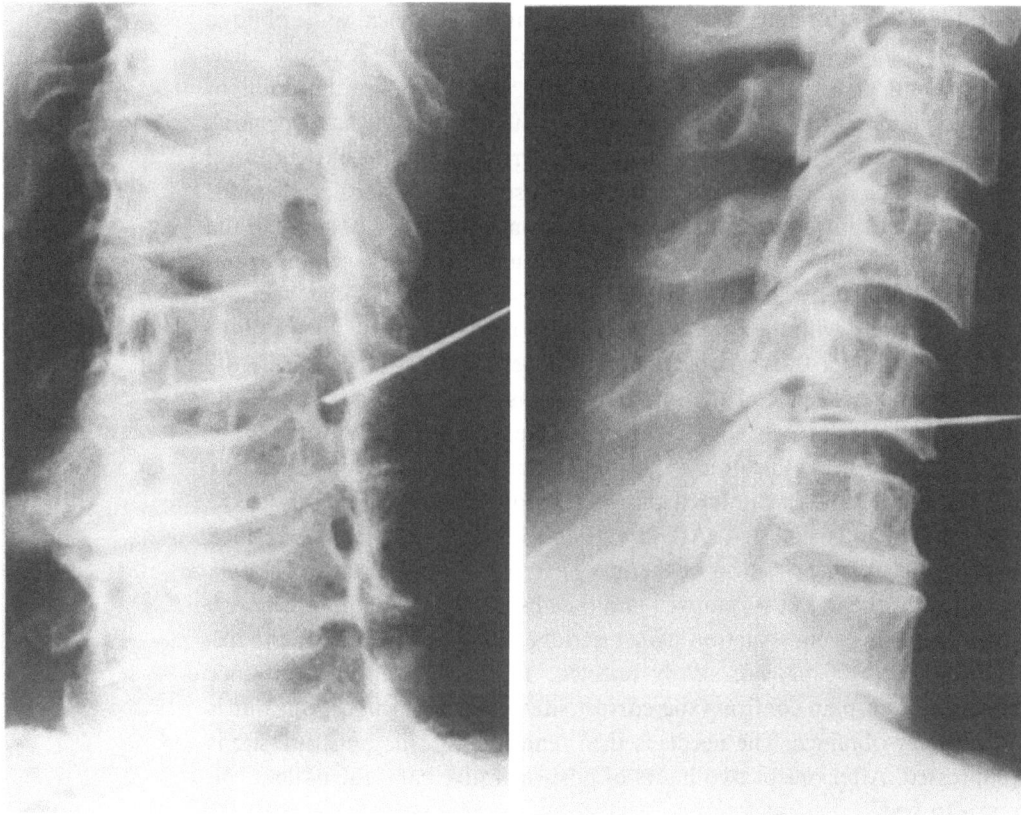

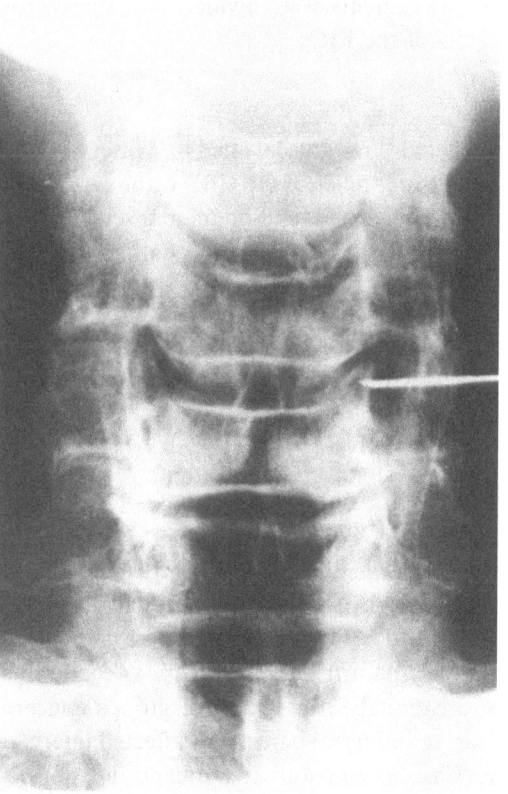

Fig. 3. Oblique (**a**), lateral (**b**), and frontal (**c**) views of the cervical spine with the C 5–6 disc seen in profile. The needle tip is placed at the anterior aspect of the foramen, against the uncus process

to profile the cervical intervertebral foramina, as for a cervical spine oblique view. The X-ray tube is tilted in a cephalad direction until the X-ray parallels the vertebral plates at the level to be treated. To determine the point of puncture, a metallic marker is placed on the patient's skin so that it projects on the relevant foramen on the fluoroscopic screen. The light of the central ray is used as a direction guide. Under usual aseptic conditions, a 21-gauge spinal needle is introduced at the point of the metallic mark, exactly in the direction of the light of the central ray, through the sternocleidomastoid muscle, while the trachea, esophagus, carotid artery and jugular vein are pushed medially with the fingers (Fig. 1). The needle is gently advanced to the foramen under fluoroscopic control. Some operators place the needle tip at the posterior part of the foramen, against the anterior aspect of the superior articular process (Fig. 2). Others place it at the posterior aspect of the uncus (Fig. 3). At this time a sharp radicular pain may occur, indicating that the needle has encountered the nerve root. In such cases, the needle position is slightly adjusted. AP, lateral and oblique views are performed to verify the correct position of the needle tip within the foramen (Fig. 2). The stylet of the needle is removed and the absence of blood or CSF return is checked using gentle suction with a syringe. Two to three ml of predni- solone acetate (50 mg) are slowly injected. This injection can reproduce radicular pain; pain confirms the correct site of corticoid injection, but is inconstantly obtained. The needle is then removed and the puncture site is compressed. After one or two hours of post-operative care, the patient can return to bed.

Results

The 51 patients were divided into three groups according to the clinical results of the FIC:

Group 1

Satisfactory results, obtained in 7 cases (14%), were defined as complete and long-lasting relief of the CBP without additional treatment. Relief was ob- tained between several hours and several days after the FIC.

Group 2

Fair results, obtained in 31 cases (61%), were defined as incomplete or transitory improvement of CBP.

Group 3

Poor results, obtained in 13 cases (25%), were defined as minimal change in CBP.

The mean age of the seven patients (all men) of group 1 was 57 years. Their CBP had lasted from ten days to four months (mean: two months) before FIC. In six of these cases, a single FIC was sufficient to obtain complete pain relief; in one case a second injection was necessary. A com- parative analysis of the three groups concerning the type of onset of the CBP, radiological patterns of affected intervertebral discs and foramen, and response to previous treatments, did not reveal a characteristic pattern among the seven cases. Neither disc narrowing nor uncarthrosis was found to be significantly correlated with results. The only significant features found in 6 of the 7 cases were the rather short duration of pain (two months) prior to FIC, and a positive response to the first FIC.

Complications

In one case an isolated headache occurred following FIC, but it lasted only a few hours and disappeared spontaneously. In another case, dizziness arose after the procedure, lasted 24 hours, and disappeared completely without treatment.

Discussion

The principle of FIC is the same as that for juxtadural injections of steroids at the lumbar level. Nevertheless, cervical anatomic conditions (spinal cord, vertebral artery, carotid artery and jugular vein) make FIC a more delicate and complicated operation, necessitating fluoroscopic control and hospitalization.

Overall results are unsatisfactory (14% satisfactory and 86% fair or poor results). Results are impossible to predict insofar as no particular clinical or radiological pattern was significantly correlated with satisfactory results.

Conclusion

Of 73 FIC procedures in 51 patients, satisfactory results, defined by rapid and complete disappearance of brachial pain, were obtained in 14% of cases.

In view of the relatively invasive nature of this procedure and the overall unsatisfactory results of our experience, we cannot recommend this technique as a routine treatment of CBP. However, a single FIC should be tried in some stubborn cases of CBP of relatively short duration.

References

1. Baleriaux D, Noterman J, Ticket L (1983) Recognition of cervical soft disc herniation by contrast enhanced CT. AJNR 4: 607–608
2. Chaouat Y, Paquet J (1968) Les névralgies cervico-brachiales trainantes. Tentatives thérapeutiques par les injections intra-rachidiennes de corticoides. Rev Rhum 26: 657–660
3. Coin CG, Coin JT (1981) CT of cervical disc disease. Technical considerations with representative case reports. J Comput Assist Tomogr 5: 275–280
4. D'Angelo CM, Zimmerman RD, Czervionke LF, Huckman MS (1984) Cervical disc herniation: CT demonstration after contrast enhancement. Radiology 152: 703–712
5. Lucherini T (1954) Premiers résultats sur l'emploi de l'hydrocortisone par voie intra-rachidienne. Rev Rhum 12: 809–816
6. Massare C, Bard M, Tristant H (1974) Discographie cervicale. Réflexions spéculatives au plan de la technique et des indications dans notre expérience. J Radiol 55: 396–399
7. Morvan G, Busson J, Massare C, Bard M, Seguy E (1984) Exploration tomodensitométrique des névralgies cervico-brachiales avec injection intra-veineuse de produit de contraste. J Radiol 65. 159–164

Aspiration of tendinous calcific deposits

C. Normandin[1], E. Seban[1], J.-D. Laredo[2], Dominique N'Guyen[2], D. Kuntz[1], and M. Bard[2]

Departments of [1] Rheumatology (Centre Viggo-Petersen) and of [2] Bone and Joint Radiology, Hôpital Lariboisière, Paris, France

Needle aspiration of tendinous calcific deposits under fluoroscopic control may be a valuable tool in selected cases of painful shoulders with chronic and debilitating pain resistant to medical treatment. This procedure is most frequently performed on the shoulder. However, it may also be carried out in certain other locations, such as tendons and joint capsules of the hip, elbow and wrist.

Background

Aspiration of calcific deposits was first performed on acutely painful shoulders with evacuation of calcification into the sub-acromial bursa [7, 18]. In these cases, superficial puncture is sufficient to accelerate the process of resorption of the deposits. Comfort et al. used this same technique primarily for acute episodes; however, they also performed the procedure in some cases of chronic painful shoulders and first stressed the value of fluoroscopic control for localization of the calcific deposits and needle guidance [6]. Comfort et al. also established that it is not mandatory to evacuate the calcific deposits in totality, since spontaneous resorption of the residual calcium frequently occurs in the weeks following needle aspiration. In 1981, Gross and Siegrist reported their experience of repeated needle aspiration and irrigation of the shoulder at two-week intervals [8]. They observed 31 complete recoveries in 60 cases of both acute and chronic painful shoulders.

In the last few years (1983–1987), this technique has primarily been used in chronically painful shoulders with intratendinous calcific deposits of the rotator cuff which have little or no spontaneous tendency to evacuate into the sub-acromial bursa [3, 14, 16].

Indications and contra-indications

The procedure has 3 goals:
- evacuation of a maximum of calcium;
- fragmentation of the residual calcific deposits, in order to facilitate resorption during the following weeks;
- reduction of inflammation secondary to the presence and migration of residual calcific deposits by in situ injection of corticosteroids.

Finally, the aim of this procedure is to avoid surgery in selected patients with chronic painful shoulders associated with tendinous calcific deposits.

Indications

Needle aspiration is indicated in selected patients with chronic and debilitating shoulder pain resistant to medical treatment and associated with persistant calcific deposits within the rotator cuff tendons (Table 1).

Table 1. Indications and prognostic factors

Indications	Criteria of good prognosis	Criteria of bad prognosis
Calcification of the rotator cuff with chronic debilitating pain	Large amount of calcium aspirated	Calcium deposity of hard consistency
No tendency to spontaneous evacuation	Secondary resorption of residual calcium	Intratendinous calcification with a striated appearance
Diameter 5 mm	Absence of associated lesions of rotator cuff tendons	

This condition usually develops over several months or years and may be complicated by occasional acute episodes. In the late stage, it interferes with sleep and sports. Professional activities may be considerably curtailed. The inefficacy of prior medical treatment such as non-steroid anti-inflammatory agents, physiotherapy and cortisone injections must be verified.

Needle aspiration should be performed if the clinical and radiologic evaluation suggest that pain is at least partially related to the presence of calcific material within the rotator cuff. Since the finding of tendinous calcifications is frequently clinically insignificant, an attempt must be made to relate the painful symptoms to the presence of the calcific deposits. This includes careful clinical and radiological evaluation to eliminate those patients with rotator cuff tear. Impingement syndrome of the rotator cuff tendons is another cause of shoulder pain which is difficult to differentiate from calcific deposit-related symptoms. However, impingement syndrome may be associated with tendinous calcification of the rotator cuff and is not a contra-indication to the technique if symptoms are likely to be partly related to the calcific deposit. Furthermore, the calcified material may play a role in the persistance of an inpingement syndrome.

Clinically, pain at night with intermittent acute exacerbation and accentuated by all shoulder motions, suggest that symptoms are related to the calcific deposit. When present, decrease in joint mobility involves all shoulder movements. Conversely, absence of pain at night, pain caused by a particular kind of motion within a given arc, and limitation of a specific motion suggest that pain is related to tendon lesions.

Radiologic assessment of calcific deposits

Prior to needle-aspiration, a careful radiologic evaluation of the calcific deposits and related lesions is required, including:

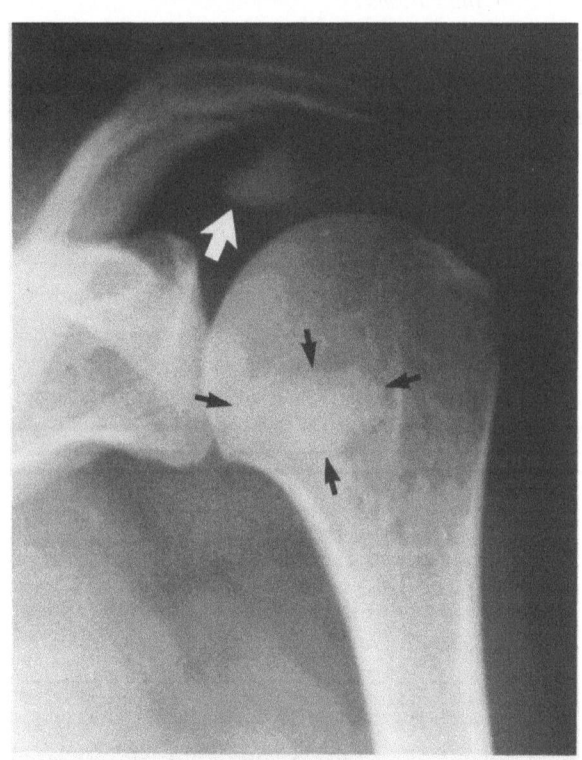

Fig. 1. Association of two separate calcifications in supraspinatus (large arrow) and subscapular tendons (small arrows)

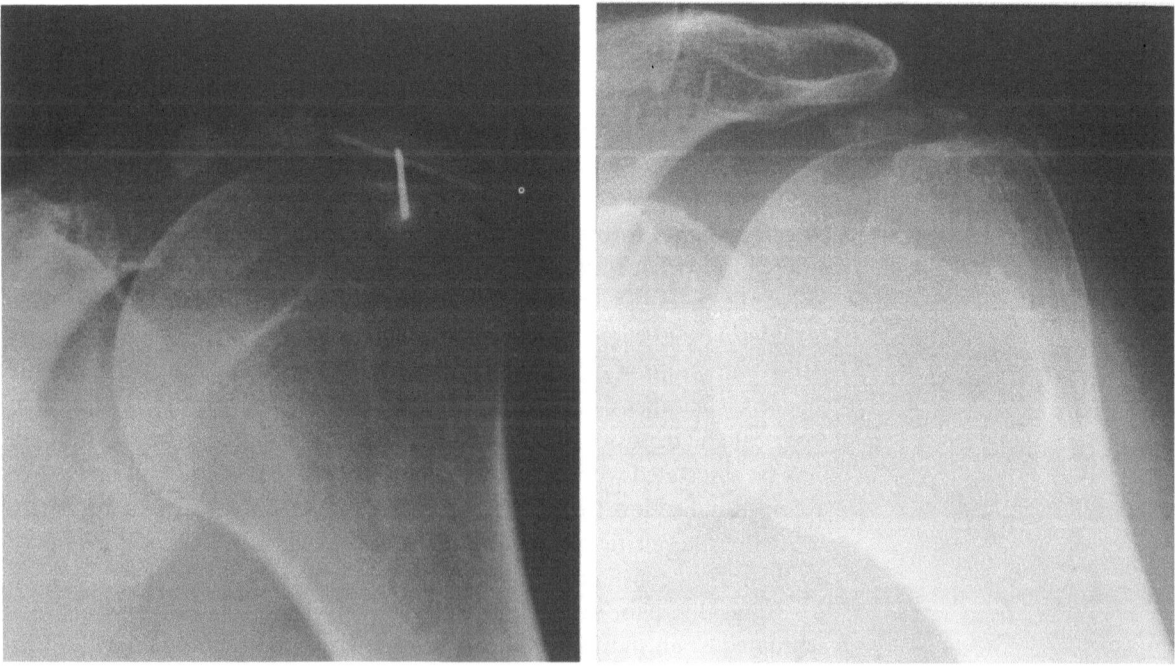

a   b

Fig. 2. Lobulated calcification. Aspiration with two needles (**a**). Residual calcific deposits at the end of the procedure (**b**)

Plain films of the shoulder

AP views with the arm in neutral position and external and internal rotation; an axillary view; a profile of the scapula which demonstrates the subacromial space in a projection orthogonal to the AP view.

The structure of the calcific deposits and their relation to the rotator cuff tendons and subacromial bursa must be determined. Other features to evaluate are number, size, density, contour, homogeneity, or tendency to evacuate into the subacromial bursa. The absence of changes on successive X-rays suggests an indication for needle aspiration. The presence of multiple calcific deposits is frequent: for example, supraspinatus calcifications and sub-scapular calcifications frequently co-exist (Fig. 1). In practice, a calcific deposit with a minimal diameter of 5 mm is accessible. Ultimately, however, the composition of the deposit will determine the success or failure of the aspiration. In this regard, both the density and contours of the deposit aid in predicting the outcome of needle aspiration. Faint calcifications with vague contours are usually associated with liquid consistency (Fig. 1). On the contrary, rock-hard solid deposits are common in the case of very dense appearing images with clearly defined margins. Distribution of the calcific deposits also influences the success of the procedure. Multiple needle punctures are usually necessary in lobulated calcifications (Fig. 2). Aspiration may be unsuccessful in scattered deposits (Fig. 3).

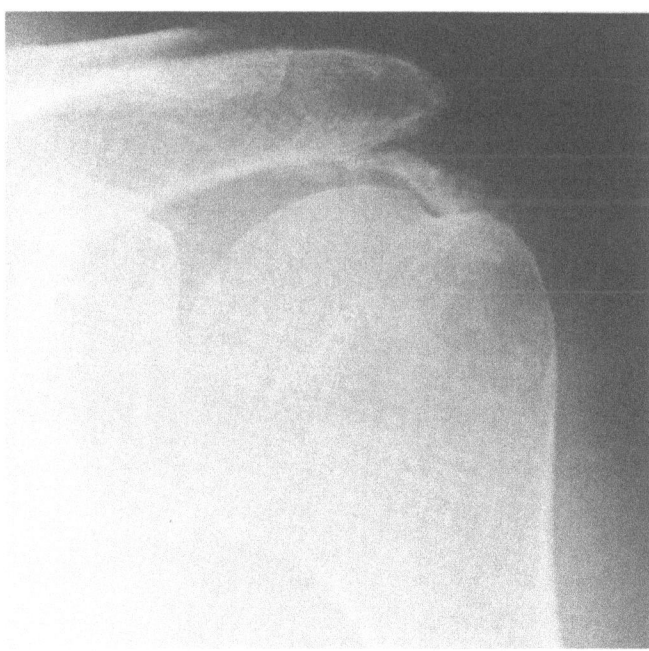

Fig. 3. Calcific deposits with striated appearance. Calcifications with this appearance are usually located within tendons fibers and cannot be significantly evacuated by needle-aspiration

Echography

Echography allows direct visualization of muscles and tendons. The thickness of the rotator cuff tendons can be accurately measured. In most cases, complete rupture of the rotator cuff can be evaluated, thus avoiding arthrography. Echography also provides easy measurement and localization of calcifications within the rotator cuff.

Arthrography

Arthrography is indicated in clinically suspected tear of the rotator cuff when echography is inconclusive. Arthrography will demonstrate the magnitude of the tendon tear and help in making a surgical decision. In addition, it allows verification of joint capacity in clinically suspected adhesive capsulitis.

Contra-indications

Concurrent infection and confirmed allergy to local anesthetics are formal contra-indications to needle aspiration.

Symptoms related to tendon or joint lesions such as complete and incomplete tears of the rotator cuff, adhesive capsulitis and abnormalities of the biceps tendon are not amenable to needle-aspiration. When these conditions are suspected on clinical grounds, radiologic examination including echography and, when necessary arthrography, must be carried out prior to needle-aspiration to confirm the diagnosis. However, the coexistence of complete rotator cuff tear and tendinous calcific deposits is a rather uncommon finding. In our personal experience, we found the average age of patients with tendinous calcific deposits who are candidates for needle-aspiration to be inferior by about ten years to that of patients with rotator cuff tears (47 versus 56 years old).

Not all calcific deposits are amenable to needle-aspiration. As stated above, the calcification must have a diameter of at least 5 millimeters. Calcifications with a milky homogenous appearance are usually located at

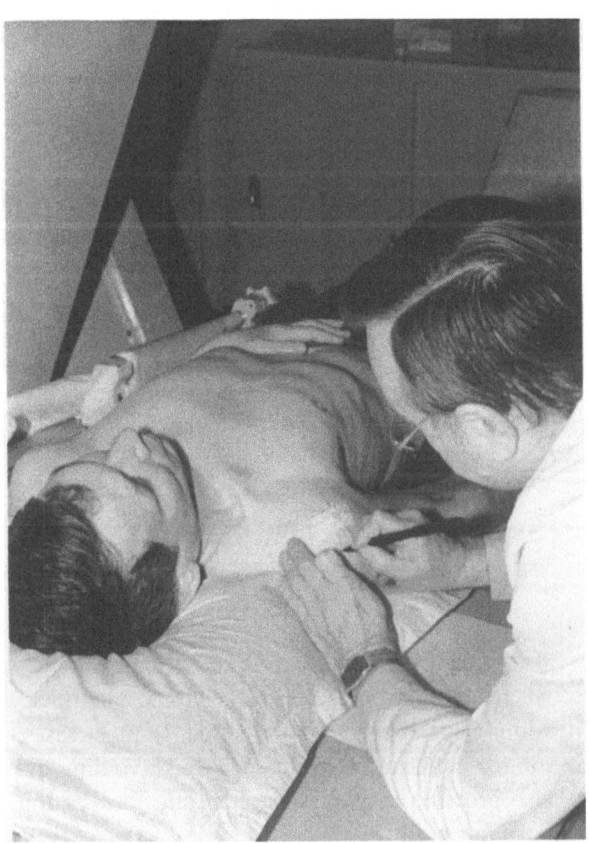

Fig. 4. Determination of point of skin puncture: a lead marker is placed on the skin so that it projects in the center of the calcific deposit on the fluoroscopic screen

the external surface of the tendons and represent good indications for needle-aspiration technique (Fig. 1). Those with a heterogeneous striated appearance usually correspond to diffuse calcific incrustation of the tendon fibers, which cannot be aspirated (Fig. 3).

Technique

The technique described concerns the shoulder, which is the most frequent site of tendinous calcific deposits. We use a direct antero-posterior approach. Some calcific deposits are accessible through a lateral or posterior route. However in our experience, the technique is more difficult using these approaches. The patient is placed in supine position on the radiographic table (Fig. 4). The X-ray beam is centered vertically to the shoulder or slightly tilted if this allows better separation of the calcium deposit from the underlying bone on the image intensifier screen. Arm position is chosen according to the location of the calcification within the rotator cuff (Table 2).

Table 2. Summary of data for localization of calcific deposits*

Location of calcification	Direction of movement with rotation of humerus		Tendon
	Internal	External	
1. Upper third, greater tuberosity	med.	lat.	supraspinatus
2. Middle third, greater tuberosity	lat.	med.	infraspinatus
3. Lower third, greater tuberosity	lat.	med.	teres minor
4. Superior rim of glenoid	none	none	bicep-long head
5. Inferior rim of glenoid	none	none	bicep-short head
6. Lesser tuberosity	med.	lat.	subscapularis

* From [21]

Superficial calcifications within the subacromial bursa are well visualized with the arm in neutral position. Clear visualization of deep tendinous calcific deposits requires a variable degree of arm rotation in order to avoid superposition of the underlying humeral head. External rotation is needed for supraspinatus tendon deposits, while internal rotation allows good visualization of infraspinatus and teres minor calcifications (Fig. 5). Calcifications within the long portion of the biceps are usually located close to the upper margin of the glenoid and are not affected by rotation of the humeral head. Calcifications within the subscapular tendon are aspirated anterior to the humeral head with slight external rotation of the arm. Once the optimal tilt of the X-ray beam and position of the arm have been determined under fluoroscopic control, the puncture point is chosen. A lead marker is placed on the skin so that it projects in the center of the calcific deposit on the image-intensifier screen [15, 16] (Figs. 4 and 5).

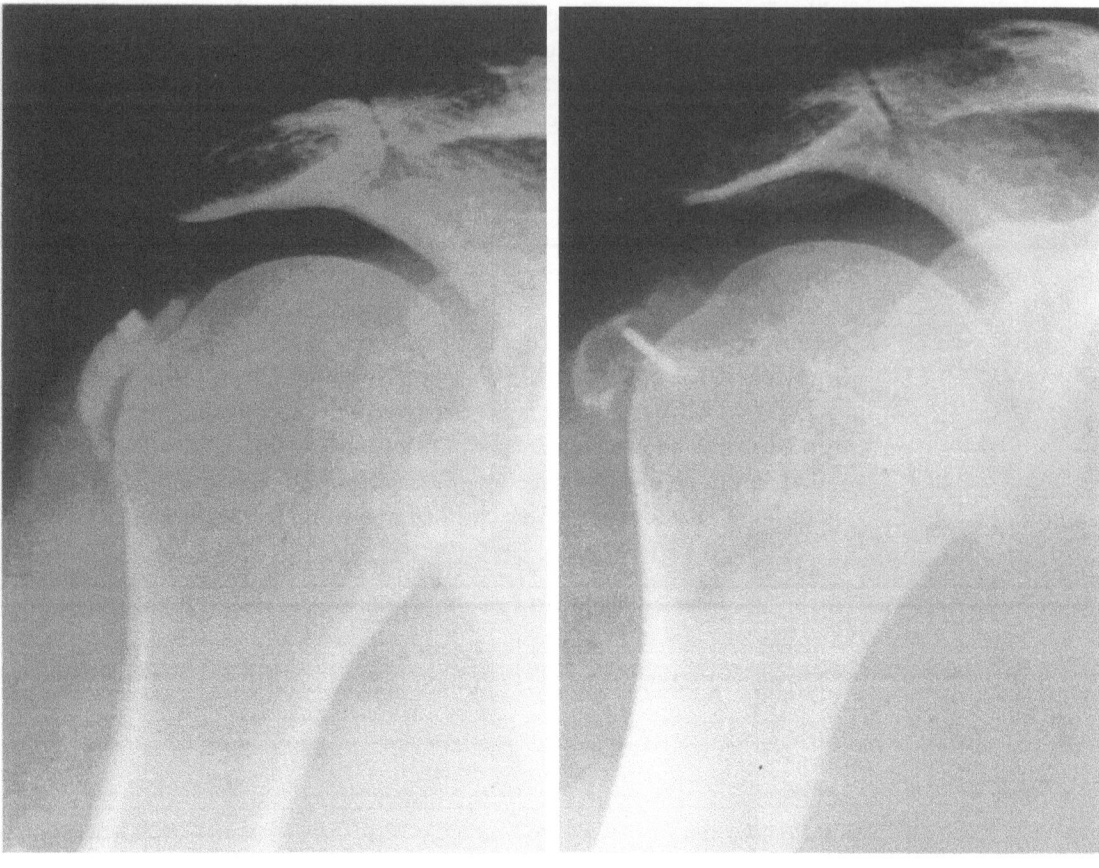

a   b

Fig. 5. Infraspintaus calcific deposits before (**a**) and after (**b**) needle aspiration. The arm is in internal rotation

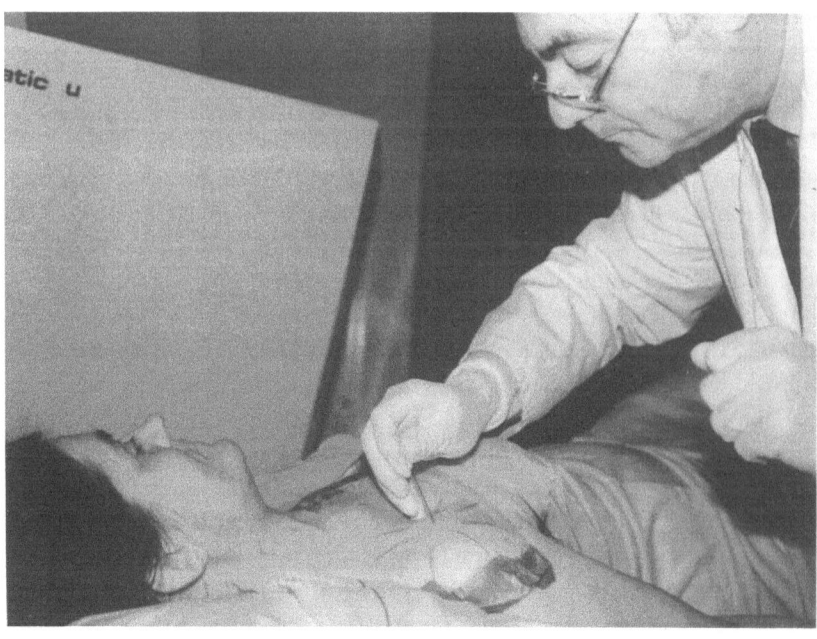

Fig. 6. Needle insertion following the direction of the X-ray beam

Puncture

Aseptic conditions are mandatory. The skin is shaved and disinfected with an alcohol-iodine solution. Sterile gloves, sterile compresses and a sterile field are used. The skin and superficial planes are anesthetized with 1 percent Xylocaine. A 19-gauge needle with stylet is inserted at the pre-determined puncture point and advanced under fluoroscopic guidance to the center of the calcification, following a direction parallel to the X-ray beam (Fig. 6) [15, 16]. During the entire approach the needle appears on the image intensifier screen as a single point in the center of the calcification (Fig. 7a). A firm sensation is obtained when the calcification is reached. At this stage of the procedure, the X-ray beam is successively tilted to two oppositive perpendicular angles to confirm on the image-intensifier screen that the needle tip has indeed penetrated the calcification (Fig. 7b, c). Calcium aspiration is performed using a syringe containing sterile water or saline solution adapted to the aspiration-needle. A succession of propulsions and suctions with the syringe piston is done. If necessary, the aspiration needle is moved back and forth as well. Aspirated calcium appears in the syringe as a white cloudy return (Fig. 8). Calcium is evacuated in a test-tube and the syringe is again filled with liquid to aspirate additional calcium. This

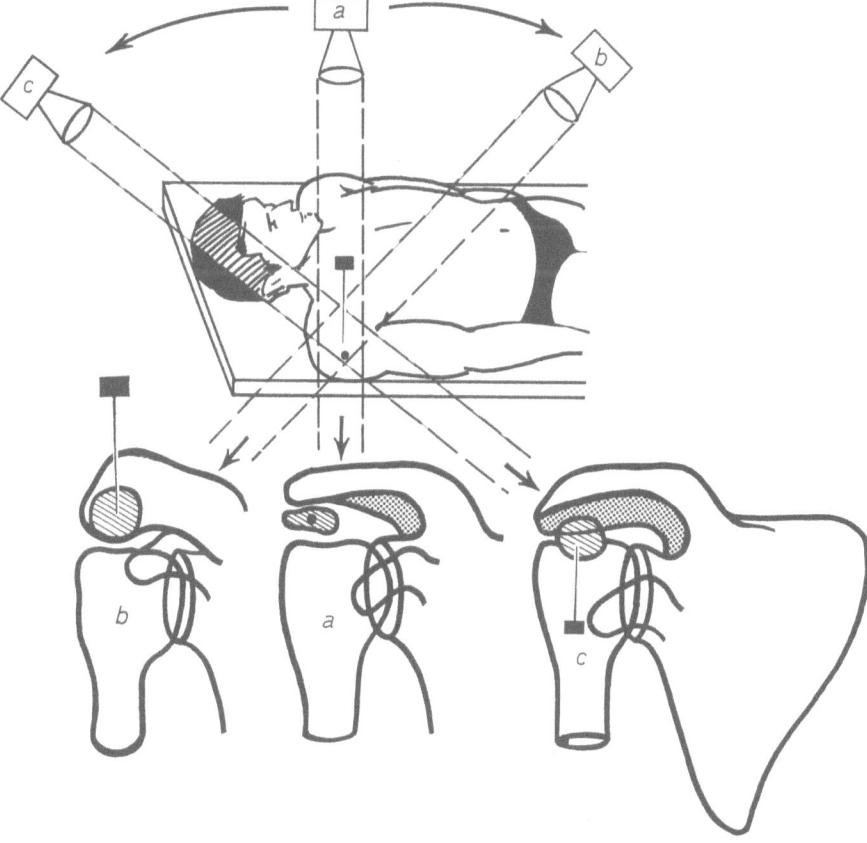

Fig. 7. Fluoroscopic procedure to check that the needle tip is within the calcific deposits. When the X-ray beam is directed along the needle axis, the needle appears as a dot ₃in the center of the calcification (**a**). The X-ray beam is then successively tilted in maximal cephalad (**b**) and caudad (**c**) directions. If correctly placed the needle tip will remain within the calcification

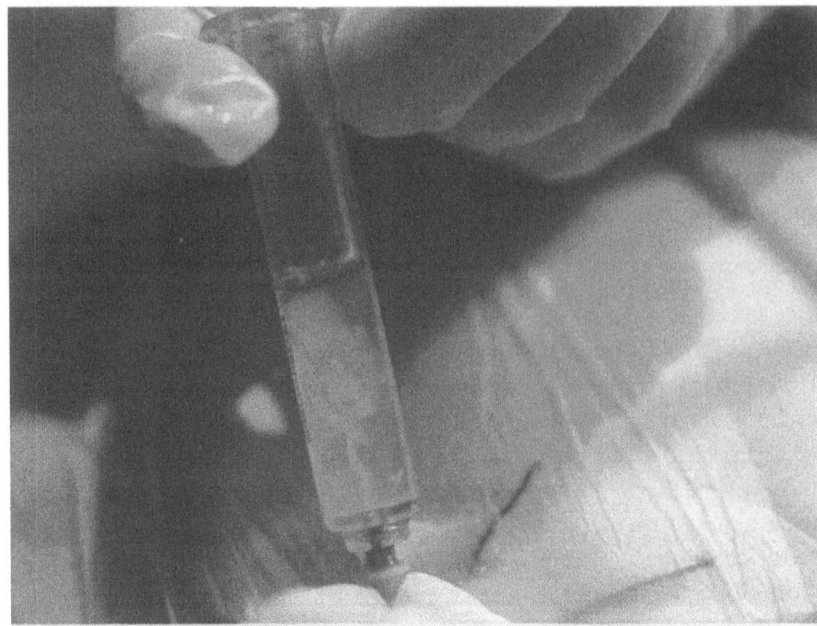

Fig. 8. Aspirated calcium appearing as a cloudy return in the syringe

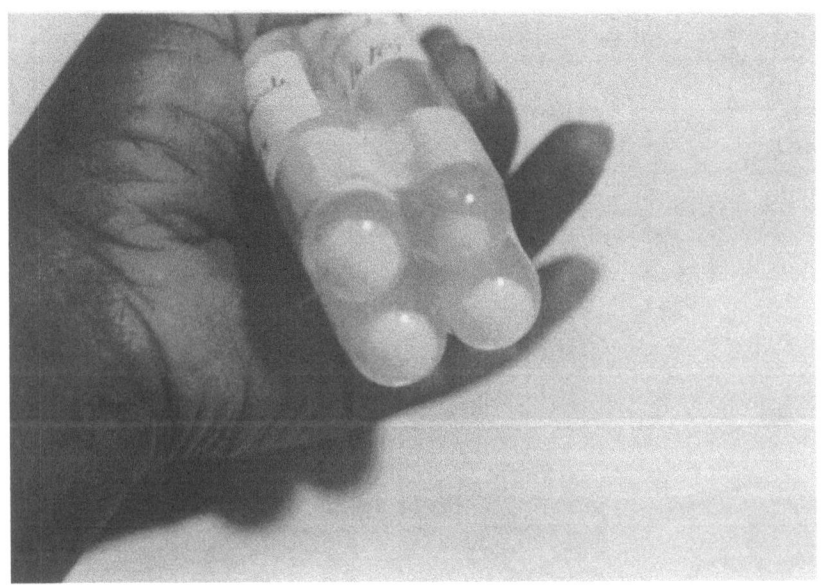

Fig. 9. Calcium aspirated at the end of a successful procedure

procedure is repeated until maximal aspiration of calcium has been obtained (Fig. 9). In large and lobulated calcific deposits, insertion of 2 parallel needles may be useful to complete the needle irrigation. The amount of calcium aspirated at the end of the procedure is variable and always incomplete (10–80%). In some cases, the calcification has a hard consistency and no calcium can be aspirated. However, grinding of the calcific deposit with the needle may in itself accelerate the process of spontaneous resorption [3, 6, 8, 14, 15, 16] and is as important as aspiration. Once maximal calcium aspiration has been performed, 2 to 3 ml of prednisolone acetate (50 to 75 mg) are injected in situ [1, 14]. Radiographs are performed at the end of the procedure to assess and document evacuation of the deposits. In the great majority of cases, the procedure is well-tolerated and painless.

Immediate follow-up

Keeping the shoulder at rest for 5 or 6 days is recommended. No bandage or orthopedic material is used. This also applies to other articulations: hip, elbow and wrist (Figs. 10 and 11). One third of patients, usually those presenting with local inflammation, have a painful reaction rarely lasting more than 2–4 days. This is managed with intermittent application of ice and prescription of antalgics and non-steroid anti-inflammatory agents [14,

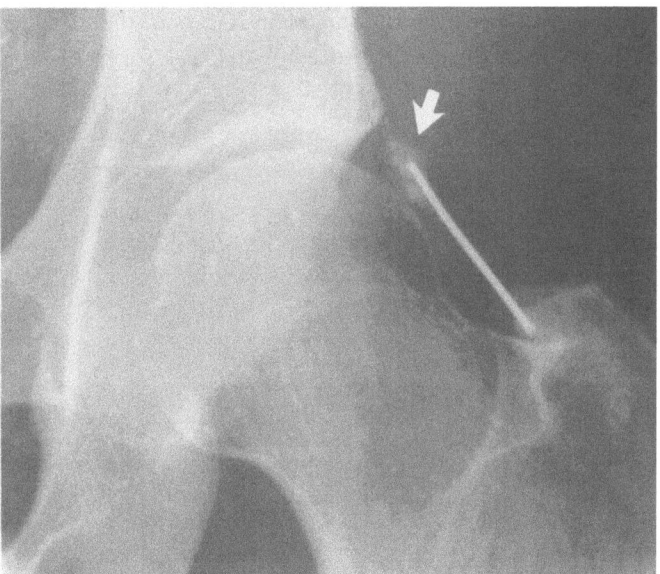

Fig. 10. Aspiration of a calcific deposit near the thip joint (arrow)

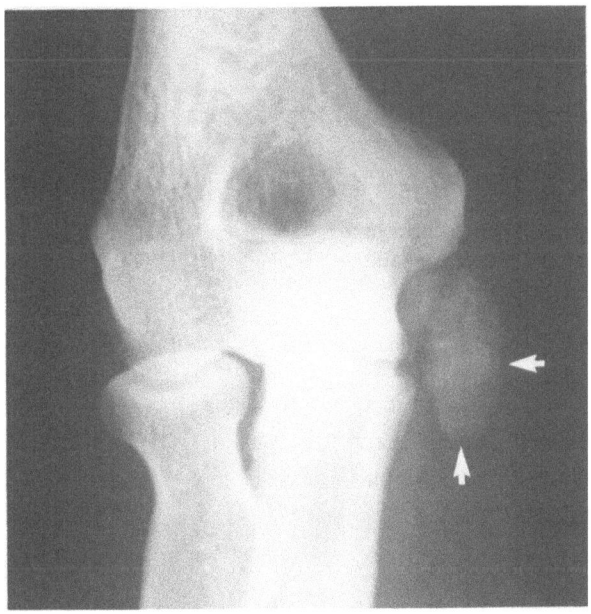

a

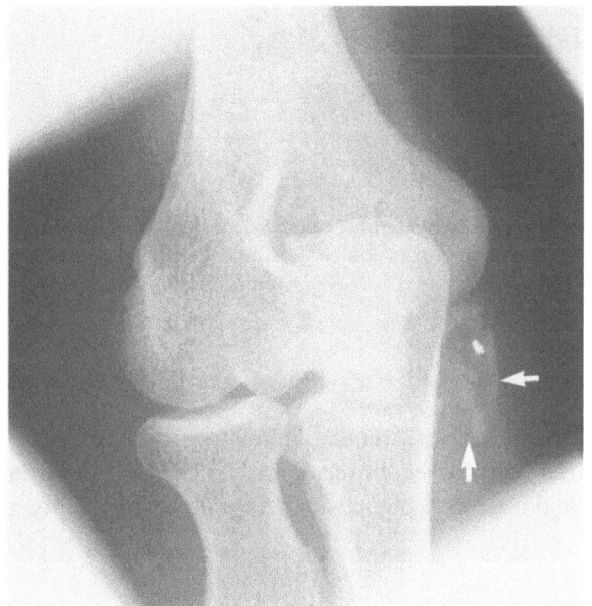

b

Fig. 11. Calcific deposit at the internal aspect of the elbow before (**a**) and after (**b**) needle aspiration (arrows)

15, 16]. Some physicians administer Colchicine 1–2 mg per os for several days in order to prevent a secondary reaction to calcific deposit fragmentation [15, 16].

The first sign of clinical improvement is usually the disappearance of pain at night, with recovery of sleep. Then, in 1–3 weeks, pain is also relieved during the day and, in cases with good and excellent results, the ability to carry out normal activities is recovered.

An untimely early return to arduous professional activity, housework, sports or vigorous rehabilitation exercises can result in relapse and sometimes in treatment failure. Additional plain X-rays of the shoulder are performed in the weeks following the procedure to check for resorption of the residual calcium (Fig. 12).

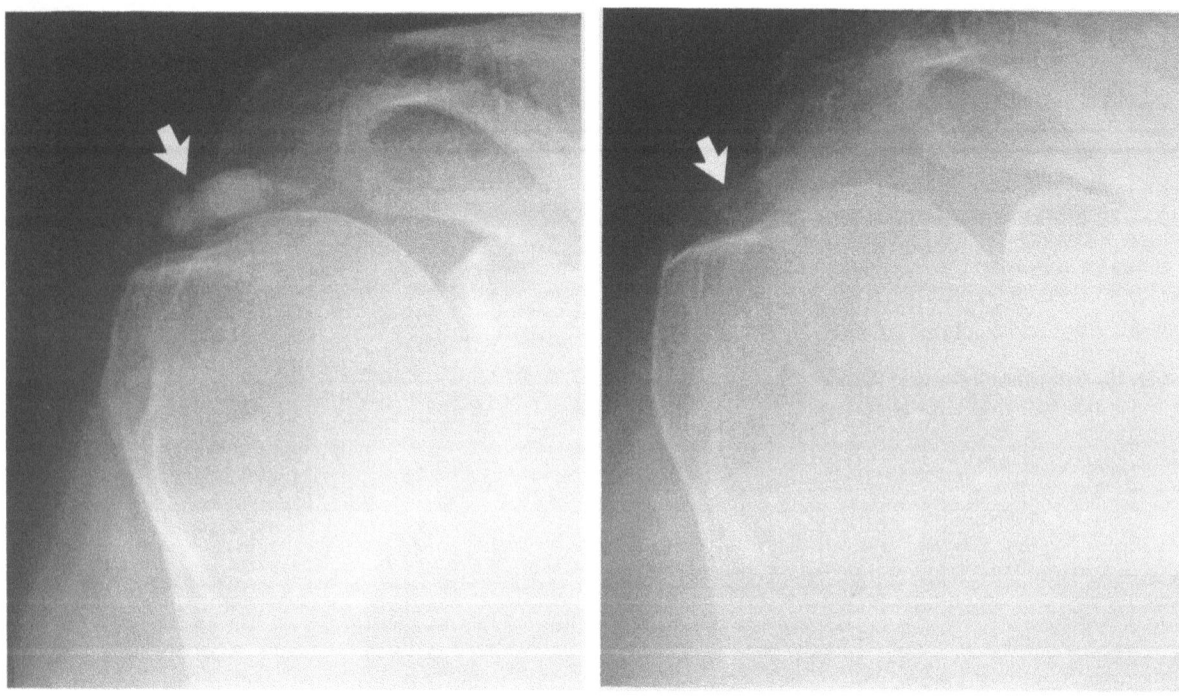

a   b

Fig. 12. a Supraspinatus calcific deposits before needle aspiration (arrow); the arm is in external rotation. **b** Residual calcific deposits after needle aspiration and secondary resorption (arrow)

Crystallographic and chemical studies

Studies of the aspirated granulations identify Type B carbonated apatites rich in hydrogenphosphate ions $HPO4_2$ [19], consisting of an organic fraction and a mineral phase with a calcium to phosphore ratio of 1, 67.

Results

Four criteria were used to evaluate results of the needle-aspiration technique in our series: recovery of sleep, percentage of subjective improvement, recovery of joint mobility and pain relief, and ultimate resorption of calcific deposits [15, 16]. Results were classified as excellent (fulfillment of four criteria), good (fulfillment of three criteria), fair (fulfillment of two criteria) and poor (fulfillment of one or no criteria). In a personal experience on 69 cases (mean age of patients 47 years), clinical result obtained at 11 to 45 months follow-up (average 24 months) was excellent or good in 42 cases (60.9%), fair in 8 cases (11.6%) and poor in 19 shoulders (27.5%) [16]. Eight patients with a poor result were subsequently operated on. Aspiration of large amounts of calcium and secondary resorption of residual deposits in the weeks following the needle aspiration were significantly associated with satisfactory results. By contrast, negative needle aspiration and the absence of secondary resorption were associated with a significant rate of failure.

In previous reports, the percentage of good and excellent results varied from 49 to 100% (Table 3) [3, 6, 8, 11, 14, 18]. However, these reports do not always specify the clinical condition of the patients treated and the length of follow-up.

Conclusion

Needle aspiration of tendinous calcific deposits is a well-tolerated conservative procedure which should be attempted after failure of medical treatment in chronically painful shoulders associated with rotator cuff deposits. Good and excellent results vary from 49 to 70% of patients. In these patients,

Table 3. Results of aspiration of tendinous calcific deposits of the shoulder

Author	No. of shoulders	Clinical condition	Follow-up	% of good and excellent results
Patterson and Darrach (1937) [18]	63	acute painful shoulders (76%)	N.S.*	90
Marchetti et al. (1968) [12]	63	N.S.*	N.S.*	66
Comfort and Arafiles (1978) [6]	9	N.S.*	9 years	100
Gross and Siegrist (1981) [8]	60	N.S.*	N.S.*	70
Cabanel (1983) [3]	58	chronic painful shoulders (93%)	10 months minimum	58
Moutonnet et al. (1984) [14]	41	N.S.*	3 months minimum	49
Mansat et al. (1984) [11]	30	N.S.*	N.S.*	63
Normandin and Ortiz-Bravo (1986) [15]	54	chronic painful shoulders (90%)	12–36 months	61
Normandin et al. (1987) [16]	69	chronic painful shoulders (90%)	11–45 months	61

* *N. S.* Not stated

dramatic and durable improvement is obtained without surgery. Findings related to a better response rate are a large amount of calcium aspirated during the procedure and secondary resorption of residual calcification in the following days or weeks.

References

1. Anglejan G de, Dupuich C, le Meignen P, Lermusiaux JL, Masse JP, Normandin C, Teyssedou JP (1983) Cortisone et tendon. In: de Seze S et al (eds) L'actualité rhumatologique 1983. Expansion Scientifique, Paris, pp 293–297

2. Bosworth BM (1941) Calcium deposits in the shoulder and subacromial bursitis. A survey of 12, 122 shoulders. JAMA 116: 2477–2482

3. Cabanel G (1983) La ponction aspiration lavage ou trituration: une modalité thérapeutique des tendinites calcifiantes rebelles de l'épaule. Mémoire CES de Rhumatologie, Paris

4. Caroit M, Patte D (1977) Faut-il opérer et quand doit on opérer les calcifications tendineuses de l'épaule? In: de Seze S et al (eds) L'actualité rhumatologique 1976. Expansion Scientifique, Paris, pp 209–216

5. Caroit M (1983) Quelques aspects particuliers de la pathologie de la coiffe des rotateurs. L'épaule douloureuse simple chronique. Rhumatologie 25: 333–339

6. Comfort TH, Arafiles RP (1978) Barbotage of the shoulder with image-intensified fluoroscopic control of needle placement for calcific tendinitis. Clin Orthop 135: 171–178

7. Flint JM (1913) Acute traumatic subdeltoid bursitis. A new and simple treatment. JAMA 60: 1224–1228

8. Gross D, Siegrist H (1981) Lavage des calcifications chez la périarthrite de l'épaule. Comm Congr Int Rhumatologie. Paris Abstract 0853, Rev Rhum Mal Osteoartic

9. Harmon PH (1958) Methods and results in the treatment of 2580 painful shoulders. With special reference to calcific tendinitis and the frozen shoulder. Am J Surg 95: 527–543

10. Kozin F (1985) Painful shoulder (calcific tendinitis). In: McCarthy DJ (ed) Arthritis and allied conditions, 9th edn. Lea and Febiger, Philadelphia, pp 1330–1333

11. Mansat C, Duboureau L, Andrieu D, Remy D, Cha P, Bordes D (1984) Les calcifications de l'épaule: traitement par ponction-irrigation. In: Simon L, Rodineau J (eds) Epaule et médecine de rééducation. Masson, Paris, pp 353–355

12. Marchetti PG, Angeletti P, Jacchia GE (1969) Trattamento delle periartriti calcificate di spalla mediante lavaggio a due vie. Minerva Ortop 20: 485–490

13. Matsen MLA, Kilcoyne RF, Davies PK, Sickler ME VS (1985) Evaluation of the rotator cuff. Radiology 157: 205–209

14. Moutonnet J, Chevrot A, Godefroy D, Horreard P, Zenny JC, Auberge T, Laoussadi S (1984) Ponction infiltration radioguidée dans le traitement des périarthrites calcifiantes rebelles d'épaules. J Radiol 65: 569–572

15. Normandin C, Ortiz-Bravo E (1987) Ponctions de calcifications de l'épaule. Concours Méd 109: 2559–2564

16. Normandin C, Seban E, Laredo JD, Bard M, Kuntz D (1987) Aspirations-triturations de calcifications tendineuses douloureuses d'épaules après échec des traitements médicaux. A propos de 69 cas. Communication, Onzième Congr Eur Rhumatologie, Athens, Abstract F 255. Rheumatology 5: 76

17. Patte D, Goutallier D (1983) Possibilités de la chirurgie (L'épaule douloureuse). Rhumatologie 35: 345–348

18. Patterson RL, Darrach W (1937) Treatment of acure bursitis by needle irrigation. J Bone Joint Surg 19: 993–1002

19. Saez-Clavere L, Legros R, Arlet J, Bonel G (1980) Étude cristallo-chimique de deux calcifications sous deltoidiennes. Rev Rhum Mal Osteoartic 47: 383–392

20. Seze S de (1974) Les épaules douloureuses et les épaules bloquées. Diagnostic et traitement. Concours Méd 96: 5328–5357

21. Vigario GP, Keats TE (1970) Localization of calcific deposits in the shoulder. Am J Roentgenol 108: 806–811

22. Welfing J (1964) Les calcifications de l'épaule. Diagnostic clinique. Rev Rhum Mal Osteoartic 31: 265–271

Subject index

G. B. J. Andersson / T. W. McNeill

Lumbar Spine Syndromes
Evaluation and Treatment

1988. 108 figures. Approx. 250 pages.
ISBN 3-211-82070-1

To be published in Fall 1988

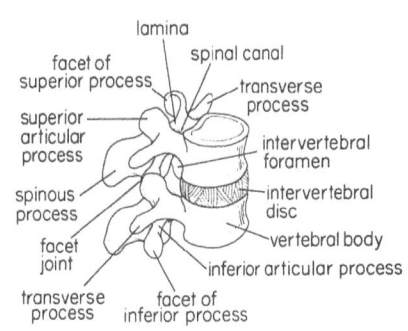

Introduction
Investigations
Treatment Modalities
The Patient with Acute Low Back Pain
The Patient with Recurrent Low Back Pain
The Patient with Pain Radiating down the Leg (Sciatica?)
The Patient with Severe, Unremitting Low Back Pain
The Patient with Bilateral Sciatica who is Unable to Void
The Patient with Bilateral Leg Pain when Walking
The Patient with Chronic, Moderate Low Back Pain
The Patient with Spine Deformity and Low Back Pain
The Child with Back Pain
The Patient with Functional Back Pain/Malingering
The Patient with a Spine Tumor
The Patient with a Spine Infection
The Patient with Inflammatory Diseases of the Spine
The Patient with a Metabolic Spine Condition
Glossary
References

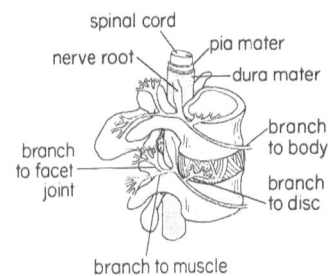

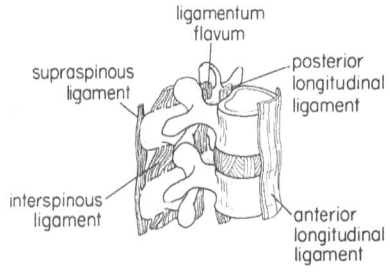

The title expresses most succinctly the subject of this book. No previous publication has attempted to present the differential diagnosis of low back pain, starting with the clinical presentation of the patient to the physician and carrying the discussion through to the conclusion of care in the correct sequential manner. The discussion of each of the syndromes center upon a decision analysis format which is also a new approach to this difficult subject. Readers will be able to develop the matrix for coherent thought and ease of future recall. Their patient care should then follow easily.

Springer-Verlag Wien New York

Moelkerbastei 5, A-1010 Wien · Heidelberger Platz 3, D-1000 Berlin 33
175 Fifth Avenue, New York, NY 10010, USA · 37-3, Hongo 3-chome, Bunkyo-ku, Tokyo 113, Japan

Vascular Anatomy of the Spinal Cord

Neuroradiological Investigations and Clinical Syndromes

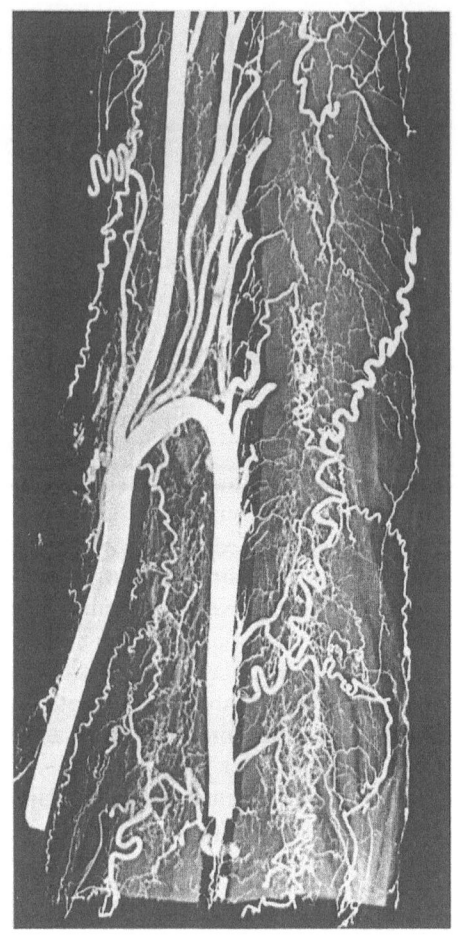

By **Armin K. Thron**
with collaboration of Ch. Rossberg
and A. Mironov

1988. 74 partly colored figures. VII, 114 pages.
Cloth DM 98,–, öS 690,–
ISBN 3-211-82015-9

Prices are subject to change without notice

The book summarizes the anatomic guidelines of external blood supply to the spinal cord. The basic principles of arterial supply and venous drainage are illustrated by explicit schemes for quick orientation.

In the first part of the book, systematic radiologic-anatomic investigations of the superficial and deep vessels of all segments of the spinal cord are introduced. The microvascular morphology is portrayed by numerous microradiographic sections in all three dimensions without overshadowing. The excellent three-dimensional representation of the vascular architecture illustrates elementary outlines and details of arterial territories, anastomotic cross-linking as well as the capillary system, particularly the hitherto unknown structure of the medullary venous system with its functionally important anastomoses and varying regional structures. These often new radiologic-anatomic findings are discussed as to their functional and pathophysiologic impact and constitute the basis on which to improve our modest understanding of vascular syndromes of the spinal cord. The neurosurgeon as well as the neuroradiologist familiar with endovascular techniques are offered information on microvascular morphology necessary for interventions at the spinal cord.

The second part of the book focuses on clinical syndromes and illustrates the present diagnostic contribution of spinal angiography with special reference to the diagnosis of spinal vascular anomalies and arteriovenous malformations. Though widely underrated in the past, the pathogenetic role of the spinal venous system in direct or indirect circulatory disorders of the spinal cord is emphasized. The book fills an obvious void both from a pathologic-anatomic and a clinical point of view, and should therefore attract every physician remotely interested in neurology.

Springer-Verlag Wien New York

Moelkerbastei 5, A-1010 Wien · Heidelberger Platz 3, D-1000 Berlin 33
175 Fifth Avenue, New York, NY 10010, USA · 37-3, Hongo 3-chome, Bunkyo-ku, Tokyo 113, Japan